Communications in Computer and Information Science   962

Commenced Publication in 2007
Founding and Former Series Editors:
Phoebe Chen, Alfredo Cuzzocrea, Xiaoyong Du, Orhun Kara, Ting Liu,
Dominik Ślęzak, and Xiaokang Yang

More information about this series at http://www.springer.com/series/7899

Antonia Moropoulou · Manolis Korres
Andreas Georgopoulos · Constantine Spyrakos
Charalambos Mouzakis (Eds.)

Transdisciplinary Multispectral Modeling and Cooperation for the Preservation of Cultural Heritage

First International Conference, TMM_CH 2018
Athens, Greece, October 10–13, 2018
Revised Selected Papers, Part II

Springer

Editors
Antonia Moropoulou
National Technical University of Athens
Athens, Greece

Constantine Spyrakos
National Technical University of Athens
Athens, Greece

Manolis Korres
National Technical University of Athens
Athens, Greece

Charalambos Mouzakis
National Technical University of Athens
Athens, Greece

Andreas Georgopoulos
National Technical University of Athens
Athens, Greece

ISSN 1865-0929   ISSN 1865-0937 (electronic)
Communications in Computer and Information Science
ISBN 978-3-030-12959-0   ISBN 978-3-030-12960-6 (eBook)
https://doi.org/10.1007/978-3-030-12960-6

Library of Congress Control Number: 2019930855

This Springer imprint is published by the registered company Springer Nature Switzerland AG
The registered company address is: Gewerbestrasse 11, 6330 Cham, Switzerland

Preface

Innovative scientific methodologies and challenging projects marking future trends in the protection of cultural heritage have initiated a universal conversation within a holistic approach, merging capabilities and know-how from the scientific fields of architecture, civil engineering, surveying engineering, materials science and engineering, information technology and archaeology, as well as heritage professionals and stakeholders in cultural heritage. The advanced digital documentation permits the data fusion of interdisciplinary innovative modeling, analytical and non-destructive techniques, supporting the emergence of a transdisciplinary field for multispectral sustainable preservation and management of cultural heritage.

The project of the rehabilitation of the Holy Sepulchre's Holy Aedicule, as a pilot multispectral, multidimensional, novel approach based on transdisciplinarity, holistic digital reconstruction and cooperation in the protection of monuments, motivated the organization of this conference. As discussed in the homonymous panel with the NTUA Interdisciplinary Team (Prof. Antonia Moropoulou, NTUA School of Chemical Engineering- Coordinator, Prof. Manolis Korres, Emeritus, Academy of Athens, Prof. Andreas Georgopoulos, NTUA School of Rural and Surveying Engineering, Prof. Constantine Spyrakos and Asst. Prof. Charalambos Mouzakis, NTUA School of Civil Engineering), National Geographic Society (Fredrik Hiebert, Archaeologist), World Monuments Fund (Yiannis Avramides, Civil Engineer), European Commission (Albert Gauthier), as well as the representatives of the three Christian Communities, Guardians of the Holy Tomb and Asst. Prof. Stefanos Athanasiou, University of Bern, Faculty of Theology, the opening of the scientific community towards society, which managed to reach out to all humanity, was successfully achieved through the cooperation of National Geographic with NTUA, as characteristically demonstrated through its contribution in highlighting and transmitting the ecumenical values related to the opening of the Holy Tomb.

The conference also explored the sustainable preservation and management of monuments, sites, and historic cities as a prerequisite, which must be met in order for them to remain within the listed Monuments of World Heritage recommended by UNESCO, such as the Medieval City of Rhodes, for which the strategic management plan proposed by the NTUA was discussed in detail. In this direction, new funding tools, schemes, and mechanisms are sought, based on the re-use of heritage assets at the service of local communities and in order to enhance their value, aiming to increase social participation and awareness and at the same time generate revenue for a sustainable preservation within the framework of circular economy, as stressed by Mrs. Bonnie Burnham, President Emeritus of the World Monument Fund, Mr. George Kremlis, E.C.-D.G. Environment, Responsible for the Circular Economy, Mrs. Le Minaidis, Secr. Gen. OWHC (Org. of World Heritage Cities) – Coordinator, Prof. Ismini Kriari, Rector at the Panteion University of Social and Political Sciences, Prof. Michael Turner, UNESCO Chair in Urban Design, Conservation Studies/Bezalel

Academy of Arts and Design, Jerusalem, Mr. Teris Chatziioannou, Deputy Mayor of Rhodes, as well as NTUA Professors Antonia Moropoulou, President of the TCG General Assembly, Sophia Avgerinou-Kolonia, ICOMOS CIVVICH, President of the Board of Directors of *Eleusis 2021* - European Capital of Culture, Eleni Maistrou, Society for the Environmental and Cultural Heritage (ELLET), Council of Architectural Heritage.

In the panel "Novel Education Approach for the Preservation of Monuments," the education of both society and members of the scientific community was discussed. The Interdisciplinary NTUA Post-Graduate MSc Program "Protection of Monuments," 20 years after its founding, offered by the NTUA School of Architecture, with the cooperation of the NTUA Schools of Chemical Engineering, Civil Engineering and Rural and Survey Engineering, represented in the panel by Prof. Irene Efesiou, NTUA, Post-Graduate MSc Program "Protection of Monuments" Em. Prof. Eleni Maistrou, President of the Council of Architectural Heritage of the Society for the Environmental and Cultural Heritage, ELLET, and Antonia Labropoulou, NTUA, the Post-Graduate Program "Protection, Conservation and Restoration of Cultural Monuments Interdepartmental Program of Post Graduate Studies" offered by the Aristotle University of Thessaloniki, represented in the panel by Prof. Maria Arkadaki, the University of the Peloponnese, represented by Prof. N. Zacharias, the School of Pedagogical and Technological Education (ASPETE), represented by Prof. Anastasia Sotiropoulou and the Frederick University of Cyprus, represented by Prof. Marios Pelekanos, recognized the significant educational value of problem-based learning and the hands-on approach in educating new scientists, while the value of the holistic digital fusion of data was highlighted by Professor Peter De Vries, Delft, European Society for Engineering Education (SEFI). The significant contribution of multimedia in the education of not only scientists, but the whole of society was discussed with Marcie Goodale, Product Director of National Geographic Learning and J. J. Kelley, Emmy-nominated film-maker of National Geographic.

In order to achieve a multilateral approach for the preservation of cultural heritage, heritage stakeholders, science, and industry must be brought together. This issue was discussed in the relative panel by Amalia Androulidaki, Dir. Gen. of Anastylosis, Hellenic Ministry of Culture and Sports – Coordinator, Yuval Baruch, Israeli Antiquity Authorities, Michalis Daktylides, PEDMEDE, European Construction Industry Federation (FIEC), Kyriakos Themistokleus, Cyprus Scientific and Technical Chamber (ETEK), member of the Executive Committee, Prof. Panagiotis Touliatos, Frederick University of Cyprus, Stefano Dellatore, Politecnico di Milano, Athina Chatzipetrou, President Hellenic Archaeological Fund, Dr. Evgenia Tzannini, Law Lecturer NTUA - T.C.G., Özgür Turan Aksan, Architect, Istanbul, Dr. Ekaterini T. Delegou, NTUA, representing the Hellenic Construction Technology Platform - T.C.G., NTUA, and Elena Korka, ICOMOS, member of the International Board.

The innovative panel procedures of the conference formatted the future landscape for the preservation of cultural heritage. The First TMM_CH conference attracted researchers from all over the world, as 237 contributions were submitted for oral presentation from 22 countries. In addition, the participants of the conference, including stakeholders, industry, and academy, especially scholars and young researchers, who came from all over the world, highlighted the fact that it was an international conference, which successfully achieved bringing together a vast number of brilliant specialists who represent the aforementioned fields to share information about their latest projects and scientific progress.

Striving to ensure that the conference presentations and proceedings are of the highest quality possible, we only accepted papers that presented the results of various studies focused on the extraction of new scientific knowledge in the area of transdisciplinary multispectral modeling and cooperation for the preservation of cultural heritage. Hence, only 73 papers were accepted for publishing (i.e., 32% acceptance rate). All the papers were peer reviewed and selected by the international Steering and Scientific Committees, comprising 105 and 155 reviewers, respectively, and from over 90 institutions. Each submission was reviewed by three reviewers. We would like to express our deepest gratitude and appreciation to all the reviewers for devoting their precious time to thoroughly review the submissions and provide feedback to the authors.

Subsequently, we are happy to present this book, *Transdisciplinary Multispectral Modeling and Cooperation for the Preservation of Cultural Heritage*, which is a collection of the papers presented at the First International TMM_CH Conference held during October 10–13, 2018, at the Eugenides Foundation of Athens in Greece. This book consists of 14 chapters, which correspond to the 14 thematic areas covered during the conference, as follows: Opening Lecture; The Project of the Rehabilitation of Holy Sepulchre's Holy Aedicule as a Pilot Multispectral, Multidimensional, Novel Approach Through Transdisciplinarity and Cooperation in the Protection of Monuments; Digital Heritage; Novel Educational Approach for the Preservation of Monuments; Resilience to Climate Change and Natural Hazards; Conserving Sustainably the Materiality of Structures and Architectural Authenticity; Interdisciplinary Preservation and Management of Cultural Heritage; Sustainable Preservation and Management Lessons Learnt on Emblematic Monuments; Cross-Discipline Earthquake Protection and Structural Assessment of Monuments; Cultural Heritage and Pilgrimage Tourism; Reuse, Circular Economy and Social Participation as a Leverage for the Sustainable Preservation and Management of Historic Cities; Inception – Inclusive Cultural Heritage in Europe Through 3D Semantic Modeling; Heritage at Risk; Advanced and Non-Destructive Techniques for Diagnosis, Design and Monitoring.

We would like to acknowledge all who made the conference possible and successful, i.e., the organizers: National Technical University of Athens, Technical Chamber of Greece, National Geographic Society, World Monuments Fund; the ministries: Ministry of Culture and Sports of the Hellenic Republic, Ministry of Digital Policy, Telecommunications and Media of the Hellenic Republic, and Ministry of Foreign Affairs who collaborated; the President of the Hellenic Republic and the Hellenic Parliament and His Beatitude Archbishop Hieronymus II of Athens and All Greece and the representatives of the three Christian communities, Guardians of the Holy Tomb who blessed the conference. The conference was held within the framework of the European Year of Cultural Heritage 2018.

We would also like to acknowledge the following for the inauguration addresses of the conference: NTUA Rector Prof. Ioannis Golias, National Geographic Society Archaeologist Dr. Fredrik Hiebert, European Commission D.G. Connect Dr. Albert Gauthier, World Monuments Fund Yiannis Avramides, ICOMOS Dr. Elena Korka,

Technical Chamber of Greece (TCG) President Mr. Giorgos Stasinos, President of the Hellenic Parliament Mr. Nikos Voutsis, President of the Standing Committee of the Cultural and Educational Affairs of the Hellenic Parliament NTUA Prof. Dimitris Sevastakis, Deputy Minister of Foreign Affairs of the Hellenic Republic Mr. Markos Bolaris, Minister of Digital Policy, Telecommunications and Media of the Hellenic Republic Mr. Nikos Pappas, and the Secretary General for Telecommunications and Media Mr. Georgios Florentis.

The conference would not have been possible without the full cooperation and commitment of the NTUA professor Ioannis Golias and the NTUA Interdisciplinary Team, Prof. Manolis Korres, Emeritus, Academy of Athens, Prof. Andreas Georgopoulos, NTUA School of Rural and Surveying Engineering, Prof. Constantine Spyrakos and Asst. Prof. Charalambos Mouzakis, NTUA School of Civil Engineering, the commitment of the Technical Chamber of Greece (TCG), president and administrative council, as well as without the valuable assistance of Leonie Kunz, Aliaksandr Birukou, Amin Mobasheri, Natalia Ustalova, Miriam Costales, all from Springer, to whom we are utmost grateful. We are very proud of the result of this collaboration and we believe that this fruitful partnership will continue for many more years to come.

TMM_CH for the year 2020 has already been announced.

December 2018   Antonia Moropoulou

Organization

General Chair

Antonia Moropoulou   National Technical University of Athens, Greece

Program Committee Chairs

Manolis Korres	National Technical University of Athens, Greece
Andreas Georgopoulos	National Technical University of Athens, Greece
Constantine Spyrakos	National Technical University of Athens, Greece
Charis Mouzakis	National Technical University of Athens, Greece

Steering Committee

Antonia Moropoulou	National Technical University of Athens, Greece
Manolis Korres	National Technical University of Athens, Greece
Andreas Georgopoulos	National Technical University of Athens, Greece
Constantine Spyrakos	National Technical University of Athens, Greece
Charis Mouzakis	National Technical University of Athens, Greece
Kountouri Elena	Directorate of Prehistoric and Classical Antiquities, Greece
Florentis Georgios	Ministry of Digital Policy, Telecommunications and Media of the Hellenic Republic, Greece
Keane Kathryn	National Geographic Society, USA
Hiebert Fredrik	National Geographic Society, USA
Burnham Bonnie	World Monuments Fund, USA
Avramides Yiannis	World Monuments Fund, USA
Korka Elena	International Council of Monuments and Sites (ICOMOS), France
Minaidis Lee	Organization of World Heritage Cities (OWHC), Canada
Potsiou Chryssy	International Federation of Surveyors (FIG), Denmark
Rodriguez-Maribona Isabel	European Construction Technology Platform (ECTP), Belgium
Forest Emmanuel	European Construction Technology Platform (ECTP), Belgium
Murphy Mike	European Society for Engineering Education (SEFI), Belgium
De Vries Pieter	European Society for Engineering Education (SEFI), Belgium

Caristan Yves	European Council of Academies of Applied Sciences, Technologies and Engineering Euro-Case, France
Prassianakis Ioannis	Academia NDT International, Italy
Massué Jean Pierre	Council of Europe, France
D'Ayala Pier Giovanni	International Scientific Council for Island Development (INSULA), France
Ronchi Alfredo	European Commission-MEDICI Framework, Belgium
Di Giulio Roberto	University of Ferrara, Italy
Ioannides Marinos	Cyprus University of Technology, Cyprus
Gravari-Barbas Maria	Paris 1 Panthéon-Sorbonne University, France
Efesiou Irene	National Technical University of Athens, Greece
Alexopoulou Aleka	Aristotle University of Thessaloniki, Greece
Tzitzikosta Aikaterini	Hellenic National Commission for UNESCO, Greece
Achniotis Stelios	Cyprus Scientific and Technical Chamber (ETEK), Cyprus
Baruch Yuval	Israel Antiquities Authority, Israel
Groysman Alec	Association of Engineers, Architects and Graduates in Technological Sciences, Israel
Nakasis Athanasios	Hellenic Section of ICOMOS, Greece
Lianos Nikolaos	Democritus University of Thrace, Greece
Avgerinou-Kolonias Sofia	National Technical University of Athens, Greece
Hadjinicolaou Teti	International Council of Museums Hellenic National Committee, Greece
Pissaridis Chrysanthos	ICOMOS, Cyprus
Mavroeidi Maria	International Committee for the Conservation of Industrial Heritage (TICCIH), Greece
Maistrou Eleni	National Technical University of Athens, Greece
Tournikiotis Panayotis	National Technical University of Athens, Greece
Corradi Marco	University of Perugia, Italy
Forde Michael	University of Edinburgh, UK
Aggelis Dimitris	Free University of Brussels, Belgium
Anagnostopoulos Christos	University of the Aegean, Greece
Asteris Panagiotis	School of Pedagogical and Technological Education (ASPETE), Greece
Athanasiou Stefanos	University of Bern, Switzerland
Balas Kostas	Technical University of Crete, Greece
Botsaris Pantelis	Democritus University of Thrace, Greece
Boutalis Ioannis	Democritus University of Thrace, Greece
Castigloni Carlo	Polytechnic University of Milan, Italy
Chetouani Aladine	University of Orleans, France
Chiotinis Nikitas	University of West Attica, Greece
Christaras Basile	Aristotle University of Thessaloniki, Greece
Coccossis Harry	University of Thessaly, Greece

D'Ayala Dina	University College London, UK
Dimakopoulos Vassilios	University of Ioannina, Greece
Distefano Salvatore	Kazan Federal University, Russia
Erdik Mustafa	Bogazici University, Turkey
Foudos Ioannis	University of Ioannina, Greece
Ioannou Ioannis	University of Cyprus, Cyprus
Kallithrakas-Kontos Nikolaos	Technical University of Crete, Greece
Katsifarakis Konstantinos	Aristotle University of Thessaloniki, Greece
Koufopavlou Odysseas	University of Patras, Greece
Koutsoukos Petros	University of Patras, Greece
Kyriazis Dimosthenis	University of Piraeus, Greece
Kyvellou Stella	Panteion University of Social and Political Sciences, Greece
Leissner Johanna	Fraunhofer Institute, Germany
Osman Ahmad	Saarland University of Applied Sciences, Germany
Mouhli Zoubeïr	Association de Sauvegarde de la Médina de Tunis, Tunisia
Loukas Athanasios	University of Thessaly, Greece
Mataras Dimitris	University of Patras, Greece
Mavrogenes John	Australian National University, Australia
Nobilakis Ilias	University of West Attica, Greece
Paipetis Alkiviadis	University of Ioannina, Greece
Papageorgiou Angelos	University of Ioannina, Greece
Pappas Spyros	European Commission, Belgium
Philokyprou Maria	University of Cyprus, Cyprus
Prepis Alkiviadis	Democritus University of Thrace, Greece
Providakis Konstantinos	Technical University of Crete, Greece
Sali-Papasali Anastasia	Ionian University, Greece
Saridakis Yiannis	Technical University of Crete, Greece
Skianis Charalambos	University of the Aegean, Greece
Skriapas Konstantinos	Network PERRAIVIA, Greece
Sotiropoulou Anastasia	School of Pedagogical and Technological Education (ASPETE), Greece
Spathis Panagiotis	Aristotle University of Thessaloniki, Greece
Stavroulakis Georgios	Technical University of Crete, Greece
Tokmakidis Konstantinos	Aristotle University of Thessaloniki, Greece
Touliatos Panagiotis	Frederick University, Cyprus
Tsatsanifos Christos	International Society for Soil Mechanics and Geotechnical Engineering, UK
Tsilaga Evagelia-Marina	University of West Attica, Greece
Tucci Grazia	University of Florence, Italy
Xanthaki-Karamanou Georgia	University of the Peloponnese, Greece
Zacharias Nikos	University of the Peloponnese, Greece
Zendri Elisabetta	Ca'Foscari University of Venice, Italy

Scientific Committee

Antonia Moropoulou	National Technical University of Athens, Greece
Manolis Korres	National Technical University of Athens, Greece
Andreas Georgopoulos	National Technical University of Athens, Greece
Constantine Spyrakos	National Technical University of Athens, Greece
Charis Mouzakis	National Technical University of Athens, Greece
Abdal Razzaq Arabiyat	Jordan Tourism Board, Jordan
Abuamoud Ismaiel Naser	University of Jordan, Jordan
Achenza Maddalena	University of Calgary, Canada
Adamakis Kostas	University of Thessaly, Greece
Aesopos Yannis	University of Patras, Greece
Agapiou Athos	Cyprus University of Technology, Cyprus
Aggelakopoulou Eleni	Acropolis Restoration Service, Greece
Andrade Carmen	International Centre for Numerical Methods in Engineering, Spain
Argyropoulou Bessie	University of West Attica, Greece
Asteris Panagiotis	School of Pedagogical and Technological Education ASPETE, Greece
Athanasiou Stefanos	University of Bern, Switzerland
Avdelidis Nikolaos	University of Thessaly, Greece
Babayan Hector	National Academy of Sciences of Armenia, Armenia
Badalian Irena	Church of the Holy Sepulchre, Israel
Bakolas Stelios	National Technical University of Athens, Greece
Benelli Carla	Church of the Holy Sepulchre, Israel
Bettega Stefano Maria	Superior Institute of Artistic Industries (ISIA), Italy
Betti Michele	University of Florence, Italy
Biscontin Guido	Ca'Foscari University of Venice, Italy
Boniface Michael	University of Southampton, UK
Boyatzis Stamatis	University of West Attica, Greece
Bounia Alexandra	University of the Aegean, Greece
Bozanis Panayiotis	University of Thessaly, Greece
Caradimas Constantine	National Technical University of Athens, Greece
Cassar JoAnn	University of Malta, Malta
Cassar May	University College London, UK
Cavaleri Liborio	University of Palermo, Italy
Chamzas Christodoulos	Democritus University of Thrace, Greece
Chiotinis Nikitas	University of West Attica, Greece
Chlouveraki Stefania	University of West Attica, Greece
Correia Marianna	School Gallaecia of Portugal, Spain
Daflou Eleni	National Technical University of Athens, Greece
De Angelis Roberta	University of Malta, Malta
Delegou Aikaterini	National Technical University of Athens, Greece

Dellatore Stefano	Polytechnic University of Milan, Italy
Demotikali Dimitra	National Technical University of Athens, Greece
Dimitrakopoulos Fotios	National and Kapodistrian University of Athens, Greece
Don Gianantonio Urbani	Studium Biblicum Franciscanum in Jerusalem - Diocese of Vicenza, Israel
Doulamis Anastasios	National Technical University of Athens, Greece
Drdácký Miloš	Academy of Sciences of the Czech Republic, Czech Republic
Dritsos Stefanos	University of Patras, Greece
Economopoulou Eleni	Aristotle University of Thessaloniki, Greece
Economou Dimitrios	University of Thessaly, Greece
Exadaktylos George	Technical University of Crete, Greece
Facorellis Yorgos	University of West Attica, Greece
Farouk Mohamed	Bibliotheca Alexandrina, Egypt
Firat Diker Hasan	Fatih Sultan Mehmet Vakıf University, Turkey
Frosina Annamaria	Centre for Social and Economic Research in Southern Italy (CRESM), Italy
Fuhrmann Constanze	Fraunhofer Institute for Computer Graphics Research IGD, Germany
Ganiatsas Vasilios	National Technical University of Athens, Greece
Gavela Stamatia	School of Pedagogical and Technological Education (ASPETE), Greece
Ghadban Shadi Sami	Birzeit University, Birzeit, Palestine
Gharbi Mohamed	Institut Supérieur des Etudes Technologiques de Bizerte, Tunisia
Gómez-Ferrer Bayo Álvaro	Valencian Institute for Conservation and Restoration of Cultural Heritage, Spain
Hamdan Osama	Church of the Holy Sepulchre, Israel
Iadanza Ernesto	University of Florence, Italy
Ioannidis Charalambos	National Technical University of Athens, Greece
Izzo Francesca Caterina	Ca'Foscari University of Venice, Italy
Kaliampakos Dimitrios	National Technical University of Athens, Greece
Kapsalis Georgios	National University of Ioannina, Greece
Karaberi Alexia	National Technical University of Athens, Greece
Karellas Sotirios	National Technical University of Athens, Greece
Karoglou Maria	National Technical University of Athens, Greece
Katsioti-Beazi Margarita	National Technical University of Athens, Greece
Kavvadas Michael	National Technical University of Athens, Greece
Kioussi Anastasia	Fund of Archaeological Proceeds, Greece
Kollias Stefanos	National Technical University of Athens, Greece
Konstanti Agoritsa	National Technical University of Athens, Greece
Konstantinides Tony	Imperial College London, UK
Kontoyannis Christos	University of Patras, Greece
Kourkoulis Stavros	National Technical University of Athens, Greece

La Grassa Alessandro	Centre for Social and Economic Research in Southern Italy (CRESM), Italy
Lambropoulos Vasileios	University of West Attica, Greece
Lambrou Evangelia	National Technical University of Athens, Greece
Lampropoulos Kyriakos	National Technical University of Athens, Greece
Lee-Thorp Julia	Oxford University, UK
Liolios Asterios	Democritus University of Thrace, Greece
Liritzis Ioannis	University of the Aegean, Greece
Lobovikov-Katz Anna	Technion Israel Institute of Technology, Israel
Lourenço Paulo	University of Minho, Portugal
Lyridis Dimitrios	National Technical University of Athens, Greece
Maietti Federica	University of Ferrara, Italy
Mamaloukos Stavros	University of Patras, Greece
Maniatakis Charilaos	National Technical University of Athens, Greece
Maravelaki Pagona-Noni	Technical University of Crete, Greece
Marinos Pavlos	National Technical University of Athens, Greece
Marinou Georgia	National Technical University of Athens, Greece
Mataras Dimitris	University of Patras, Greece
Matikas Theodoros	University of Ioannina, Greece
Mavrogenes John	Australian National University, Australia
Milani Gabriele	Polytechnic University of Milan, Italy
Miltiadou Androniki	National Technical University of Athens, Greece
Mitropoulos Theodosios	Church of the Holy Sepulchre, Israel
Mohebkhah Amin	Malayer University, Iran
Neubauer Wolfgang	Ludwig Boltzmann Institute for Archaeological Prospection and Virtual Archaeology, Austria
Nevin Saltik Emine	Middle East Technical University, Turkey
Oosterbeek Luiz	Polytechnic Institute of Tomar, Portugal
Ortiz Calderon Maria Pilar	Pablo de Olavide University, Spain
Osman Ahmad	Saarland University of Applied Sciences, Germany
Pagge Tzeni	University of Ioannina, Greece
Panagouli Olympia	University of Thessaly, Greece
Pantazis George	National Technical University of Athens, Greece
Papagianni Ioanna	Aristotle University of Thessaloniki, Greece
Papaioannou Georgios	Ionian University, Greece
Papi Emanuele	University of Siena, Italy
Pérez García Carmen	Valencian Institute for Conservation and Restoration of Cultural Heritage, Italy
Perraki Maria	National Technical University of Athens, Greece
Piaia Emanuele	University of Ferrara, Italy
Polydoras Stamatios	National Technical University of Athens, Greece
Psycharis Ioannis	National Technical University of Athens, Greece
Rajčić Vlatka	University of Zagreb, Croatia
Rydock James	Research Management Institute, Norway
Saisi Antonella Elide	Polytechnic University of Milan, Italy

Santos Pedro	Fraunhofer Institute for Computer Graphics Research IGD, Germany
Sapounakis Aristides	University of Thessaly, Greece
Sayas Ion	National Technical University of Athens, Greece
Schippers-Trifan Oana	DEMO Consultants, The Netherlands
Siffert Paul	European Materials Research Society, France
Smith Robert	Oxford University, UK
Stambolidis Nikos	University of Crete, Greece
Stavrakos Christos	University of Ioannina, Greece
Stefanidou Maria	Aristotle University of Thessaloniki, Greece
Stefanis Alexis	University of West Attica, Greece
Tavukcuoglu Ayse	Middle East Technical University, Turkey
Theoulakis Panagiotis	University of West Attica, Greece
Thomas Job	Cochin University of Science Technology, India
Triantafyllou Athanasios	University of Patras, Greece
Tsakanika Eleutheria	National Technical University of Athens, Greece
Tsatsanifos Christos	International Society for Soil Mechanics and Geotechnical Engineering, UK
Tsoukalas Lefteris	University of Thessaly, Greece
Turner Mike	Bezael Academy of Arts and Design Jerusalem, Israel
Tzannini Evgenia	National Technical University of Athens, Greece
Van Grieken René	University of Antwerp, Belgium
Van Hees Rob	Delft University of Technology, The Netherlands
Varum Humberto	University of Porto, Portugal
Varvarigou Theodora	National Technical University of Athens, Greece
Vesic Nenad	ICOMOS International Scientific Committee for Places of Religion and Ritual (PRERICO), Greece
Vintzilaiou Elissavet	National Technical University of Athens, Greece
Vlachopoulos Andreas	University of Ioannina, Greece
Vogiatzis Konstantinos	University of Thessaly, Greece
Vyzoviti Sophia	University of Thessaly, Greece
Ward-Perkins Bryan	Ertegun Graduate Scholarship Programme in Oxford, UK
Yannas Simos	Architectural Association School of Architecture, UK
Zacharias Nikos	University of Peloponnese, Greece
Zachariou-Rakanta Eleni	National Technical University of Athens, Greece
Žarnić Roko	University of Ljubljana, Slovenia
Zervakis Michael	Technical University of Crete, Greece
Zervos Spyros	University of West Attica, Greece
Zouain Georges	GAIA-heritage, Lebanon

Executive Organizing Committee of the Steering Committee

Lampropoulou Antonia	National Technical University of Athens, Greece
Psarris Dimitrios	National Technical University of Athens, Greece
Directorate of European Affairs & International Relations, Department of International Relations	Technical Chamber of Greece, Greece

Organizational Support of the Steering Committee

Kolaiti Aikaterini	National Technical University of Athens, Greece
Keramidas Vasileios	National Technical University of Athens, Greece
Kroustallaki Maria	National Technical University of Athens, Greece
Alexakis Emmanouil	National Technical University of Athens, Greece
Skoulaki Georgia	National Technical University of Athens, Greece

Technical Editing

Psarris Dimitrios	National Technical University of Athens, Greece
Kroustallaki Maria	National Technical University of Athens, Greece
Kolaiti Aikaterini	National Technical University of Athens, Greece

Kind Support

Stavridi Christina	Tsomokos S.A., Greece
Georgia Vlachou	Tsomokos S.A., Greece
George Markantonatos	Tsomokos S.A., Greece

The venues of the conference are a kind offer of:

Eugenides Foundation

Sponsors

1. Aegean Airlines

2. DALKAFOUKIOIKOS

3. Sintecno

post scriptum

4. post scriptum

5. NEOTEK

NeoMech
www.shop3d.gr

6.NeoMech

With the kind support of:

SGT SYMEON G. TSOMOKOS S.A.

Contents – Part II

**Sustainable Preservation and Management Lessons Learnt
on Emblematic Monuments**

Improving the Geometric Documentation of Cultural Heritage: Combined
Methods for the Creation of an Integrated Management Information
System in Greece 3
   *Eleftheria Mavromati, Eleni Stamatiou, Leonidas Chrysaeidis,
and Konstantinos Astaras*

The Temple of Apollo at Corinth. Observations
on the Architectural Design 22
   Dimitra Andrikou

Anastylosis of Roman Temples in the West End of the Roman Forum, as
Part of a Management Plan of the Archaeological Site of Ancient Corinth . . . 36
   Evangelos D. Kontogiannis

The Restoration of the Katholikon of Daphni Monastery: A Challenging
Project of Holistic Transdisciplinary Approach, Novel Methodologies
and Techniques and Digital Modelling. 59
   Androniki Miltiadou-Fezans and Nikolaos Delinikolas

Restoration and Consolidation of the Byzantine Church of Palaia
Episkopi in Tegea, Arcadia 78
   Konstantina Siountri

Decryption, Through the Collaboration of Historical Accounts and
Technology, of the Hidden Phases of Construction of the Church
of the Holy Monastery of Kykko in Cyprus. A Preliminary Report 88
   Nasso Chrysochou

The Oslo Opera House – Condition Analysis and Proposal for Cleaning,
Protection and Maintenance of Exterior Marble. 104
   *Pagona-Noni Maravelaki, Lucia Toniolo, Francesca Gheraldi,
Chrysi Kapridaki, and Ioannis Arabatzis*

Modern Architecture and Cultural Heritage. 117
   Agnes Couvelas

Cross-Discipline Earthquake Protection and Structural Assessment of Monuments

Post-seismic Restoration Project of Basilica Churches in Kefallonia Island . . .   131
Themistoklis Vlahoulis, Apostolia Oikonomopoulou, Niki Salemi, and Mariliza Giarleli

Investigation of the Structural Response of Masonry Structures.   143
Georgios A. Drosopoulos, Jan Phakwago, Maria E. Stavroulaki, and Georgios E. Stavroulakis

The Timber-Roofed Basilicas of Troodos, Cyprus (15th–19th Cen.).
Constructional System, Anti-seismic Behaviour and Adaptability   157
Marios Pelekanos

Alternative Dome Reconstruction Method for Masonry Structures   177
Argyris Fellas

Cultural Heritage Structures Strengthened by Ties Under Seismic
Sequences and Uncertain Input Parameters: A Computational Approach.   188
Angelos A. Liolios

Masonry Compressive Strength Prediction Using Artificial Neural
Networks   200
Panagiotis G. Asteris, Ioannis Argyropoulos, Liborio Cavaleri, Hugo Rodrigues, Humberto Varum, Job Thomas, and Paulo B. Lourenço

Nominal Life of Interventions for Monuments and Historic Structures   225
Constantine C. Spyrakos and Charilaos A. Maniatakis

Cultural Heritage and Pilgrimage Tourism

Restoration and Management of World Heritage Christian Cultural Sites:
Problems of Religious Tourism – Challenges - Perspectives   239
Alkiviadis Prepis

On the Greatest Challenge in the Management of Living Religious
Heritage: Linking the Authenticity of Heritage and the Authenticity
of Tourist Experiences to the Authenticity of Religious Tradition   262
Ioannis Poulios

Reuse, Circular Economy and Social Participation as a Leverage for the Sustainable Preservation and Management of Historic Cities

Towards a New Heritage Financing Tool for Sustainable Development   275
Bonnie Burnham

Public Built Cultural Heritage Management: The Public-Private
Partnership (P3)   289
    Cristina Boniotti

A Programme for Sustainable Preservation of the Medieval City of Rhodes
in the Circular Economy Based on the Renovation and Reuse of Listed
Buildings   299
    Antonia Moropoulou, Nikolaos Moropoulos, George Andriotakis,
    and Dimitrios Giannakopoulos

Inception – Inclusive Cultural Heritage in Europe through 3D Semantic Modeling

Advanced 3D Survey and Modelling for Enhancement and Conservation
of Cultural Heritage: The INCEPTION Project   325
    Roberto Di Giulio, Federica Maietti, and Emanuele Piaia

INCEPTION: Web Cutting-Edge Technologies Meet Cultural Heritage   336
    Ernesto Iadanza, Peter Bonsma, Iveta Bonsma, Anna Elisabetta Ziri,
    Federica Maietti, Marco Medici, Federico Ferrari,
    and Pedro Martín Lerones

Cultural Heritage Sites Holistic Documentation Through Semantic
Web Technologies.   347
    Anna Elisabetta Ziri, Peter Bonsma, Iveta Bonsma, Ernesto Iadanza,
    Federica Maietti, Marco Medici, Federico Ferrari,
    and Pedro Martín Lerones

In Situ Advanced Diagnostics and Inspection by Non-destructive
Techniques and UAV as Input to Numerical Model and Structural
Analysis - Case Study   359
    Vlatka Rajčić, Mislav Stepinac, and Jure Barbalić

Current and Potential Applications of AR/VR Technologies in Cultural
Heritage. "INCEPTION Virtual Museum HAMH: A Use Case on BIM
and AR/VR Modelling for the Historical Archive Museum
of Hydra Greece"   372
    Dimitrios Karadimas, Leonidas Somakos, Dimitrios Bakalbasis,
    Alexandros Prassas, Konstantina Adamopoulou, and George Karadimas

Heritage at Risk

Reconstructing the Hellenistic Heritage: Chemical Processes, Devices
and Products from Illustrated Greek Manuscripts—An Interdisciplinary
Approach   385
    Dimitrios Yfantis

Glittering on the Wall: Gildings on Greek Post-Byzantine Wall Paintings. . . .   397
Georgios P. Mastrotheodoros, Dimitrios F. Anagnostopoulos,
Konstantinos G. Beltsios, Eleni Filippaki, and Yannis Bassiakos

Advanced and Non-Destructive Techniques for Diagnosis, Design and Monitoring

Kinematic Analysis of Rock Instability in the Archaeological Site of Delphi
Using Innovative Techniques   407
Kyriaki Devlioti, Basile Christaras, Vasilios Marinos,
Konstantinos Vouvalidis, and Nikolaos Giannakopoulos

Monitoring and Mapping of Deterioration Products on Cultural Heritage
Monuments Using Imaging and Laser Spectroscopy   419
Kostas Hatzigiannakis, Kristalia Melessanaki, Aggelos Philippidis,
Olga Kokkinaki, Eleni Kalokairinou, Panagiotis Siozos,
Paraskevi Pouli, Elpida Politaki, Aggeliki Psaroudaki,
Aristides Dokoumetzidis, Elissavet Katsaveli, Elissavet Kavoulaki,
and Vassiliki Sithiakaki

Radar Interferometer Application for Remote Deflection Measurements
of a Slender Masonry Chimney.   430
Georgios Livitsanos, Antonella Saisi, Dimitrios G. Aggelis,
and Carmelo Gentile

Multispectral and Hyperspectral Studies on Greek Monuments,
Archaeological Objects and Paintings on Different Substrates.
Achievements and Limitations   443
Athina Alexopoulou, Agathi Anthoula Kaminari, and Anna Moutsatsou

Infrared Hyperspectral Spectroscopic Mapping Imaging from 800
to 5000 nm. A Step Forward in the Field of Infrared "Imaging"   462
Stamatios Amanatiadis, Georgios Apostolidis, and Georgios Karagiannis

Fusion of the Infrared Imaging and the Ultrasound Techniques to Enhance
the Sub-surface Characterization   472
Stamatios Amanatiadis, Georgios Apostolidis, and Georgios Karagiannis

Moisture Climate Monitoring in Confined Spaces Using
Percolation Sensors   482
Helge Pfeiffer, Charlotte Van Steen, Els Verstrynge, and Martine Wevers

Author Index   495

Contents – Part I

Opening Lecture

The Project of the Rehabilitation of Holy Sepulchre's Holy Aedicule
as a Pilot Multispectral, Multidimensional, Novel Approach Through
Transdisciplinarity and Cooperation in the Protection of Monuments   3
    Antonia Moropoulou, Manolis Korres, Andreas Georgopoulos,
    Constantine Spyrakos, Charalambos Mouzakis,
    Kyriakos C. Lampropoulos, and Maria Apostolopoulou

**The Project of the Rehabilitation of Holy Sepulchre's Holy Aedicule
as a Pilot Multispectral, Multidimensional, Novel Approach
Through Transdisciplinarity and Cooperation in the Protection
of Monuments**

The Holy Sepulchre as a Religious Building.   29
    Stefanos Athanasiou

Preliminary Assessment of the Structural Response of the Holy Tomb
of Christ Under Static and Seismic Loading   44
    Constantine C. Spyrakos, Charilaos A. Maniatakis,
    and Antonia Moropoulou

Corrosion Protection Study of Metallic Structural Elements
for the Holy Aedicule in Jerusalem   58
    Eleni Rakanta, Eleni Daflou, Angeliki Zacharopoulou, George Batis,
    and Antonia Moropoulou

Innovative Methodology for Personalized 3D Representation and Big
Data Management in Cultural Heritage   69
    Emmanouil Alexakis, Evgenia Kapassa, Marios Touloupou,
    Dimosthenis Kyriazis, Andreas Georgopoulos, and Antonia Moropoulou

Governance and Management of the Holy Edicule Rehabilitation Project   78
    Antonia Moropoulou and Nikolaos Moropoulos

Digital Heritage

Branding Strategies for Cultural Landscape Promotion: Organizing Real
and Virtual Place Networks   105
    Konstantinos Moraitis

3D Survey of a Neoclassical Building Using a Handheld Laser Scanner
as Basis for the Development of a BIM-Ready Model   119
  Dimitrios-Ioannis Psaltakis, Katerina Kalentzi,
  Athena-Panagiota Mariettaki, and Antonios Antonopoulos

Exploring the Possibilities of Immersive Reality Tools in Virtual
Reconstruction of Monuments   128
  Panagiotis Parthenios and Theano Androulaki

Reconstruction and Visualization of Cultural Heritage Artwork Objects   141
  Anastasia Moutafidou, Ioannis Fudos, George Adamopoulos,
  Anastasios Drosou, and Dimitrios Tzovaras

Vandalized Frescoes' Virtual Retouching   150
  Melina Aikaterini Vlachou, Dimitrios Makris, and Leonidas Karampinis

Idea: Ancient Greek Science and Technology   171
  Panagiotis Ioannidis, Angeliki Malakasioti, and Maria Mavrokostidou

Cross-Sector Collaboration for Organizational Transformation: The Case
of the National Library of Greece Transition Programme to the Stavros
Niarchos Foundation Cultural Center (2015–2018)   184
  Julia Elmaloglou, Georgia Angelaki, and Stephania Xydia

Volos in the Middle Ages: A Proposal for the Rescue of a Cultural Heritage   203
  Konstantia Triantafyllopoulou

Information Technology, Smart Devices and Augmented Reality
Applications for Cultural Heritage Enhancement: The Kalamata
1821 Project   222
  Vayia V. Panagiotidis, George Malaperdas, Eleni Palamara,
  Vasiliki Valantou, and Nikolaos Zacharias

DiscoVRCoolTour: Discovering, Capturing and Experiencing Cultural
Heritage and Events Using Innovative 3D Digitisation Technologies
and Affordable Consumer Electronics   232
  Constantin Makropoulos, Dimitra Pappa, René Hellmuth,
  Alexander Karapidis, Stephan Wilhelm, Vassilis Pitsilis,
  and Florian Wehner

The Digitization of the Tangible Cultural Heritage and the Related
Policy Framework   250
  Konstantina Siountri, Evangelia Vagena, Dimitrios D. Vergados,
  Joseph Stefanou, and Christos-Nikolaos Anagnostopoulos

Novel Educational Approach for the Preservation of Monuments

Authentic Learning to Better Prepare for Preservation Work   263
  Pieter de Vries and Antonia Moropoulou

20 Years of the N.T.U.A. Interdisciplinary Post Graduate Programme
"Protection of Monuments"   273
  Irene Efesiou, Eleni Maistrou, Antonia Moropoulou, Maria Balodimou,
  and Antonia Lampropoulou

Education and Training for the Preservation of Cultural Heritage,
as a Strategic Aim of the Department of Architecture, Frederick
University Cyprus   285
  Marios Pelekanos and Byron Ioannou

Cultural Heritage and Education/Training/Occupational Activity
of Engineers in Greece   309
  Stamatia Gavela and Anastasia Sotiropoulou

The Historic Centre of Vimercate: Investigation, Education,
Community Involvement   319
  Stefano Della Torre, Rossella Moioli, and Lorenzo Cantini

Connections with the Cultural Heritage in Formal and Informal Learning:
The Case of an Interactive Visual Game of School Life Museum in Chania . . .   329
  Maria A. Drakaki

From Discovery to Exhibition - Recomposing History: Digitizing a Cultural
Educational Program Using 3D Modeling and Gamification   337
  Maria Xipnitou, Sofia Soile, Ioannis Tziranis, Michail Skourtis,
  Alcestis Papadimitriou, Athanasios Voulodimos, Georgios Miaoulis,
  and Charalabos Ioannidis

Resilience to Climate Change and Natural Hazards

Climate Information for the Preservation of Cultural Heritage: Needs
and Challenges   353
  Lola Kotova, Daniela Jacob, Johanna Leissner, Moritz Mathis,
  and Uwe Mikolajewicz

Heritage Resilience Against Climate Events on Site - HERACLES Project:
Mission and Vision   360
  Giuseppina Padeletti and and HERACLES Consortium Staff

Resilient Eco-Smart Strategies and Innovative Technologies to Protect
Cultural Heritage.   376
    Anastasios Doulamis, Kyriakos Lambropoulos, Dimosthenis Kyriazis,
    and Antonia Moropoulou

Interventions on Coastal Monuments Against Climatic Change   385
    George Alexandrakis, Georgios V. Kozyrakis, and Nikolaos Kampanis

Increasing the Resilience of Cultural Heritage to Climate Change
Through the Application of a Learning Strategy   402
    Elena Sesana, Chiara Bertolin, Arian Loli, Alexandre S. Gagnon,
    John Hughes, and Johanna Leissner

Conserving Sustainably the Materiality of Structures and Architectural Authenticity

Taxonomy of Architectural Styles and Movements Worldwide Since
8500 BC   427
    Christos Floros

Monastery of Kimisis Theotokou, Valtessiniko, Arcadia,
Greece - Restoration of the Temple and Integration of New Structures   449
    Vobiri Julia

Deterioration of Monument Building Materials: Mechanistic Models
as Guides for Conservation Strategies   456
    Dimitra G. Kanellopoulou, Aikaterini I. Vavouraki,
    and Petros G. Koutsoukos

Interdisciplinary Preservation and Management of Cultural Heritage

Technological Innovations in Architecture During Antiquity.
The Case of Cyprus   473
    Maria Philokyprou

TLS Survey and FE Modelling of the Vasari's Cupola of the Basilica
dell'Umiltà (Italy). An Interdisciplinary Approach for Preservation of CH . . .   487
    Grazia Tucci, Gianni Bartoli, and Michele Betti

Interdisciplinary Approaches in Cultural Heritage Documentation:
The Case of Rodakis House in Aegina.   500
    Georgiadou Zoe, Alexopoulou Athina-Georgia, and Ilias Panagiotis

Assessment of Masonry Structures Based on Analytical Damage Indices   513
    Athanasia Skentou, Maria G. Douvika, Ioannis Argyropoulos,
    Maria Apostolopoulou, Antonia Moropoulou, and Panagiotis G. Asteris

A Historical Mortars Study Assisted by GIS Technologies   532
  Panagiotis Vryonis, George Malaperdas, Eleni Palamara,
  and Nikos Zacharias

Towards a Blockchain Architecture for Cultural Heritage Tokens   541
  Aristidis G. Anagnostakis

Protection and Highlighting of a Waterfront Zone Disposing Strong
Cultural Characteristics   552
  Dimitris Psychogyios and Helen Maistrou

Author Index   567

Sustainable Preservation and Management Lessons Learnt on Emblematic Monuments

Improving the Geometric Documentation of Cultural Heritage: Combined Methods for the Creation of an Integrated Management Information System in Greece

Eleftheria Mavromati[1](✉) , Eleni Stamatiou[2] ,
Leonidas Chrysaeidis[3] , and Konstantinos Astaras[4]

[1] Senior Investigator at the Greek Ombudsman – Quality of Life
Department/Archaeologist MA/Aristoteleion University of Thessaloniki,
Conservator of Antiquities and Pieces of Art Technological University
of Athens/Professor of Archaeology and Conservation at Vocational
Institution I.IEK. Enosi, Street Art Conservator,
3 Paridi Street, Polygono, 114 76 Athens, Greece
teripieri@gmail.com

[2] Senior Investigator at the Greek Ombudsman – Architect Engineer NTUA,
MS Regional Development, Dr Regional & Town Planner (PhD, PPhD),
Professor Consultant at the Hellenic Open University,
7 Areos, 175 62 Paleo Faliro, Greece
elestamlac@gmail.com

[3] Surveyor Engineer of Products and Services Department - Hellenic
Cadastre/Rural and Surveyor Engineer (M.B.A.)/National Technical University
of Athens, 3 Paridi Street, Polygono, 114 76 Athens, Greece
lchrysai@gmail.com

[4] Surveyor Engineer GEODAITIKI ASTARAS - Surveyor Engineer
Technological Educational Institution of Athens, 15 Papandreou Andrea Street,
Agia Marina, 151 27 Melissia, Greece
mail@astaras.gr

Abstract. The complete registration of areas of cultural interest, monuments and public immovable cultural property in a single and systematic way requires the creation of an integrated information system.

The architecture of the integrated digital registration system should include the descriptive incorporation of all Archaeological Sites and Historic Sites, Monuments, Protected Areas, Regional Zones and their wider environments, the full geographic recording of their data and relevant environmental data.

The architecture of the Integrated Management Information System (MIS) should include a Geographical Information System (GIS) in a structured, modern, easy-to-use, unified platform, along with the development of a Relational Database (RDB) of the above-mentioned cultural protection thematic areas and their related elements, analyzed in the article.

Methods of scientific approach are promising, since the choice of using the appropriate topographic, photographic and software tools is entirely consistent with the categorization and cultural significance of the recording objects corresponding to the individual fields of the information system. In addition, they

© Springer Nature Switzerland AG 2019
A. Moropoulou et al. (Eds.): TMM_CH 2018, CCIS 962, pp. 3–21, 2019.
https://doi.org/10.1007/978-3-030-12960-6_1

provide data on existing damages and enable detection and assessment of areas of construction requiring conservation interventions.

The extent and importance of the Greek cultural heritage requires the modification and optimization of the methodology of spatial identification and recording of real estate belonging to the Ministry of Culture, protection areas of the cultural environment and immovable monuments, so as to ensure the sustainable management and protection of the cultural object on Greek Territory, the design of their future conservation activities, access to cultural information and floating geospatial and cartographic data.

Keywords: Cultural heritage · Hellenic Archaeological Cadastre · The Greek Ombudsman · Hellenic Cadastre · Management Information System · Photogram-metrical methods · Geospatial data · Archaeological sites protection

1 Introduction

The proper registration of the areas of protection of cultural heritage for monuments and more generally of the public immovable cultural property in a uniform and systematic way requires the development of a Monitoring Information System (MIS).

Its operation should include full geospatial recording and relevant environmental data, and it is essential to collect historical data on the recording of archaeological data. The article refers to the way of recording the cultural objects, the archaeological legislation and its evolution in Greece, the alternative methodology on the basis of the modern topographic and photogrammetric survey methods and the optimization of the methodologies for the complete and uniform creation of the Greek Archaeological Cadastre. As it is already evident from previous considerations, approaches and practices and their results and as determined by the relevant legislation, the recording and mapping of the Greek cultural heritage, given its overriding importance, requires a number of modifications and improvements.

The examination of Greek reality, the findings, needs and priorities of the last decades of the 20th century have highlighted the urgency for the creation of a single, systematic and complete inventory of real estate and areas of cultural heritage, and the need to frequently update the information with the help of a new organized, methodical, reliable and simple to use information system (MIS). This information system is a unified and constantly updated system of information, and, besides the recording of new archaeological findings data of cultural heritage, it will also be used to document the public property managed by the Ministry of Culture and map the cultural objects under protection throughout the Greek Territory.

The article refers to the contents of the Greek Archaeological Cadastre (project scope, implementation plan, scanning, georeferencing and vectoring of the archive material of the involved entities) and the corresponding problems that it resolves. The proposals that arise from the archaeological and Surveying Engineering experience, from the development of the Hellenic Cadastre which aims the creation of a modern, fully automated real estate property record compilation and the research of the reports to the Independent Authority of the Greek Ombudsman focus on (a) the archaeological work and the necessary actions (categorization by importance, historical description, categorization in thematic units, clear delineation) to be combined with field mapping, and (b) methods for the improvement of topographical or photogrammetric surveying and photographic documentation accuracy and precise results. Finally, possibilities and prospects are analyzed, focusing on the expected benefits after the completion of the Archaeological Cadastre, which are summarized in the securing of the archaeological real estate of the state, in the facilitation of the public officers and the academic and research institutions, as well as the ability of citizens-owners to monitor and control their property (especially in cases of imminent expropriation, archaeological engagement or imposing financial burdens, etc.).

2 Methods Regarding the Recording of Archaeological Data. Historical Information

The research of archaeological data in the past[1] was based on the macroscopic survey of topography and other sciences through field research, archaeological findings, Greek and foreign literature review, travelers' testimonies, etc.

Researchers used to collect archaeological data from sources, maps, travelers' testimonies, and excavations. They used the existing limited number of data without having the full capability of linking archaeological finds with wider environmental factors or scenery, which in many cases was restrictive, forcing archaeologists into smaller divisions of the site due to the terrain [2, 3].

The survey methods to locate archaeological sites were occasionally carried out, often repeated because of geological access restrictions and climatic conditions and required a large number of staff or were based on summer programs involving students or researches conducted by foreign archaeological schools. The positioning was made on extracts of maps of the Army Geographic Service (GHS), usually using the scale 1:5,000. This scale has been widely used for the purposes of demarcation of cultural sites, either for archaeological research or for the construction of infrastructures and major projects in which old and new archaeological sites emerged [3, 7].

The absence of geospatial data[2] for the demarcation and Geo-referencing of archaeological sites and monuments in Greece until now has caused many problems, both in the protection of the monuments and in the protection of the properties around them. The Greek Ombudsman, an Independent Authority, has been involved, for

---

[1] Since the middle of the 17th century, interest has begun to protect cultural goods from the negative actions of opponents. The first law to protect national monuments was Swedish and was issued in 1666. With regards to Greece, Adamantios Korais first expressed interest in the preservation and protection of Greek antiquities with his memoirs to the Ecumenical Patriarch. Already in 1825, the Minister of Interior and Police in the government of Kountouriotis-Grigorios Dikaios (Papaflessas) issued a decree on the collection of antiquities and their preservation in the school buildings [13: 60] [19: 60).

[2] Topographic, cadastral or other types of charts that do not accompany the properties and the declarations of their protection and do not exist in the archives of the involved entities, as established by the already three completed pilot projects of the National Cadastre project.

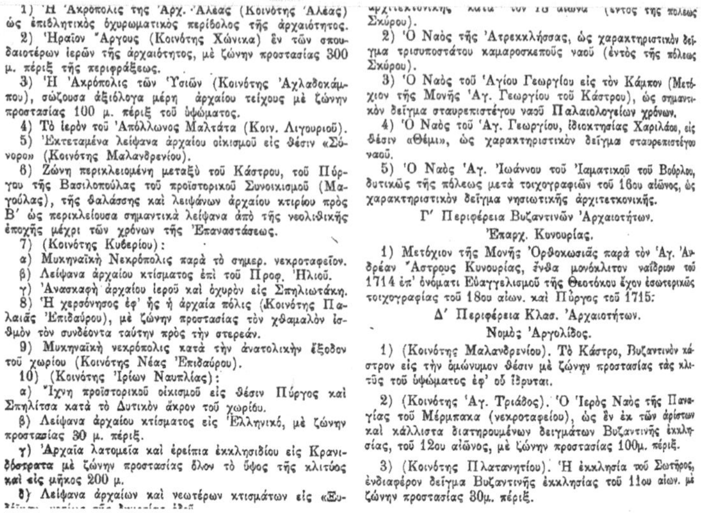

Fig. 1. Extract from the 1965 Government Official Journal, which lists unclear declarations (in Greek) (Government Journal 605/B/1965)

example, over many years in the non-demarcated archaeological areas and the problems arising from the inability to exploit private property due to inadequate localization through the system of state cores and maps of the boundaries declarations of archaeological sites and the delimitation of Zones A and B within them[3] (see Fig. 1) [15, 22].

Developments in the science of topography and the requirements of archaeology have led to the collaboration of scientists in both fields in order to draw conclusions based on various methods, to record archaeological data and to reconstitute the environment [5, 25, 26].

As early as in the beginning of the 20th century, aerial methods were used for archaeology in various countries worldwide, and since 1980 the evolution of topographic and photographic systems has introduced new fields of science and we now refer to aerial archaeology and archaeological topography, which, combined with the current climatic conditions and modern methods such as the Airborne lidar (light detection and ranging)[4] (see Fig. 2) reveal hidden or lost antiquities, which are impossible to be perceived, they are not observable from ground level, and are aimed at protecting and highlighting the cultural heritage [1, 2, 6–8, 11].

---

[3] Indicatively, Mavromati (2013) (22.01): Mediation summit on "Implications of the Unclear Limits of Archaeological Sites, An example of the redefinition of the declared archaeological site of the Muses Valley of Voiotia", which was posted on the website of the Ombudsman, http://www.synigoros.gr/?i=quality-of-life.el.loipa_adeiodotisi.83170 and Ombudsman, Life Cycle, Protection Zones and Uses, https://www.synigoros.gr/?i=quality-of-life.el.zwnes_prostasias).

[4] Airborne lidar measures the height of the ground surface and other features in large areas of landscape with a very high resolution and accuracy. «Therefore, it provides the ability to collect very large quantities of high precision three-dimensional measurements in a short time. This facilitates very detailed analysis of a single site, or data capture of entire landscapes» [6, 7].

Fig. 2. Prize winning archaeological KAP image in print. CHRISTY LAWLESS http://arch.ced. berkeley.edu/kap/discuss/index.php?p=%2Fdiscussion%2F5828%2Fprize-winning-archaeological-kap-image-in-print

3 Legislation and Methods of Protection of Cultural Heritage in Greece. Findings, Needs and Perspectives

3.1 Legislation and Methods of Protection of Cultural Heritage in Greece

According to Article 24 of the Constitution, individual and social rights include the protection of the cultural environment, *"Protection of the natural and cultural environment is a duty of the State and the right of everyone. In order to preserve it, the State has an obligation to take particular preventive or repressive measures within the framework of the principle of sustainability"*.

According to Article 18 of the Constitution, *"Special laws regulate the ownership and disposal of archaeological sites [...] and any other deprivation of the free use and appropriation of property, required by particular circumstances [...] Monuments, traditional areas and traditional elements are protected by the State"*.

Therefore, the Constitution enshrines the protection of the natural and cultural environment as a state matter and as an individual right of citizens. In implementing this specific provision of the Constitution, the cultural environment and, in particular, the monuments, traditional areas and traditional elements must be constantly protected throughout the year – at this point comes the concept of sustainability, which defines the way of the state protection of cultural goods[5] [17].

---

[5] Skouris Panagiotis and Trova Eleni, *The protection of Antiquities and Cultural Heritage*, Athens Thessaloniki, 2003, pp. 31–32.

The Law 3028/2002 introduces the archaeological legislation, the framework for the protection of the Greek cultural heritage, the modern legislative tool for the protection of the protected goods. Monuments before 1453 belong to the State, they cannot be sold or seized and they are not the object of any transaction. Monuments after this historic date can be owned by individuals, with restrictions and conditions imposed by the local Archaeological Authority in order to ensure their protection[6].

Great importance is now given to the protection of the monuments as a whole with the surrounding area and relevant provisions are established to be used as the highest tool in their protection. Article 10 of Law 3028/2002, on actions permitted in immovable monuments and their environment, is widely used by Archaeologists Special Scientists in the Greek Ombudsman, Independent Authority. This article refers to the basic categories of protection of archaeological and historical sites and provides for the control of any activity near the ancient ruins, thus protecting cultural goods and areas that may not have been declared under protection[7] (see Fig. 3) [14].

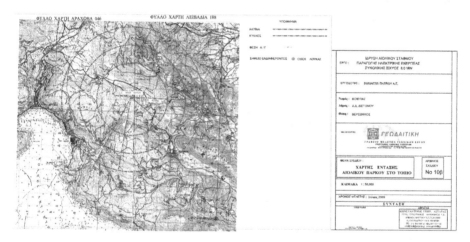

Fig. 3. A map of wind farm integration in the landscape and distance control to see if there is a legal distance from the Osios Loukas monastery of Viotia and if there is a visual nuisance in the monument. File Geodaitiki Astaras.

In the archaeological sites located within or outside settlements, where the State may be able to designate Protection Zones, such as Zone A' Absolute Protection or Zone B' Protection with the possibility of building on terms and restrictions, if the settlements are not inhabited and are archaeological sites, the construction is forbidden and only the conservation of buildings and the demolition of buildings that are considered to be dilapidated is allowed. If the settlements are inhabited and part of them is an archaeological site, it is forbidden to intervene in a way that alter the character and

---

[6] According to the Article 7 of Law 3028/2002 for the ownership of monuments and the Article 9 for the maintenance of ancient monuments.

[7] Mavromati E., Chrysaeidis L. (2017).

the already existing urban plan or disrupt the relationship between the buildings and the open spaces. Before any intervention, the permission of the Minister of Culture and the opinion of the competent Council is presented[8] [14, 15, 20–22].

In cases where archaeological sites, settled in areas outside the city plan or outside the legal settlements, have been declared, the possibility of agricultural or other activities, as well as the rebuilding of existed buildings is given by Decision of the Minister of Culture. In certain areas, designated as Zones B', specialists apply conditions of building and activation. The properties, in which ancient ruins are visible and is decided by specialists to be of great importance, are usually included in Protected Area A', which means that the area is a Zone banned from construction and other activities and properties must be obligatory expropriated[9] [22].

3.2 Findings, Needs and Priorities

Moving towards the end of the 20th century and in the twilight of the 21st century, it was considered necessary to have a uniform, systematic and complete registration of the Real Estate and the Areas of Protection of Cultural Interest, which would be continuously updated and organized. It was also considered necessary to document the public properties managed by the Ministry of Culture and to map the cultural heritage protected throughout the Territory (according to Law 3028/2002, these areas include the Archaeological Sites, Protected Areas A' and B', Historic Sites, Regional Areas for the Protection of Monuments and Sites, the Sites surrounding the Monuments, etc.), as well as the data concerning the Real Estate Monuments in Greece.

In 2012, the international open competition of the Ministry of Education, Religious Affairs, Culture and Sport was proclaimed, based on Directive 2004/18/EC[10], on the establishment of the Greek Archaeological Cadastre. The Archaeological Cadastre falls within the definition of archaeological work, within the meaning of Presidential Decree 99/92, according to Law 3207/03, as supplemented by Law 4049/12. The Archaeological Cadastre is a unified and constantly updated system of information, that contains the scientific, systematic and continuous collection, documentation, digitization, exploration, codification, material digitization and archaeological content production, data input, spatial identification and mapping, storage and management, using a Management Information System (M.I.S.), design and implementation of electronic data management applications of all the (a) real estates acquired by the Ministry of the Environment and Energy; (b) areas for the protection of the cultural environment

---

[8] According to the Article 14 of Law 3028/2002, for the archaeological sites in settlements or settlements-archaeological sites.

[9] According to the Article 13 of Law 3028/2002 for archaeological sites at rural areas and areas of protection.

[10] With the title *"Implementation of the Archaeological Cadastre: Digitization Works (Scanning, Georeferencing, Vectorization), Real Estate Locations/Places/Monuments and Field Works (Declaration on the physical object of subproject 3 of the approved amended technical bulletin No. 186090 edition 5)"* for Greece.

(archaeological sites, historical sites, Protection Zones, etc.) where restrictions/ prohibitions apply to the anthropogenic activities of immovable monuments of all historical periods throughout the Territory.

The strong and permanent objective is the integrated, quantitative and qualitative performance of the cultural, historical, archaeological and socio-economic reality of the cultural repository and the promotion of this repository on the internet. The objective is to integrate descriptive/geospatial data concerning: (a) approximately 7,500 public properties (inside/outside cities and settlements, buildings, etc.) acquired by the State; and (b) approximately 4,200 areas of protected cultural environment and 18,000 immovable monuments. According to articles, there were only 2345 properties in the whole of the Greek territory that belonged to the management jurisdiction of the Ministry of Culture until they were recorded in the directories of the Real Estate Public Authority. After the establishment of the Archaeological Cadastre, entries were found to be 27,300. It is estimated that the total number of entries will be around 30,000[11].

The aim of the project is also the creation of a computational infrastructure for the provision of Electronic Services to the citizen and the public institutions, which will constitute a platform for the interconnection of additional systems of the Ministry of the Environment and Energy, the upgraded system of the National Monument Register. In the future, the Archaeological Cadastre will be able to incorporate other information as appropriate, such as the results of archaeological investigations, archaeological findings and other excavation data within archaeological sites and public buildings after an appropriate extension study.

The work of the Greek Archaeological Cadastre includes: (1) Scope of the project, with a Plan of Implementation; (2) Scanning, Georeferencing and Distribution of the archival material of the involved entities; (3) Localization of objects of cultural interest all over the Territory; (4) Field and Office mapping and Surveying Works for Uninterrupted Records, Diagramming and Correlation with Existing Cadastral diagrams and cadastral registrations of the National Cadastre, (5) Input of digital and geospatial data in Management Information System (MIS), (6) Quality controls[12].

---

[11] Article "YPPO… disappearance of the Archaeological Cadastre", *To Vima tis Kiriakis* newspaper, 30.09.2018, p. 55.

[12] According to the no. ΥΠΑΙΘΠΑ/ΓΓΠ/ΓΔΑΠΚ/ΔΑΑΠ/ΕΣΠΑΑΚ/142047/43861/1739-19.12.2012 document of the Ministry for the Education, Religious Affairs, Culture and Sport: "(1) Study of the implementation of the project:" The Ministry of Education, Religious Affairs, Culture and Sports": The Contractor will prepare and submit a Scope Plan, which will be the detailed guide for the implementation of the project. The Project will be the development guide and the reference base for monitoring the progress of work throughout the project. (2) Scanning, Georeferencing and Distribution of the archive material of the involved entities, to be collected by the Contracting Authority and will be available at the headquarters of the agencies within the Attica Region. (3) Terrestrial Localization Works: It concerns all objects of cultural interest, namely the identification of the State Property, the Protection Areas and the Monuments of Real Estate throughout the Territory. (4) Field and Office Mapping for Non-identified Records: It will include field and office work as well as the drawing up of Cadastral diagrams and the drawing of conclusions in relation to existing cadastral diagrams and the cadastral registrations of the National Cadastre (where these exist). (5) Introduction of Digital Archives and Geospatial Data in the Integrated Information System (M.I.S.). (6) Quality Controls.

4 Proposals and Desired Results

Although antiquity has a special importance in Greece, the full extent of our archaeological/historic environment is still unknown. With the use of remote sensing we try to identify and record our archaeological sites, historical landscapes and monumental ruins, as well as to improve the understanding thereof. Cultural documentation requires precision, visual quality of final products, and combined use of different pioneering methodologies[13]. In order to create a Management Information System (MIS), we propose the following fields be provided for and included:

Archaeological Works
In order to achieve the establishment of the Hellenic Archaeological Cadastre with reliable data, it is necessary to have ensured in advance the organization and the provision of the necessary archaeological information on the field mapping by the competent Public Authority and the Regional Services, such as the categorization by area of the significant monuments and the desirable demarcation of the archaeological sites and the Protection Zones, identifying elements etc. This information is the basis for the completeness of the cultural objects mapping.

More particularly, taking into consideration that some settlements have been wrongly included in Zone A', some areas protecting the cultural objects are larger that the ones needed (See Footnote 14), that some sites are only indicative, using point detection, and that the extent of the ruins to be protected is not known, all the above should be reconsidered by the Ministry of Culture in order to define the monuments sites and the boundaries of the archaeological sites and Protection Zones, and then, or at the same time, to complete the surveying engineering on the field (See Footnote 16)

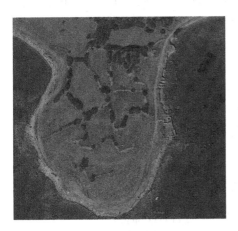

Fig. 4. In ancient antiquities, the use of modern survey methods becomes a necessity. Plytra Laconia, Source: Google Earth.

Fig. 5. Plytra Laconia. Inability of mapping with terrestrial survey methods. File Geodaitiki Astaras.

---

[13] Tsingas V. (2010), where these methods about Erechtheion are described.

and to give the exact coordinates in accordance with the GNSS (Greek National State geographic coordinate System) of 1987[14] [14, 15, 24].

In particular, considering that in the archaeological Zones A' and B', some settlements which should not have been included in Zone A' were mistakenly been part of them, or areas larger than those actually needed for the protection of cultural objects[15] are declared or that some sites are only indicative, by point detection and the extent of the ruins to be protected are not known, they should be re-examined by the Ministry of Culture and the locations of the monuments, the boundaries of archaeological sites and Protection Zones, so that, in parallel with or in conjunction with the specific workshops to complete the surveying engineering of the field[16] and to give the exact coordinates in the GNSS (Greek National State geographic coordinate System) of 1987[17] [14, 15, 24].

For cultural objects with unclear declarations in the Official Journal of the Hellenic Republic (see Fig. 3) or without cadastral diagrams[18], or without a new correct optimized location, two types of topographical survey work have to be completed at the same time: (a) fully mapping of the archaeological sites extent and the Protection Zones as declared; and (b) provision of the correct limits of the declared area to be mapped [22]. In order to identify the actual location of cultural remains, it is proposed to use data from the Regional Services of the Ministry of Culture, to carry out surface surveys and mapping for accessible areas, with a multidisciplinary group of archaeologists, archaeological guardians and surveying engineers, or, to use balloons,[19] the Unmanned Aerial Vehicle (U.A.V.) modern technology [6, 23] and satellite imagery for inaccessible areas and monuments of interest. Finally, photo-interpretation studies will be set up by surveyor engineers (see Figs. 4 and 5). At the same time, the already declared areas will be mapped again, since the existing declarations have resulted in administrative acts with legal rights of third parties. The declarations of the archaeological sites, which have been published in the Official Journal of the Hellenic Republic, will

---

[14] Greek Geographical Reporting System of 1987 or EGSA 87. This is a geodetic reference system that has been used in Greece since 1990. The reference system focuses on the framework of parameters and coordinate systems associated with a particular region or a specific area in the Greek Territory and in which the points and objects on the surface of the Country.

[15] Mavromati E.2013/Tsakirakis E.2013, https://www.synigoros.gr/resources/docs/344700.pdf.

[16] Those works are expected to take place either under the steering assistance of the archaeologists or at the suggestions of the guardians of antiquities serving at the Regional Services of the Ministry of Culture who, at the present stage of the preparation of the Archaeological Cadastre, are insufficient for this work, due to of the financial difficulties of the State.

[17] Greek Geographical Reporting System of 1987 named also EGSA 87. This is a geodetic reference system that has been used in Greece since 1990. The reference system focuses on the framework of parameters and coordinate systems associated with a particular region or a specific area in the Greek Territory and in which the points and objects on the surface of the Country.

[18] There are declarations of archaeological sites and Protection Zones that are not accompanied by a topographic map or the topographical of the declaration, the boundaries of the archaeological sites are not clear or the declaration is accompanied by a topographic map with incorrect coordinates and boundaries of the space.

[19] For monuments of special cultural importance to Greece and the world cultural heritage, such as the Acropolis of Athens, the U.A.V. fly above the monument is not allowed. In these cases, the imprint is done with pampering instruments such as sun balloons (see Tsingas V., 2012).

be corrected, via the correct boundaries re-publication of the aforementioned archaeological sites.

The photogrammetric survey/studies of the cultural objects are one of the pillars for the completion of the Archaeological Cadastre. The parallel contribution of the Ministry of Culture and the creation of a Working Group that will deal exclusively with the Hellenic Cadastre and will carry out a series of tasks including the research, collection and review of the existing files and other items[20] are necessary for the identification and recognition of monuments and sites of cultural interest, so that the single database will include all distinct information about the property - ownership, urban, cultural and geographic interest data, etc.

The multiple levels of information of the MIS system that will be created, accessed by criteria[21], we suggest to include, among others, a brief description of the known findings, a link to the historical events and the sources in which they are recorded, the construction materials and the conservation status for different periods of time, the classification in thematic units per historical period, area and location in Greece, the importance of the monument, the identity and character of the monument, the connection with other monuments in the area, the analysis of their role in the development of the region, etc.

The proposed "GAIA" e-services platform should be interactive and allow its visitors to record their comments on the cultural object (the popularity of the monument, comments on its status, staff, accessibility, etc.).

Surveying Engineering

For the completeness of the Archaeological Cadastre mapping, it is necessary to collect all the metric information of the monument or the monumental ruins and the areas of cultural interest, their location in relation to the site, their possible past extent, their use, their function and their significance or their symbolism, each building element of the monument in relation to the environment, the form of the broader natural environment, as well as the recording of the detailed qualitative characteristics of their construction and materials. The ultimate aim serves the creation of an integrated unified information system, easy to access, with easy-to-use and valid metadata. For the holistic mapping of cultural objects, the following are necessary: (a) the identification of the best access route; (b) the production of the appropriate travel accessibility trail; (c) the recording of proposed alternative routes; and (d) the paths inside the archaeological areas and to the monuments, including paths for people with limited mobility and vision ability.

Based on the technical specifications of the Project, the exact location of the entities to be designated is indicated in situ, to the person responsible for the preparation of the

---

[20] Historical, archaeological and other archives (e.g. archives of the Geographical Service of the Army, planning services), historical and geological maps, surveys and publications, data from the Regional Services of the Ministry of Culture should be collected and registered in the single database of the Greek Archaeological Cadastre.

[21] Access has been reported with criteria for the protection of antiquities in the Workshop on "*The Integrated Information System of the Archaeological Cadastre and the presentation of the web portal Archaeological Cadastre*", organized by the Ministry of Culture, Education and Religious Affairs, 20.03.2015, at the Amphitheater of the Acropolis Museum in Athens, Greece.

survey, by the staff of the Regional Ephorates of Antiquities[22]. The Employer has the ability to increase or remove the required number of entities. The mapping credibility of the monument or cultural area can be assured only if the financial provision for the mapping is sufficient to provide a great number of entities density and, as a result, to give the exact location of the object of interest, while, the mapping of the enclosure of the site/monument, or even a point spatial object detection of the monument's location doesn't constitute a reliable collection of data metrics.

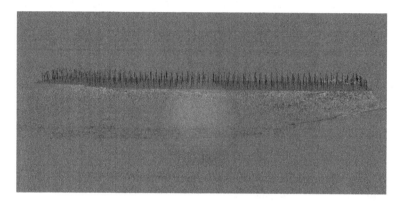

Fig. 6. Snap from U.A.V. Create point cloud, Astypalaia Island. File Geodaitiki Astaras

Given the increased importance of the protection of cultural heritage in the Greek Territory, scanning with scanners (laser scanners and hand-held scanners)[23] is proposed, based on the significance of each monument and given its accessibility. With regard to large-scale geometric documentation, it is proposed to automate the collection of large volumes of data using the technique of three-dimensional laser scanning, which enables the processing of data through modern computing systems. In this way, the problem of limited scanning points is overcome, as this technique allows the detection of monuments and the collection of data with controlled parameters, such as coordinate knowledge, desired density and evenly distributed points. The result that is produced is a cloud of points[24] (point cloud) with a high capability of topographic rendering of cultural objects (see Fig. 6) [5, 23, 25, 26].

Furthermore, the importance of monuments and archaeological sites can be a criterion for the creation of three-dimensional (3D) model[25] for monuments and objects of cultural heritage, of various dimensions, sizes and complexity, or color-coded models, for the visible antiquities, archaeological and historical sites and landmarks, which are

---

[22] The suggestion can be made by the archaeologists of the Ephorate of Antiquities of the area, the staff of the site or the excavation team.

[23] Tsingas V., 2012.

[24] L. Barazzetti et al., (2011).

[25] Tsingas V., 2012.

Fig. 7. Example of an aerial photo turned to a dimensional geographic data (source: https://www.opendronemap.org/odm/)

of high detail and give the maximum detail as there is the possibility of collecting many points in a minimum of time[26] (see Fig. 7). For non-accessible cultural areas, it is proposed to use modern photogrammetric methods and software [23, 25, 26].

Generalizing, the aerial method[27] (UAV, helicopter or airplane) for all non-city entities would reduce the time and cost of mapping objects and would bring closer to the completion of the Hellenic Archaeological Cadastre, while, at the same time, would provide ample information - photos of building materials, monument conservation status, damages, unidentified locations, creation of photo mosaic and digital models [Digital Elevation Model (DEM), Digital Surface Model (DSM), Digital Terrain Model (DTM)] for the buildings and the relief of the surrounding area[28]- data necessary for archaeologists, scholars, conservators of antiquities and other cultural heritage stakeholders (see Fig. 8).

Geo-referencing and Photographic Documentation

Any method of modern photographic documentation[29] provides additional information about the geographical location of the objects they represent. The selection and composition of photographic or other information allows the design and implementation of a spatial database so that each area of the orthophoto map appears in the correct geographical location[30] (see Figs. 8 and 9). The necessity of using the above

---

[26] There are open sources tools for the process of aerial images for example, that turn simple point-and-shoot camera images into two- and three- dimensional geographic data that can be used in combination with other geographic data sets. Other programs take images as an input and produce a variety of georeferenced assets as output, such as maps and 3D models.

[27] The aerial observation was already used in the 1980s in the archaeological site of Dion Pieria, where the ancient sacred city of Dion is located, The Aristotle University, responsible for the systematic excavation at the area, conducted a research for the identification of the Zeus temple. On low helicopter flights, observation of vegetation and scars on the ground, photographing, and then by surface scanning of the area with a Geo radar determined the depth and the most accurate location of the monument.

[28] Tsingas V., (2010), Tsingas V., (2012).

[29] Cameras or specific devices/applications/software programs for the production of photos 360°, scanned aerial photos, satellite sensors or scanned maps.

[30] For the photogrammetric mapping see, among other things, the Erechtheion modeling documentation, Tsingas V. (2012) and the Acropolis Restoration Service website for the development of geographic information systems for the Acropolis of Athens, http://acropolis-gis.ysma.gr.

Fig. 8. Drone flight charts on the island of Astypalaia. File Geodaitiki Astaras

mentioned and the creation of a structured MIS would lead to the accurate description and identification of the spatial object (code, name and location) in the background maps[31] which will be made available.

Furthermore, photographic documentation of cultural objects must be produced with united specifications and standards and provided with the necessary information relevant to their categorization and importance. The above must include: the geolocation and orientation of the photos, the facades of the subject, and the 360° panorama. The produced photos are suggested to be created with high resolution tens of megapixels[32] (see Fig. 10).

The Greek Archaeological Cadastre includes the development of an M.I.S., namely a platform that will incorporate vector and raster geospatial data, concerning public properties that have been acquired by the State in various ways, archaeological sites and other areas of protection, as well as immovable monuments (ancient and newer) managed by the Ministry of Culture [13, 22].

It is proposed to create an easy-to-use free Geographic Information System (GIS) on a platform of a digital true orthophotomosaic, with the precision of image resolution 25 cm, created between the years 2015–2016, for the LSO25 project of the

---

[31] Point visualization of cultural objects in the background maps of the Hellenic Cadastre (http://gis. ktimanet.gr/wms/ktbasemap/default.aspx) or other free applications such as the Geodata portal, Google map etc.

[32] For example, new generation smartphones, digital cameras, drone cameras.

Fig. 9. Signalized of antiquities by geodetically measured points using G.P.S. File Geodaitiki Astaras

Fig. 10. U.A.V. drone capture from 50 m height. Plan of the Holy Monastery of Agnoudos, Prefecture of Argolida, with an image resolution size 4.000 pixels × 3.000 pixels. Derived information of structure, construction, materials and environment. File Geodaitiki Astaras

Hellenic Cadastre. At the same time, it is advisable to allow the simultaneous use of existing historical backgrounds, such as the historical maps of the Greek Cadastre, created between 1945 and 1960, and the digital background of Google or other free institutions worldwide. The data set of the Greek Archaeological Cadastre, derived from the public documents (expropriation, archaeological zones, building restrictions, etc.) should be included in the free information database[33]. The experience of (a) the Acropolis of Athens Monument Conservation Authority, (b) other authorities of the Ministry of Culture, (c) the Hellenic Cadastre and (d) possibly other authorities, should be taken into account and cooperation between them should be achieved.

5   Discussion and Conclusions

The creation of an Archaeological Cadastre has been, for years, a key issue for the Ministry of Culture. The establishment of a unique and systematic digital recording of public property managed by the Ministry of Culture aims at: (a) protecting and encouraging healthy business, once specific conditions and proper criteria for development and nomination are implemented; and (b) setting up the National Monument Archives, which will include the archive of all Hellenic monuments [9, 10].

The systematic recording and mapping of the archaeological sites and immovable monuments in the country and the completion of the information system compilation [12, 13] by using modern technologies and software, is a fundamental condition for fulfilling the multifaceted project of adequate protection of cultural heritage.

The development and implementation of the Archaeological Cadastre introduces Greece into a new era of digital documentation of all archaeological and architectural wealth, with the ultimate aim of knowing the qualitative and quantitative elements of the country's public cultural reserve, its conservation, restoration and promotion, as well as the proper management and the assurance of its integrity.

The expected benefits resulting from the perfect, consistent, modernized implementation and completion of the project are as follows:

- As far as the protected entities are concerned, detection of errors, deficiencies in the clear identification of archaeological and historical sites, cultural objects not designated as protected monuments, or mistakes and omissions in the declaration of monuments or the detection of illegal demolition of monuments, etc.
- Elimination of incorrect documentation issues that sometimes lead to inappropriate declaration or to the dual declaration of the same monument.
- Reconstruction of space, location, city, region and anthropogenic environment, which is an active historical background element of the social and economic, political and historical status.

---

[33] Therefore, the Ministry of Culture should include the vector data, the entire archive material and the existing spatial identities in the unique base of the Archaeological Cadastre, in cooperation with the Hellenic Cadastre.

- Recording of the history of the cultural heritage from the foundation of the Greek state to the present day, achieved through selection, focus and systematic classification of a large amount of information on the indirect or distant past.
- Reorganization and systematization of the size and range of the cultural wealth archive and creation of a centralized database (until now, this information has been scattered).
- Temporal assurance of the archival material and information through digitization, with simultaneous access to the archives of all Ephorates of the Ministry of Culture and the co-competent bodies [15–17].
- Exact knowledge of the cultural repository, its protection and management status, with controlled access by the Ministry of Culture [22–24].
- Reduction of the average time for handling cases, completing pending consultations, and exporting safe conclusions.
- Transformation of digital management framework of cultural heritage, removal of bureaucratic malfunctions for the direct and friendly service to the citizen and any interested user [15–17].
- Supply of information on cultural heritage, accessible to various users (public, private and citizens) via the Gaia web portal, with the prospect of continuous information and enrichment, in order to create an integrated digital encyclopedia on the country's cultural map [5–7]. Citizens information for the condition of property status and location, by searching into GAIA, and the ability of interested persons to register, for personalized information regarding cases of compulsory expropriation due to the existence of antiquities, monuments, traditional buildings and traditional settlements etc.

6 Epilogue

The Greek Archaeological Cadastre is a vital project for the mapping, protection and promotion of the cultural wealth of Greece [14]. It is considered as a unique multi project, with high definition requirements and performance of precision products. For this reason, it requires a combination and pioneering methodology, the use of a variety of modern software, an appropriate legislative framework for cultural protection and, finally, cooperation between operators and services.

Since the integrity of Greece's cultural heritage is being hampered due to the country's unfavorable economic situation and the desire of foreign operators and private companies to acquire or exploit monuments, antiquities and land in Greek Territory[34], the Greek Archaeological Cadastre, in combination with the Hellenic Cadastre and the Forest Maps completion, is expected to become a powerful tool [6–8] for the institutional defense and protection of archaeological/historical areas, cultural sites and monuments of the ancient and modern history of the country [14–19].

---

[34] Legislative arrangements are expected for the complete exclusion of the monuments from their transfer to the Greek Privatization Superfund. Critical proposals are also expected to be included for the monuments built after 1453 and with the ultimate goal of their complete protection according to the Constitution of the Country.

References

1. Aerial investigation and mapping. https://historicengland.org.uk/research/methods/airborne-remote-sensing/aerial-investigation
2. Alexakis, D., Sarris, Ap., Astaras, Th., Albanakis, K.: Integrated GIS, remote sensing and geomorphologic approaches for the reconstruction of the landscape habitation of Thessaly during the Neolithic period. J. Archaeol. Sci. **38**, 89–100 (2011)
3. Georgia, A.G.: Contribution to Archeology and Topography of Azania (Northern Arcadia) Province of Kalavrita, Ph.D. thesis, vol. A, text, Patras (2014). (in Greek)
4. Archaeological Cadastre (2012). www.digitalplan.gov.gr/portal/recource/ARCHAIOLO-GIKO-ΚΤΗΜΑΤΟΛΟΓΙΟ-087595c3-f581-4ada-95d5-ad0221cb5f95
5. Barazzetti, L., Binda, L., Scaioni, M., Taranto, P.: Photogrammetric survey of complex geometries with low-cost software: application to the 'G1 temple in Myson', Vietnam. J. Cult. Herit. **12**, 253–262 (2011)
6. English Heritage: 3D Laser Scanning for Heritage Advice and Guidance to Users on Laser Scanning in Archaeology and Architecture, David M. Jones. English Heritage Publishing, Swindon (2007)
7. English Heritage: The Light Fantastic, Using Airborne Lidar in Archaeological Survey, David M. Jones. English Heritage Publishing, Swindon (2010)
8. Kaimaris, D., Georgoula, O., Patia, P.: Aerial and Satellite Archeology, Excavation, 10 (2009). https://anaskamma.files.wordpress.com/2009/10/kaimaris_etal.pdf. Accessed 18 July 2018
9. Lacroix, R.N., Lagos, D.: Competitive advantage of computer science and telecommunications in the tourism industry. J. WSEAS Trans. Energy Environ. Ecosyst. Sustain. Dev. **2** (5), 659–666 (2006). WSEAS International Conference, Vouliagmeni, Greece, 11–13 July 2006. ISSN 1790-5079
10. Lacroix, R.N., Ladias, Chr.: Digital conversion & upgrading of Greece's public administration services. In: WSEAS International Conference on Energy, Environment, Ecosystems and Sustainable Development, Vouliagmeni, Athens, Greece, 11–13 July 2006 (paper 535-151, 6 pages on Proceedings CD) (2006)
11. Lidar (Light Detection and Ranging). https://historicengland.org.uk/research/methods/airborne-remote-sensing/lidar/. Accessed 18 July 2018
12. Helen, M.: Prehistoric Crete, Topography and Architecture, from Neolithic to Prehistoric Times. Kardamitsa Publications (2002). (in Greek)
13. Marmaras, E., Raptis, St., Stamatiou, E.: Protection of Cultural Heritage, YPEPTH - Pedagogical Institute, Athens (2000). (in Greek)
14. Mavromati, E., Chrysaeidis, L.: The protection of cultural heritage and the protection of property in Hellas country: problems and prospects at the beginning of the 21st century, Heritage and Society, 4th Heritage forum of Central Europe, 1–2 June 2017, International Cultural Centre, Krakow (2017)
15. Mavromati, E.: Mediation Summary on "Impacts from the Unclear Limits of Archaeological Sites, An example of the redefinition of the declared archaeological site of the Muses Valley of Voiotia", which was posted on the website of the Ombudsman (2013). http://www.synigoros.gr/?i=quality-of-life.el.loipa_adeiodotisi.83170. Accessed 15 July 2018
16. Mavromati, E., Stamatiou, E.: Interventions in the public space of historical cities and settlements in Greece. Administrative pathogenesis, weaknesses, possibilities and prospects. The experience of the Ombudsman, Scientific Conference Scientific Support in Decision-Making for Sustainable and Compatible Materials and Maintenance and Protection of Cultural Heritage, 28–29/9.2015, National Technical University of Athens (2015). (in Greek)

17. Skouris, P., Trova, E.: The protection of Antiquities and Cultural Heritage, Athens Thessaloniki (2003)
18. Stamatiou-Lacroix, E., Sapounaki-Drakaki, L.: Evolution de la Législation et de la Politique Urbaine en Grèce, Discussion Paper Series, 9(20), 447–488, University Press of Thessaly, Volos, Greece
19. Stamatiou, E., Tsironi, D.: From Liberation, to Reconstruction and Development in Greece - The State Officer in the field of environment, infrastructure and quality of life, Environment and Law, vol. 2/2013, 64, 225–245 (2013). (in Greek)
20. Stamatiou, E.: Cadastre-Institutional framework, administrative pathogenesis and practice in Greece - the international experience. Environment & Law 1/2010, 51 (2010) 43–95. (in Greek)
21. Stamatiou, E.: Land Tenure and Land Relations in Greece, An International Encyclopedia of Land Tenure, relations for the Nations of the world (4 volumes), Edited by: Ministry of Agriculture of the Russian Federation/Research Institute of Land Relations and Land Planning of the country-Prof. Vladimir Belenkiy/Academician, vol. II, pp. 88–199. Edwin Mellen Press, USA (2004). (in English and Russian)
22. Mavromati, et al.: Special Report, Expropriation, Deprivation, Limitations of Property and Compensation, The Greek Ombudsman (2005). http://www.synigoros.gr/?i=quality-of-life. el.sterisi_xrisis.33912 and www.synigoros.gr. Accessed 22 Sep 2018
23. Tokmakidis, K.: Impressions of monuments and archaeological sites, Open Academic Courses, Aristotle University of Thessaloniki. https://opencourses.auth.gr/modules/document/ file.php/OCRS195/%CE%A0%CE%B1%CF%81%CE%BF%CF%85%CF%83%CE%B9% CE%AC%CF%83%CE%B5%CE%B9%CF%82/%CE%95%CE%9D%CE%9F%CE%A4% CE%97%CE%A4%CE%A1_07.pdf. Accessed 23 Sep 2018
24. Tsakirakis, E.: Summary of mediation on "Restricting the zone of absolute protection of the archaeological site of Meteora, after the intervention of the Ombudsman", which was posted on the website of the Ombudsman (2013). https://www.synigoros.gr/resources/docs/344700. pdf. Accessed 25 Sep 2018
25. Tsingas, V.: Acropolis of Athens: Recording, modeling and visualizing a major archaeological site. J. Herit. Digit. Era 1, 169–190 (2012)
26. Tsingas, V.: The photogrammetric impression of the rock and the acropolis walls. In: Proceedings of "Contemporary Technologies in Acropolis Restoration 19 March 2010", News from the restoration of the Acropolis monuments, 10 (2010), 20–23 (2010). http:// www.ysma.gr/static/files/Newsletter_10_gr.pdf. Accessed 27 July 2018

The Temple of Apollo at Corinth. Observations on the Architectural Design

Dimitra Andrikou[✉]

7 Notara Str., 106 83 Athens, Greece
dimitra_andrikou@yahoo.gr

Abstract. From the second half of the 18th century onward there are detailed measurements and architectural drawings of the surviving parts of the 6th century B.C. temple of Apollo at Corinth. Scholars who studied and/or excavated the temple propose different dimensions regarding the stylobate and intercolumniations. This is due to extended material loss and disturbed axes of the standing columns. The height of the temple, as well as the dimensions of various architectural elements of the building, have not been estimated. Moreover there are different views and interpretations regarding parts of the cella. In this paper the results of previous research and scholarship are reexamined, analyzed and discussed in view of new observations and measurements conducted by the author on the field in 2007. The purpose of this work is to elucidate representation issues and explore whether there is surviving evidence or fragmental data that could extend our knowledge on parts of the temple and missing elements of the plan and elevation. One of the novel contributions of this study is to present a graphical reconstruction of the facade and side of the temple of Apollo at Corinth, as well as a reconstruction - to the extent possible - of the pronaos and longitudinal section. Finally, a statement of significance and a synopsis of the outstanding values of the temple of Apollo is attempted.

Keywords: Apollo · Archaic Doric Temple · Corinth

1 Introduction

The ca 560 B.C. temple of Apollo at Corinth is founded on the natural bedrock on a limestone ridge to the north - northwest of the Agora of Corinth. The temple has been excavated, studied and published by Dörpfeld [1] and the American School of Classical Studies at Athens [2–13]. From the various dates proposed for the chronology of the temple, ranging from 570–560 to 550 B.C., Bookidis [10] proposed the early third quarter of the 6th century B.C. as the most objective date based on ceramics from the poros chips layer formed during the construction of the temple. In a later publication Bookidis and Stroud [13] refer to the temple as of ca. 560 B.C. Winter [9] dated the terracotta roof at 550–540 B.C. and notes a possible beginning of construction ca. 560 B.C.

In the present state a small part of the temple stands still to the level of the architrave. Small scale restoration works took place since the first excavation. Dörpfeld undertook a reinforcement of the krepis at the expense of the Greek government. A reinforcement of the stylobate, krepis and restorations of members of the elevation, a column capital and

© Springer Nature Switzerland AG 2019
A. Moropoulou et al. (Eds.): TMM_CH 2018, CCIS 962, pp. 22–35, 2019.
https://doi.org/10.1007/978-3-030-12960-6_2

architraves took place during the years 1906–1907, 1937–1938 and after 1940 [14]. So in the present state parts of the stylobate and krepis are concealed. The restored part of the krepis follows the dimensions proposed by Stillwell [3].

The axes of the standing columns are disturbed. This is evident at first observation. The interaxial spacings of the five standing columns of the west facade measure - from north to south - 4.01 m, 4.03 m, 3.90 m, corner 3.84 m (measurements conducted by the author in 2007). For the south pteron the interaxial spacings are 3.82 m, corner 3.50 m. Powel [2] and Stillwell [3] published a chronicle of the history and adventures of the temple since 1436 based on accounts of travelers, depictions and drawings. They also note earthquakes that could have affected the state of the columns. The disturbed axes could also be the effect of other acts of destruction during the long life of the temple. Indeed the position of fallen columns of the south and north pteron shows that both north and south columns did not collapse towards the same direction, as one would expect in the event of an earthquake. This indicates a designed destruction or a case on an explosion. Powel [2] and Stillwell [3] mention acts of destruction, such as the one that reportedly took place around 1745 when columns were to be thrown down by the Turkish owner to provide material for the building of his villa inside the temple. Indeed at the drawings of travelers a dwelling expanding through the years (1751, 1755, 1776, 1801, 1816) is depicted inside the temple [2, 3]. Stuart and Revett [15] also noted that columns have been blasted into fragments by gunpowder. Such destructions and reuse of the building material resulted in the diminution of the standing columns from twelve until the last quarter of the 18th century to seven sometime between 1776 and 1801. Moreover, there is an extended loss of the stylobate and krepis material and from the walls of the temple only the foundation trenches cut into the bedrock survive.

2 Study

2.1 Stylobate Width

The exact width and length of the 6 × 15 column stylobate can not be defined with accuracy, but only estimated approximately. To calculate the width of the stylobate we need the "apparent column diameter", the initial interaxial spacings and the stylobate face to column distance. Stillwell [3] notes that the stylobate face to column distance seems to be 0.07 m. I was able to measure this distance only at the west face of the stylobate of the third column of the west facade. However, it could be some centimeters more, as the axis of the columns are now disturbed and the stylobate of the corner column concealed by the restoration. The "apparent diameter" of the column is 1.72 m. We will call "apparent diameter" the distance of two diametric acnes as it appears to be when the column is viewed from the front. This differs from the column diameter which is the distance of two diametric ances (in this case 1.74 m). According to the results of previous research the intercolumniations of the five standing columns of the west pteron are as follows. Blouet [16] notes central interaxial spacings 4.00 m, corner 3.85 m, (hence stylobate width 21.56 m). Dörpfeld [1] notes 4.00 m, 4.02 m, 4.00 m for the three central spacings, corner 3.70 m, stylobate width 21.36 m. Powell [2] notes "spacing of the columns from axis to axis is" 4.00 m, 4.02 m, 4.00 m, corner 3.70 m. Stillwell [3] notes 4.00 m, 4.02 m, 4.00 m, corner 3.85 m (for the corner he takes Blouet's measurement). Hence stylobate width 21.58 m. Of all the above suggestions, the stylobate 21.56 m is the most commonly used in bibliography regarding the temple of Apollo at Corinth. There are also other proposals such as stylobate width 21.49 m [17].

In the present state (author's measurements in 2007) the interaxial spacings are - from north to south - 4.01 m, 4.03 m, 3.90 m, corner 3.84 m. The average of the three central spacing is 3.98 m. Similarly, the architrave length is min. 3.95 m - max. 4.00 m, 3.98 m in average (Stillwell [3] gives 4.00 m, Powel [2] 4.00 m, Shaw in the drawing nr 051025 [18] notes 3.95 m). I believe that the guide-marks [12] on the bedrock of the foundations provide evidence for the intercolumniations. The guide-marks were discovered by J. W. Shaw in 1970 while Robinson was excavating on Temple Hill and are noted on the drawing of the plan of the temple at the publication of Pfaff [12]. They are engraved at the bedrock trenches of the foundations of the cella walls, as well as at the block beneath the stylobate of the fifth column of the west facade. Pfaff [12] noted that they are placed accurately among parallel trajectories and served as guides for laying out the foundations of the temple. The distance between the marks indicates that the pronaos and opisthodomos correspond to the axis of the second and fifth column of a facade with cental interaxial spacings of 3.96 m - max. 4.00 m. It should be noted that the axes could correspond to the stylobate of the porches or to the anta or but a few centimeters to the anta as is the case at the very similar temple of Apollo at Syracuse, the colony of Corinth.

Since the fourth spacing should be corrected from 3.90 m to max. 4.00 m, a corner spacing of max. 3.74 m arises (instead of 3.84 m as is at the present state). This is very close to 3.70 m noted by Dörpfeld [1] and Powell [2] and to the ca 3.67 m length of the corner antithema of the corner architrave, according to the drawing of Shaw [18]. Hence the corner interaxial spacing would be approximately 3.70 m - max. 3.74 m. From the above data the stylobate width could be estimated as max. 21.34 m (Fig. 1). Dörpfeld's and Powell's suggestions are in line with the above rational, instead of the 21.58 m most commonly used in bibliography. Another observation that supports the proposed width is the following. At the northeast corner of the temple is in place a marble plaque from the roman temenos pavement [5]. The plaque bordered the euthynteria of the Roman krepis of the temple [12]. The Roman euthynteria, survives formed on the rock, at the southwest corner. The perpendicular distance between the euthynteria and the plaque, the euthynteria to stylobate distance deducted, points towards the proposed stylobate width.

2.2   Stylobate Length

Three columns from the south pteron survive standing. Blouet [16] notes the intercolumniation 3.71 m, corner 3.61 m, (53.60 m stylobate length is assumed). Dörpfeld [1] notes spacing ca 3.70 m, corner 3.48, stylobate length 53.30 m. Powel [2] notes the same as Dörpfeld. Stillwell [3] used the length of the architrave, 3.75 m, for the intercolumniation and the existing spacing, 3.48 m, for the corner (stylobate length 53.82 m). The present state measurements are 3.82 m, corner 3.50 m. The architrave lengths on Shaw's [18] drawings are ca 3.72 m, corner 3.62 m. Given the fact that the spacings of the west pteron are close to the length of the architraves, we could then use

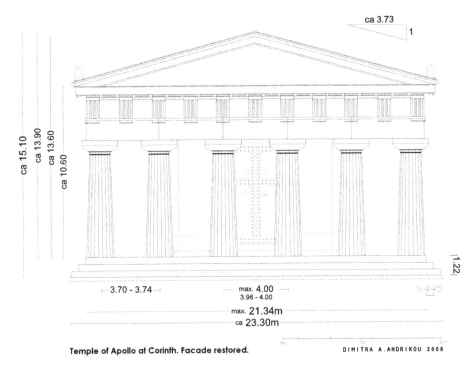

Temple of Apollo at Corinth. Facade restored.   DIMITRA A.ANDRIKOU 2008

Fig. 1. Temple of Apollo at Corinth, graphical reconstruction of the façade

the length of the architraves as indicative for the south pteron. According to this the 3.82 m spacing should be corrected to ca 3.70. Subsequently the corner 3.50 m should be corrected to ca 3.60. From this we get stylobate length ca 53.46 m. Alternatively, with spacings 3.72 m and 3.62 m add up to max. 53.74 m stylobate length. However there is not enough evidence to support a conclusive answer.

2.3   Krepis

Stillwell [3] suggests a four-stepped krepis, (stylobate height 0.44 m, steps height ca 0.36 m, ca 0.59 m wide) and an euthynteria. This wide krepis has proportions rather strange for an Archaic temple. I examined the stepped bedrock at the west facade and propose steps ca 0.44 m wide (Fig. 3). Although a four-stepped krepis appears at the earlier temple of Apollo at Syracuse, it may be that the temple at Corinth should probably be restored with a three-stepped krepis. This is because a fourth step would cover the scaffolding area in front of the west facade. There are rectangular scaffolding holes cut into the bedrock [3]. While discussing the shape of the worked bedrock, Manolis Korres made an observation that provides a better explanation of the initial plan. The height of the stylobate is ca 0.43 m. Stillwell notes 0.44 m to 0.42 m, a curvature of 0.02 m. The curvature [2, 3, 11] of the krepis seems to have been introduced at the second step, the ends measuring 0.38 m, the center 0.405 m. A third step ca 0.38 m is expected. However, the bedrock is instead cut into two parts, namely

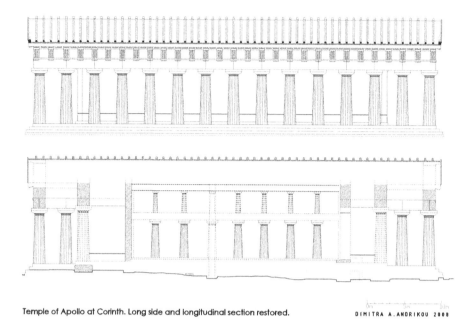

Temple of Apollo at Corinth. Long side and longitudinal section restored.   DIMITRA A.ANDRIKOU 2008

Fig. 2. Temple of Apollo at Corinth, graphical reconstruction of the long side and longitudinal section.

0.215 m and 0.17 m height, 0.385 m total height (Fig. 3a). This indicates that the bedrock has been reworked. It seems that initially a bedrock step, 0.385 m in height, formed the third Archaic krepis step (Fig. 3b). At a second phase, probably as part of the Roman renovation of the temple, the bedrock was reworked so as to receive smaller steps more fitting to the roman custom (Fig. 3c). Pfaff [12] notes the existence of Roman tooling on euthynteria blocks of the west and south side. In front of the third step are three rectangular holes for scaffoldings, cut in the bedrock among the same distance the one form the other survive. However, at least a fourth hole is expected at the northwest corner in order for the scaffolding system to be sufficient for the construction of the temple [19]. During the roman renovations the rock was lowered towards the northwest corner. This probably cleared off the now missing fourth rectangular scaffolding hole, as well as other scaffolding holes and traces that were above the level of the roman temenos floor.

2.4 Elevation

The total height of the temple is not known. The column height is 7.24 m. For the epistyle height Powel [2] gives 1.32 m, Pfaff [12] 1.34 m (1.215 m up to the regulae level, ca 0.115 m height of regulae). The regulae, guttae and taenia, as well as a triglyph fragment [3] support the existence of a Doric frieze with triglyphs and metopes. Pfaff restored the epistyle 1.337 m, the frieze ca 1.31 m, the triglyph height 62–63% of the frieze height, the metope height 85–86% of the frieze height [12]. Stillwell notes that the dimensions of the Archaic temple of Apollo at Delphi and the

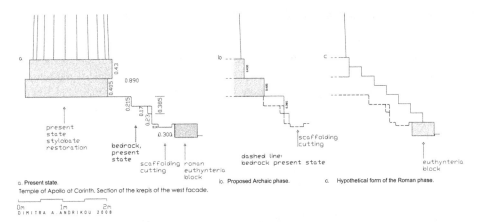

Fig. 3. Temple of Apollo at Corinth, section of the krepis of west facade: a. present state, b. proposed Archaic phase, c. hypothetical form of the Roman phase.

temple at Corinth present similarities [3]. Moreover, it seems that the two temples share similar proportions: number of columns: Delphi 6 × 15, Corinth 6 × 15, axial spacing/lower diameter: Delphi 2.28, Corinth 2.29, upper/lower diameter: Delphi ca. 0.73, Corinth 0.74; epistyle height/axial sxacing: Delphi 0.345, Corinth ca. 0.334; triglyph/metope width: Delphi estim. 0.70, Corinth 0.70; triglyph width/halph lower diameter: Delphi 0.94, Corinth 0.95. To estimate missing members I used the proportions of Delphi: frieze height/axial spacing: estim. 0.33; taenia height/regula height: 1.142; triglyph mutule/metope mutule: 1.00. Using these analogies, the regula height, measuring 0.115 m, could give ca 0.13 full taenia height. This restores the total epistyle height as ca 1.345 m. The triglyphs and metopes are 0.83 m and 1.17 m respectively (for the long sides 0.747 m and 1.12 m respectively) [3]. The height of a rectangular front metope would be ca 1.17 m, the frieze height ca 1.32 m. A large number of guttae from the site constitute the only surviving part of the horizontal geison. They provide information on the diameter, height, pitch and distance from the edges of the mutule [3]. Based on this data and on the proportions and form of horizontal Doric geison from Archaic Corinthia [12] a horizontal geison of total height ca 0.70 m, with three rows of six guttae each could be restored.

For the raking geison there is no evidence. No block of the gable tympanum survived. So, the angle of the gable - hence the total height of the temple - have not been estimated. From the tilled roof [9] one fragment belongs to the top kalypter. I measured the kalypter angle ca 154°. This should be indicative for the gable angle, which would possibly be slightly less. So assuming ca 153° for the central tympanum angle we get ca 13.5° for the corner angle, that is 1:4.16 slope (Fig. 1). Although the Archaic gable slopes range [12] from 11° to 17.10° it should be noted that the proposed slope is close to the 13.52° of the Alkmaionid Temple of Apollo at Delphi, a temple with proportions similar to the Temple of Apollo at Corinth, as already mentioned. From the above, the stylobate to gable height of the temple of Apollo, the raking geison included, could be estimated ca 13.60 m (Fig. 1). The elevation of the facade would appear more complete with acroteria, as according to Bookidis study acroteria in the form of sphinx can be associated with the temple of Apollo [10].

2.5   Pronaos and Opisthodomus

From the opistodomus some blocks of the stylobate substructure survive in situ [1–3]. From the columns in antis only one of the opisthodomus survived until the last quarter of the 18th century A.D. [2, 3]. Therefore, any depictions or drawings of the temple predating the destruction of the column are to be studied carefully. Stuart and Revett published an architectural drawing of the long side and depictions of the temple [15]. Also there is a depiction after Le Roy [20]. As Prof. Manolis Korres observed both publications show the column of the opistodomus being of the same dimensions with the pteron columns and standing at a higher level, ca 0.35 m above the pteron stylobate. This dimension is assumed by Stuart and Revett's drawing of the long side. It is fortunate that the higher level of the column appears in both publications. It could, therefore, be accepted as accurate. For the pronaos, given the absence of data, we could use the data coming from the opisthodomus (namely 1.66 m lower diameter, 1.64 m apparent diameter, 7.24 m height) (Fig. 2).

For the antae and the walls of the cella the guide-marks engraved on the foundation bedrock could be, I believe, elusive. These guides, discussed previously in this text, show the position of the cella long walls. The marks for one long wall give a width ca 1.24 m. It's too wide for the long wall of the cella. So it probably refers to the width of the wall, the projection of the anta included. The measurement fits to a wall ca 0.80–0.90 m wide (similar to the walls of the temple of Apollo at Syracuse.) with an anta projecting ca 0.20 m–0.30 m. There is one block surviving from an anta though not in situ [3]. I believe that the height of the block 0.985 m provides evidence for the height of the orthostates of the walls.

From the pronaos door only fragments of a threshold survive [3] (Fig. 4). The threshold was composed of two long blocks attached with an anathyrosis (taenia height 0.106 m) at the upper edge. On the upper surface cuttings show the position of wooden jambs, door pivots and the latch attached to the left wooden door leaf. There is also a doorstop (its height is 0.085 m) formed at the upper surface bordering with the rectangular pivot cuttings. The latch cutting is expected to be off centered. The width of the door, ca 3.03 m, can be estimated using the jamb to latch cutting distance that is 1.43 m. Based on these data and on Doric door proportions and form [21] a two leafed door, ca 6.06 m height, ca 0.12 m thick, hanging on metal sheathed pivots can be restored. Metallic nails attached to vertical stiles, top, middle and bottom horizontal rails would form each leaf. The threshold fragments reveal the width of the door wall, 1.36 m, thick enough to bear the thrust and forces caused by the weight and movements of the door (Figs. 5 and 6).

2.6   Cella

Inside the cella, the architrave or most probably the stylobate of the internal colonnade would serve as doorstop for the pronaos door, preventing its leaves from widely opening. The door would be about the height of the architrave of the internal colonnade (Fig. 2). Parts of the monolithic internal columns and capitals rest today at the South Stoa. The capital, published by Williams, [7] and the column shafts are similar to the external columns of the temple. Indeed the internal column appears as a scaled

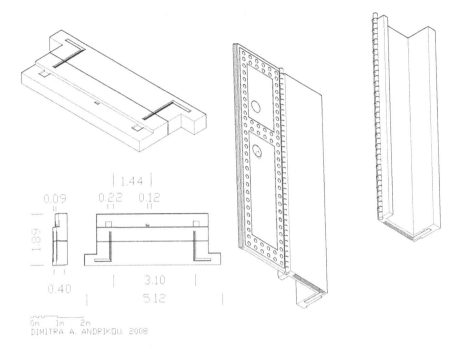

1.44

0.09   0.22   0.12

1.89

0.40

3.10

5.12

0m   1m   2m
DIMITRA A. ANDRIKOU. 2008

Fig. 4. Temple of Apollo at Corinth. Threshold restored, axonometric view, plan and section. Pronaos door and door jambs restored, axonometric view from below.

reproduction of the external one. I came to this conclusion when I scaled the external column to fit the dimensions of the capital published by Williams. This produced a column with height 4.954 m and lower diameter 1.177 m, that is identical to the actual diameter of the existing fragmental column shafts. With an additional ca 1.00 m for the architrave and some extra centimeters for the stylobate, we get a total height ca 6 m, which is very close to the proposed 6.06 m height of the door. The above calculations prove that the surviving threshold matches the overall dimensions of the Archaic temple. The threshold is of pentelic marble and could represent a renovation phase. Therefore it seams that during this renovation there were no severe alterations in the general dimensions of the door. It is possible that during the Roman renovations the cella was changed from a secos composed of two colonnaded rooms, into, most probably, a single large column-free room [7].

The cella was divided by two internal colonnades probably two storied. There is a transverse foundation trench cut on the bedrock, same in technique as the other trenches for the long and transverse walls of the temple. Dörpfeld [1], Powell [2] and Stillwell [3] restored the ground plan of the temple with a transverse wall within the cella. In the western part of the cella some foundation blocks survive, usually interpreted as the first foundation course of a statue base (Fig. 2). Pfaff [12] considers these blocks to be Archaic on grounds of tooling and finishing but questions the date of their use in their current position. He also sets the question whether the division of the cella was a Roman alteration, noting that he is inclined to accept the traditional view with the cross-wall in

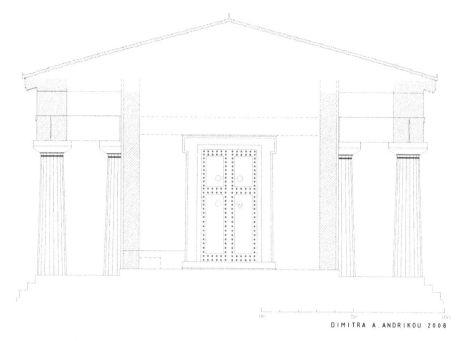

Fig. 5. Temple of Apollo at Corinth, restored section of the pronaos.

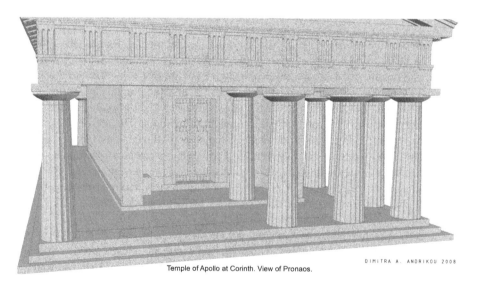

Temple of Apollo at Corinth. View of Pronaos.

Fig. 6. Temple of Apollo at Corinth, perspective view of the proposed graphical reconstruction of the pronaos.

the Archaic plan although the evidence is not decisive. Bookidis and Stroud [13] accept the existence of an Archaic wall forming a western and eastern chamber. During the examination of the transverse bedrock trench I observed that there are guide-marks engraved on the bedrock of the same kind same as the other guide-marks for the long and transverse walls of the temple. The mark has a north-south direction and was therefore intended as a guide for a transverse element of the plan. This provides an evidence for an intended transverse construction dividing the cella of the Archaic temple into two chambers as depicted at the plan of Dörpfeld [1] and Stillwell [3].

The western chamber has been subjected to different interpretations. Dörpfeld [1] initially thought that this was a case of a large treasure room complementary to the cella or a type of a cella with adyton. However he accepted the solution of an eastern and a western cella only because a transverse wall just at the east of the statue base would not allow an entrance to this room by a central opening. Stillwell [3] notes the possibility of a treasury instead of a cult statue base. The deep opisthodomus of the temple could cover the increased needs for a treasury, as in the case of the Archaic temple of Apollo in Delphi, where a great offering base has been identified. Bookidis and Stroud [13] concluded by literary evidence that a temple of Apollo with a treasury existed in the time of Periander and proposed the attribution of this function to the western chamber of the Archaic temple.

The restored plan for the temple of Apollo shows a double temple, namely a temple with double cella (eastern and western). Bookidis and Stroud [13] thoroughly examined all the evidence concerning the dedicated deity and concluded that this is a temple dedicated to Apollo and not a double deity. Moreover on the double temple type in Greek architecture Prof. Korres [23] argues that this is a feature of Parthenon and other Acropolis temples and constitutes a temple type so far identified only on the Acropolis of Athens following a local Athenian cult tradition. According to the above, in order to restore the plan of the temple, it is more secure to follow the typology most frequently met in the temples of Apollo. In Hollinshead's [22] inventory we observe that Apollo arises as the most common deity to possess a temple with inner room. Temples of Apollo and/or Artemis, cover about the one third of this inventory. Also Korres [26] has argued that the adyton - frequently with side access or accesses - constitutes an architectural element of Apollo's temples, such as Delphi, Bassae (the Archaic and the Classic temple), Sikyon, Didyma, Metropolis, Kalapodi and temples at South Italy and Sicily [17, 24–28].

In the temple of Corinth no concrete evidence survives for the access to the western chamber. Stillwell [3] proposes, although not shown in his plan, openings on the transverse wall on the axes of the side aisles. It should be noted that there is a cutting on the bedrock for a foundation bordering the north wall of the cella at the east of the transverse foundation, but the purpose of this foundation is not yet defined. Another observation is that the respective foundation trenches for the transverse structure dividing the cella from the adyton bear a resemblance with a feature met at the temple of Hera in Olympia, Apollo in Bassae and Apollo at Sikyon with walls connected to or bordering with the internal column (Fig. 7). This similarity raises the question if this old Peloponnesian architectural feature, as described by Gruben [17], was applied in the temple of Corinth. For the graphical restoration of the longitudinal section (Fig. 2) that we propose an abstract depiction of the transverse element of the cella is employed.

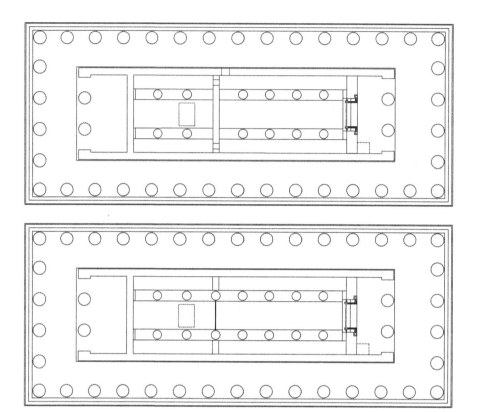

Temple of Apollo at Corinth. Possible versions of the plan.

DIMITRA A. ANDRIKOU 2008

Fig. 7. Temple of Apollo at Corinth. Possible versions of the plan. a. With a transverse wall with side accesses to the adyton. b. With transverse walls bordering the internal columns and a lighter construction dividing the adyton from the cella.

An adyton room would also provide the appropriate space for an oracular use, if there was one in this temple. Although an oracle here is not certain, Bookidis and Stroud [13] mention the connection between Corinth and the Oracle of Apollo at Klaros at the time of Hadrian. They also note that Robinson [6] linked the temple with a function of an oracle. The possibility of a deep or sunken adyton can not be excluded. There are ca 1.90 m between the estimated level of the cella to the level of the bedrock (Fig. 2). A case of a construction under the level of the stylobate survives also in the pronaos of the temple. It is a cyst, or depository for offerings. Its lower parts survive on the foundation bedrock, ca 2.20 m deeper than the pronaos level [3]. The architectural element of a crypt within the temple is also met in the temple of Zeus at the nearby Nemea. A sunken or underground adyton is met in the typology of Apollo's temples (Delphi, Didyma, the older temple of Apollo at Bassae and the temple of Apollo at Klaros). However, for the temple of Corinth, this remains a hypothesis. If there was a deep or sunken adyton, then the surviving blocks usually interpreted as part of the foundation of the statue base could be part of the crypt construction, or part of a little stair.

3   Significance of the Temple of Apollo

An integrated analysis of the architecture of the temple includes the evaluation of its significance. A synopsis of the outstanding values of the temple is attempted below. The temple of Apollo at Corinth is an emblematic monument for Doric Architecture. It bears the value of rarity and of archaeological testimony for architectural, aesthetical, constructional and technological issues. It is one of the oldest examples towards the application of the canonical Doric system in the Archaic world. Its monolithic columns present the most heavy proportions (namely the ratio column height/lower diameter) within Doric Temples in Greece, only second to the columns of the temple of Apollo at the colony of Corinth Syracuse. Also its columns are the oldest fully preserved standing Doric columns in Greece, only second again to the columns of the temple of Syracuse. The temple has the earlier known curvature at the krepis and entablature [3, 11]. Also it presents some of the earlier traces for the use of scaffoldings. To this should be added numerous holes on the bedrock around the temple that could be associated with the implementation of ancient technologies for surmounting heavy weights, such as pulleys or winches. Finally, the terracotta roof provides the earliest known example of the ovolo sima, which became the standard type for Corinthian roofs [9].

4   Conclusions

To conclude, the reexamination of the data and results of previous scholarship regarding the temple of Apollo and the observations and measurements conducted by the author on the site of the temple, along with a proportional analysis of the architectural elements of Archaic Doric temples resulted in the estimation and graphical representation of architectural features of the temple such as the krepis, the stylobate width and length, the intercolumniations, the entablature, the gable angle, the total height of the temple, the dimensions of the pronaos door and of the columns in antis, the dimensions of the cella walls. Also, observations on the foundations of the temple, the comparison with other surviving parallels and the consideration of the typology of Apollo's temples allowed the attribution of an inner chamber - probably an adyton - to the cella. There is no concrete evidence regarding the access to the adyton or the form of the transverse division of the cella. A transverse wall with side entrances on the axes of the aisles is the most probable proposal. However a lighter dividing structure, such as a grid between the columns of the internal colonnade, as well as transverse walls bordering with the columns, as in the case of Bassae or Sikyon, could also serve as an interpretation of the plan. Moreover the hypothesis of a deep or sunken adyton can not be excluded. In that case the blocks considered as foundation of the statue base could be part of a construction of this crypt, such as for a small stair. Based on the above results a graphical reconstruction of the facade, long side, longitudinal section and aspects of pronaos of the temple of Apollo at Corinth is presented.

Acknowledgments. This paper presents results of a research conducted under the supervision of Prof. Manolis Korres for the MSc thesis of the author (2008, School of Architecture, National Technical University of Athens). Results of this MSc thesis have been also orally presented in

2014 within the framework of the Lectures on the History of Architecture (National Technical University of Athens). I am grateful to Prof. Manolis Korres for his precious guidance. Our discussions on the subject and his observations substantially improved the results of my study on the temple. Any errors and/or misconceptions remain my sole responsibility. I would like to thank the American School of Classical Studies at Athens for granting me permission to study the remains temple and have access to the archives and storage areas of the ASCSA. I also thank the Ephorate of Antiquities at Corinth for granting me permission to study on the site of the temple.

References

1. Dörpfeld, W.: Der Tempel in Korinth. AM **11**, 297–308 (1886)
2. Powell, B.: The temple of Apollo at Corinth. AJA **9**, 44–63 (1905)
3. Stillwell, R.: The temple of Apollo. In: Fowler, H.N., Stillwell, R. (eds.) Corinth I.1, Cambridge, pp. 115–134 (1932)
4. Stillwell, R., Ess Askew, H.: The peribolos of Apollo. In: Stillwell, R., Scranton, R.L., Freeman, S.E., Ess Askew, E. (eds.) Corinth I.2, Cambridge, pp. 1–54 (1941)
5. Robinson, H.S.: Excavations at Corinth: temple hill, 1968–1972. Hesperia **45**, 203–239 (1976)
6. Robinson, H.S.: Temple hill, Corinth. In: Jantzen, U. (ed.) Neue Forschungen in griechischen Heiligtiimern, Tubingen, pp. 239–260 (1976)
7. Williams, C.K.: Doric architecture and early capitals in Corinth. AM **99**, 67–75 (1984)
8. Williams, C.K.: The Refounding of Corinth: some Roman religious attitudes. In: Macready, S., Thompson, F.H. (eds.) Roman Architecture in the Greek World, The Society of Antiquaries, Occasional Papers 10, London, pp. 26–37 (1987)
9. Winter, N.A.: Greek Architectural Terracottas, Oxford (1993)
10. Bookidis, N.: Corinthian Terracotta sculpture and the temple of Apollo. Hesperia **69**, 381–452 (2000)
11. Pfaff, C.: Curvature in the temple of Apollo at Corinth and in the South Stoa and classical temple of Hera at the Argive Heraion. In: Haselberger, L. (ed.) Appearance and Essence. Refinements of Classical Architecture, Philadelphia, pp. 113–125 (1999)
12. Pfaff, C.: Archaic Corinthian architecture. In: Williams, C.K., Bookidis, N. (eds.) Corinth XX. The Centenary, pp. 95–140 (2003)
13. Bookidis, N., Stroud, R.S.: Apollo and the Archaic temple at Corinth. Hesperia **73**, 401–426 (2004)
14. Mallouchou Tufano, F.: The restoration of ancient monuments in Greece 1834–1939, Athens (1998). (in Greek)
15. Stuart, J., Revett, N.: The Antiquities of Athens, vol. 3, London (1794)
16. Blouet, G.A.: Expédition scientifique de Morée, vol. 3, Paris (1838)
17. Gruben, G.: Heiligtümer und Tempel der Griechen, Athens (2000). (in Greek)
18. Corinth Excavations. http://corinth.ascsa.net. Accessed 4 Feb 2017
19. Korres, M.: From Penteli to Parthenon, Athens (1993). (in Greek)
20. Le Roy, J.D.: Les ruines des plus beaux monuments de la Grece, vol. II, Paris (1770)
21. Korres, M.: Study of the Restoration of the Parthenon, vol. 4, Athens (1994). (in Greek)
22. Hollinshead, M.B.: "Adyton", "Opisthodomos", and the inner room of the Greek temple. Hesperia **68**, 189–218 (1999)
23. Korres, M.: The architecture of Partenon. In: Tournikiotis, P. (ed.) The Parthenon and Its Impact in Modern Times, Athens (1994). (in Greek)

24. Krystalli-Votsi, K., Østby, E.: The temples of Apollo at Sikyon. In: International Congress of Classical Archaeology Meetings Between Cultures in the Ancient Mediterranean, Bollettino Di Archeologia, Roma, pp. 54–62 (2010)
25. Hellner, N.: Räumliche Führung am Beispiel der spätgeometrischen und archaischen Süd-Tempel von Abai/Kalapodi. In: Kurapkat, D., Schneider, P.I., Wulf-Rheidt, U. (eds.) Architektur des Weges gestaltete Bewegung im gebauten Raum: internationales Kolloquium in Berlin, Regensburg, pp. 289–307 (2014)
26. Korres, M.: The temple of Apollo in Eretria. In: Conference Eretria - Crossing Paths Swiss Excavations at Eretria (1964–2014), Athens (2014)
27. Intzesiloglu, B.: The Archaic Temple of Apollo at ancient Metropolis (Thessaly). In: Stamatopoulou, M., Yeroulanou, M. (eds.) Excavating Classical Culture: Recent Archaeological Discoveries in Greece, Oxford, pp. 109–115 (2002)
28. Amandry, P., Hansen, E.: Le Temple d'Apollon du IVe siècle Fouilles de Delphes II, Topographie et Architecture 14, Paris (2010)

Anastylosis of Roman Temples in the West End of the Roman Forum, as Part of a Management Plan of the Archaeological Site of Ancient Corinth

Evangelos D. Kontogiannis[(✉)]

NTUA, Athens, Greece
evagkon@hotmail.com

Abstract. After the end of excavations that took place between 1907 and the late 1940s' in the archaeological site of Ancient Corinth, it was revealed that the western boundary of the Roman Forum was organized by seven small temples, as opposed to the rest of the complex, which consisted of administrative and commercial buildings, giving a special character to the Forum. Although the temples [1] have not been fully identified, they belong to the characteristic examples of Roman architecture, and especially the ones of small temples placed on podiums or the circular monoptera temples, without an alcove like the one dedicated to the Cn. Babbius Philinus.

Presently, the absence of organized protection of the revealed architectural parts of the temples, makes them particularly vulnerable to natural deterioration. As a result, they are becoming less recognizable and thus non-visiting, although they are placed on the main course of the archaeological site to the center of the forum and the so-called step of Apostle Paul. The systematic depiction and study of the blocks that are preserved from each temple, and the current state of the West Terrace, is necessary in order to evaluate the available ancient material of each temple for a restoration proposal. The ultimate goal for the visitor is to understand the meaning of the site, the ancient route through the forum, the archaeological value of the monument and the scale of temples that previously did not exist [2].

Keywords: Ancient Corinth · West Terrace · Roman Temples

1 Ancient Corinth

1.1 Corinth in Roman Times

With the entry of Corinth in the Achaic Confederacy in 243 BC and its placement as a seat just after 200 BC, the city reached it's zenith of political power until 146 BC and it's breach with the Lacedaemonians and their Roman allies. The Roman troops under the Lucius Mummius orders were victorious against the state and conquered Corinth, where the soldiers of the Confederacy sought shelter. Ancient writers, such as Strabo, Polybius and Pausanias, speak of total destruction, smashing of artworks and devastation of all its inhabitants [3], destroying a continuous inhabitation since the 5th millennium BC according to archaeological finds.

© Springer Nature Switzerland AG 2019
A. Moropoulou et al. (Eds.): TMM_CH 2018, CCIS 962, pp. 36–58, 2019.
https://doi.org/10.1007/978-3-030-12960-6_3

After 111 BC [4] Corinth's land belongs to the Roman state and is taxed as public land (ager publicus). In 44 BC Julius Caesar reestablished the city as a colony of the Caesar and Augustus family (Gens Julia) with the establishment of Greeks, Romans and Liberals, and the new city was named Colonia Lous Julia Corinthiensis or Clara Laus Julia Corinthus or Julia Corinthus Augusta [5]. As far as urban planning is concerned, it was designed with horizontal and vertical axes, which consisted of isles in which public buildings and private monuments of wealthy Romans and Greeks were erected, as an emphatic expression of their presence. Many of the pre-existing public market structures were rebuilt in the same place during Corinth redeployment. During the period of 69–79 AD, Corinth is reestablished as a colony of the Julia and Flavia Family to which the new emperor (Colonia Iulia Flavia Augusta Corinthiensis) belonged and extends further to the wider area of the plains including the land of Isthmus. In the years that followed, the Roman emperors honored the city, that reached it's peak in the 2nd century AD, during which important public buildings were erected or renovated [5] (Fig. 1).

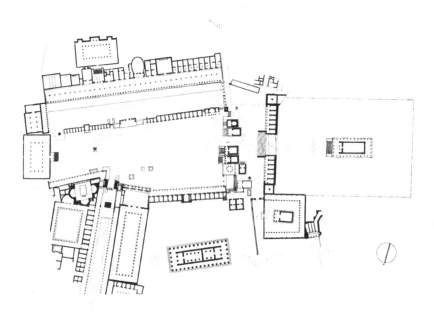

Fig. 1. Forum in Corinth. (2nd century AD). Graphic restoration of the fora with visitors' routes through the market. (Sketched by the author based on the existing bibliography)

1.2   The Roman Forum of Ancient Corinth

The center of the new city belongs to the market, which is known as Forum Romanum. With dimensions up to 200 × 100 m, it was placed close to the ancient Temple of Apollo, oriented East-West. The new Agora was divided into two levels with the medium part being defined by the Central Stores and the Roman Step (Rostra). To the east was the so-called Joulian Basilica, a building of the 1st century AD, probably dedicated to the worshipness of Gens Julia, the imperial family, while in the south behind the Step stood the South Stoa and the Southern Basilica (1st century AD). On the opposite side of the forum to the west, the boundary was defined by the West Terrace with the seven small dedicatory temples and then by the West Stores through which the visitor enters the precinct of the temple E. Peripheral Roman temple of the 1st century AD, surrounded by an early Ionic and later Doric order Stoa, was dedicated to Gens Julia and the imperial worshipness, while in another aspect Jupiter Capitolinus was worshiped in it. The main core of the city was joined by its two streams through Cardo Maximum to the south with the port of Kenchreai and to the north with the harbor of Lechaeum. Thus, on the northern side of the market, apart from the Northern Market, the Propylaea was also found with the triumphal arch and the street of the same name.

1.3   The Romans Temples at the West End

Although the research in the Ancient Corinth has been carried out since 1886 and 1892 [6], the systematic study has began with the excavation work by the American School of Classical Studies (ASCSA) in 1896. In the West Terrace, work began in 1907 with the discovery of Temple D by Oscar Broneer and continued with Temple F (Gladys Davidson). The rest of the Temples were excavated under the supervision of Robert Scranton, and in 1938 their organized study began. After the war, their registration was completed during the years 1946–1947 with the contribution of Ioannis Traylos and Elias Skrumpelos, who designed the final plans of the 1951 publication [7]. Thus, seven small temples have been registered on the formed opening and they were dedicated to the protective gods of the city who are now known by the following names (from South to the North): temple F or Venus Genetrix, temple G or Apollo Clarion, temple H or Hercules, temple J or Poseidon, the monoptera circular building of Corinthian rhythm, dedicated to the Greek descent duvir Cnaeus Babbius Philinus, the unknown temple K and the temple D or Mercury [8] (Figs. 2, 3 and 4).

Temple F
Built in the south of the Terrace, a four-columned Ionic building is accessed from the Market through stairs. Richly decorated temple with a statue inside it, dedicated to the corresponding honored god. Built after Temple G, as evidence proved by the way the foundations are constructed, with the stonework placed in trenches on the wall line, while the center of the temple rests on a stone base. With a three-terrain bastion, the stilobate ends up to a length of 5.60 m, while the walls have grown along the inner edge of the foundation, with a width of no more than 0.85 m. Many architectural parts are not saved from the Temple. Most interesting are the bases of the columns and the central block of the drum, by blue-white marble. In the center there is a raised, rough surface, circular in shape, surrounded by a series of arranged pins, possibly for fixing a copper shield or possibly a pendant of some kind while at the left end, the last three letters of early imperial times of the sign VENERI exist.

Fig. 2. Foundations of Temples H and J. Photo from ASCSA's archive, Corinth Excavations (Number of the image: bw4636, year: 1938).

Temple G

To the south of the corridor to the West Shops is placed Temple G [9]. With a strong stoneware measuring 8.50 m × 13.4 m, the foundation of the pronaos is formed by irregular porous volumes embedded in the stone, while there is no infrastructure for steps. Probably the temple originally constructed to occupy a small surface of Terrace, whereas the construction of Temple F in a later building phase made the building inaccessible, imposing the construction of steps. In a later period it was reconstructed, at least in part, by marble blocks. Some of the moldings appear to have been cut into reusable blocks of stone. The study of the capitals and the frieze is much more difficult. Two Corinthian capitals found near the temple are measured at 0.57 m in height and 0.55 m in diameter. The carving of them, carefully elaborated, lies stylistically between the Babbius monument and the work of the temples of Hercules and Poseidon. Thus it dates back to the end of the second century, a period that may coincide with the installation of the steps towards the Roman Market.

The Fountain of Poseidon

On the northern side of the corridor to the West Shops traces of the H and J temples were revealed, which belong to the last building phase of West Terrace. Originally in the same location there was an installation that hosts Fountain or Nymphaium. However, the building program for the construction of the temples led to the complete demolition of the existing structure, with the blocks that are now found to be reused and the temples have covered most of its foundations [10]. The fountain system is established by the slight relics of the drainage basins located below the temple J from white mortar and small pebbles. Traces at the foundations of Temple H and a mass of hard concrete similar to the tap in the northern part of the front foundation of the

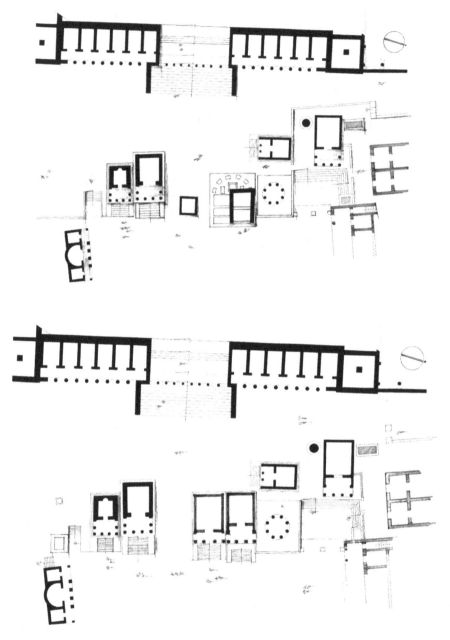

Fig. 3. The temples on the West End of Forum Romanum. The first image presents the first period (1st centuryBC–1st century AD) with the fountain in the center of the complex. The second image represents the last period (2nd century AD) which ends with the replacement of the fountain by building the temples H and J. (Graphic restoration of the terrace with visitors' routes through the temples. (Sketch of the author based on the existing bibliography-In the fountain's restoration they presented the two strong proposals for the monument)

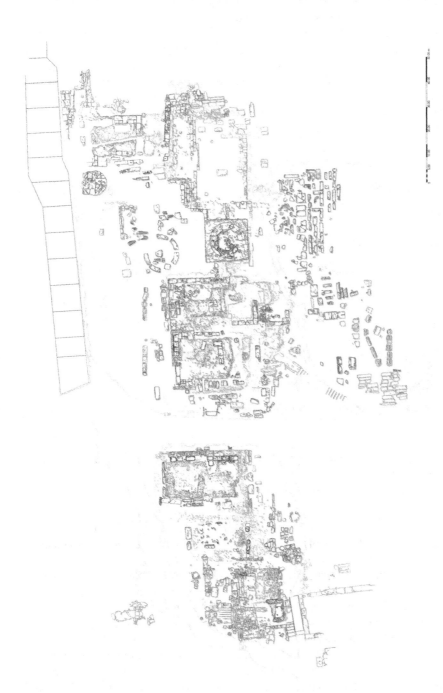

Fig. 4. The complex of Romans temples in the West End of the Forum. Graphic presentation of the present situation. (Drawing by the author for his diploma's project)

Temple contribute to the idea that the width was approximately 11.00 m [11], while its length, across the width of the structure, would be about 8.20 m. Regarding the origin of the water source for the Fountain there is no specific indication [12]. The form of the monument is not clear, although a female figure and marble bases of blue marble have been revealed with a dedicated inscription along with a dolphin and blocks of marble that mimic the natural rock. If the fusion complex is the same as that described by Pausanias [13], then in the Poseidon's fountain a dominant position in the center would have been captured by a bronze statue of this god with a foot on a dolphin, from whose mouth the water would run by filling its underlying basins (Fig. 5).

Fig. 5. Dolphin's statue with the base from Poseidon's fountain. The inscription refers the name of duvir Cnaeus Babbius Philinus (Image from author's archive)

Temple H
In the 2nd century AD, after the complete demolition of the Fountain-Nymphaium that was dedicated to Poseidon, the building of two temples took place, known as H and J [14], respectively occupying the relative position and extending to the south of the western boundary of the Roman Forum, limiting the initial passage to West Shops. Both temples, set on a foot, are of Corinthian style with a view that the visitor comes across through a large scale. From the foundations of Temple H (given by the name of Hercules) they are massive and solid, forming a base of dimensions 5.70 m in length and more elongated and higher than the floor of the Temple of Poseidon. It is noted that for the southern wall of the Temple of Hercules the seat of the wall occupies 2.00 m while for the northern wall it is only 1.00 m leading to the conclusion that the southern Temple is the last of the two temples since it is limited to the north, so it spreads freely towards the south. The remains of the foundations for the stairs are small, relatively to

the floor of the alcove set at approximately 2,70 m above the level of the forum. From the foot, three rows of porous blocks are retained on the west side, with the most remarkable feature of these pieces being that they are connected by double ties, some of which are kept in place. The final size of the temple reaches approximately 7,50 m width and 12,00 m length with an extra 4.60 m for the steps of the stair. The internal proportion of the cella will be different, with it reaching 6,00 m in width and 7,80 m in length.

Two marble fragments of the stylobate are retained, while the bases for the columns can be classified only temporarily. Additional red marble columns can be assigned for the twin temples of Heracles and Poseidon, partly because of the scale and partly because there are too many fragments of these columns to have been sent to any of the rest of the temples of the Terrace. However, capitals can be identified with a high degree of probability, due to the fact that a significant number was found in the same position as the other parts of the entablature. The most interesting of the architrave is the one on the southern front bearing an inscription with the name of the emperor Commodus that has been erased, and no other part of the frieze from the facade has been identified. From the pediment, a large part of the cornice is preserved at a height that can be attributed to the final shape, while many of the drum blocks have been identified. From the threshold there are two fragments that do not come into contact on their broken surfaces, while the central part, which has been reused upside down as a threshold in at least one medieval structure, is absent (Fig. 6).

Fig. 6. View of the foundation of Temple H with the remains blocks. Present situation (Image from author's archive)

Temple J
Although the foundation of both temples H and J occupies an area of approximately 16,50 m^2, the boundary between the two buildings is not easy to ascertain precisely,

but the differences in concrete and the traces of crossing in some positions conclude that the temple J (considered that it bears the name of god Poseidon) was first built and was about 7,50 m wide [15]. Concrete between the walls to provide foundation for the floor to the top of the foundation is estimated at a height of 2.10 m above the paved floor of the Forum. The foundations for the stairs have almost disappeared. The overall dimension of the temple, finally, is about 6.80 m wide and 12.00 m length, with an approachment about the stairs up to 4.50 m, while the cella would be about 5.50 m wide and 6.50 m length. The Temple of Poseidon offers fewer opportunities for restoration, as fewer architectural parts are preserved for which uncertainty is preserved because both temples H and J are in many ways identical. The difference in the original design is quite obvious, but the structural details are so similar that for a long time it was considered that all the blocks were from the same building. The frieze with the inscription from the right front decoration also carries part of the emperor Commodus sign as the left corner architrave of the Temple of Hercules, as the profile, style and general scale are similar. However, there are many small differences that are sufficient for scholars to accept that the second arched frieze belongs to Temple J. The design of their capitals is similar to that of Temple H. It is assumed that the temple of the Temple J abstained from its columns more distant than that of Temple H. Otherwise, the elevation of the two temples would be similar.

Tempietto of Babbius

The Babbius monument [16] was well known before it was discovered since many of its blocks, including part of the frieze bearing the name Cnaeus Babbius Philinus, were found in a second use in 1907 around the church of Agios Ioannis (12th century) built on the traces of the monument. In 1935 the foundations were revealed by the demolition of the narthex of the Christian temple. The Roman monument, a circular building with eight Corinthian columns, protruded from the main body of the blade, placed on a rectangular podium. The foundation for the actual tempietto is a core of solid concrete and broken stone 6,00 m wide on each side of the square and 2,70 m high above the market level. Taking into consideration the construction of the foot, numerous pieces of blue marble have been identified, whole and fragmentary 0.32 m in height, ranging from 0.30 m to 0.50 m in thickness and from about one meter in length to 1.75 m for the largest (broken) fragment. At the top there are traces of joint using. A series of blue marble posts were found in the same area. The preserved examples have a height of 1.30 m and about 0.30–0.35 m in thickness. Their thickness indicates that they could support the white marble coating, and the similarity of the joints and fins confirms the relation. The tips at the top indicate that they are attached not only to each other but that they are also blocked at the back, indicating that the support structure of porous material is possible. The crown element of the pedestal rose higher than the plane of the raised wall of the bend and perhaps continued as an extension behind the Babbius monument to describe a square in which the building was located.

Four blocks from the pillar are practically intact and there are fragments of the other four. It is made of blue marble with a similar construction and height, showing that the interior of the floor was made of lighter material, possibly mosaic. On the upper surface of the squares, we can trace parts of four column bases, approximately 0.75 m in diameter. The center of the column has been estimated to approximately connect with the joint between the blocks, so the area concealed is used for the joints joining the blocks and, of course, for the pins that secure the column bases from which we have only two fragments. Two well-preserved Corinthian capitals have been found, with a lower diameter of about 0.53 m and a height of 0.675 m. The finishing consists of a very low abacus measuring approximately 0.69 m^2 in the centers, not including rosettes.

Six of the eaves are kept whole, although in some of them the corners have been carved and broken. A fragment that includes more than half a seventh is also retained. The blocks are also white marble, and the construction is probably the best one found in the building. The length measured on the inner surface is approximately 1.45 m. The roof of the monument, probably monolithic, is preserved in white marble pieces cut along the wedge-shaped cross-sections of a cone surface. The upper surface is treated with a scale sculpting pattern reducing in size towards the top. The lower edges supported, as we have seen, the protrusion on the backward inner side of the shelves, with their upper edges meeting around a small circular opening in which there probably was a finishing that also served as a foundation stone. This finishing may have been done with plant decoration. There is, however, a marble cone that mimics the cone of a pine, which resembles both it's pattern, it's construction and it's material, as well as the tiles from the roof, and which may have served as its core (Fig. 7).

Fig. 7. View of the Tempietto's foundation with parts of stylobatis and its frieze. On the background the view of the south elevation of Apollo's temple. Present situation (Image from author's archive).

Temple D

It belongs to the oldest temples of the West Terrace. The four-columned temple is built of limestone during its first phase (1st century BC–1st century AD) and is placed on a plateau designated by the retaining wall of the Terrace near the northwest arcade. The entrance to the temple is through three stairs located only on the eastern side of it. The remains of limestone found on the perimeter can not be identified with the Temple, and even the foundations and the capital of the Tuscan rhythm that have been revealed, can not belong to it with certainty. The blocks are made of blue marble and the capitol decorated with four rosettes is inherent to the top of the body of the column. In the second phase (1st century BC), the temple becomes a tetra-styled Ionic temple, while the walls of the altar are coated with thin marble. Columns, architraves, pediments and drum are similarly made of marble. The level south of the temple is elevated and a large circular base is added to the southwest side for the worshiping statue, which is at the same level as the alcove floor. The remains are now more, however, no significant conclusions come out, with the exception of the frieze. The eastern section in front of temple D in the second building phase is downgraded to a level of about 1.80 m from the level of the Terrace floor, forming a platform with which one goes to the Temple through marble stairs. Three phases are identified which relate to the variation of the dimension of the plateau according to the interventions on the stairs starting initially with a dimension of 12,00 m × 7,00 m and resulting to the dimension of 10,00 m × 2,50 m in the form of being simple, without shafts and stairs made of marble and the blue marble floor, which was coated with white.

Temple K

The position of Temple K is located at the western boundary of the Terrace. From the finds of the excavations, the foundations of the building have been revealed, consisting of concrete layers for the walls, the stilt and the steps of the foot. The axis of the foundation is estimated at 11,80 m in length and 7,30 m in width, while the height for the basement according to the estimated foundation depth would reach 1,50 m. The main feature of the sanctuary is East-West orientation which is probably due to the lack of sufficient space based on the building program on the eastern boundary of the complex. From the foundations, though, it seems to be interconnected with the foundations of Temple J, but without sufficient information for the way it's done. However, the evidence of how it was built and connected to the neighboring building indicates that it has preceded it [17] (Figs. 8, 9 and 10).

2   The Roman Temples of the West End of the Roman Forum and Their Role in the Cultural Heritage Management Plan of Corinth

2.1   The Management Plan

From 1947 until today and after the relative Scranton's [18], publication in 1951, the studies on the Temples of West End of Roman Forum and generally on the western boundary, are focused on the rendition of temples and the divinity to which they were dedicated. No new excavations have been made in the archaeological site itself. Instead, there is an attempt to assign architectural parts and redefine the building phases of the temples, even their form. Thus the visitor today has approximately the same image as if he was visiting the archaeological site in the middle of the last century.

Fig. 8. Foundations of Temples D and K. View of the lower level of Forum from west. Photo from ASCSA's archive, Corinth Excavations (Number of the image: bw4634, year: 1938).

Instead, what changes dramatically in the archaeological site is it own presence. After so many years without any restoration program or restoration work, the Temples give the impression of abandonment. With the attraction of the visitor to the Apollo archaic temple and the Rostra that is connected with Apostle Paul (1st century AD), all the other buildings, including the temples of the West Terrace, are often passed by without being given the proper attention. In addition, the architectural blocks in their majority, are located in situ in the archaeological site, in front of the each temple, at the western edge of the plateau of the Forum. Thus their prolonged exposure to the environment causes alterations in their structure and form. The same thing applies to the stone-footing of foundations where native vegetation, whether small or in the form of trees, makes them more vulnerable. Finally, the development of trees around the forum contributes to the misconception of the forum, although they are often staging points especially in the summer months.

The state of the western boundary of the Ancient Corinth Forum is located not only in the specific area but also in the wider area of the Roman market, at Lechaion Road in the northeast of the area or Temple E and the Conservatory and Theater in the West and Northwest Archaeological Site. Individual rehabilitation or restoration efforts at each monument or in each section do not contribute to the implementation of a wider, larger scale program, as there are no basic principles of organization in size not of a unit, but as a whole that includes the study of urban elements or residential grids in historical landscapes. Thus, in December 2015 [19], the "Plan for the Management of the Cultural and Natural Heritage in the Greater Archaeological Site of Corinth" is being developed. A study that attempts to capture and formulate views and suggestions on how to organize the wider archaeological site of Corinth, within a wider context, despite the spatial discontinuities found in the region (Fig. 11).

Fig. 9. The complex of Romans temples in the West End of the Forum. Eastern elevation. Graphic presentation of the present situation. (Drawing by the author for his diploma's project)

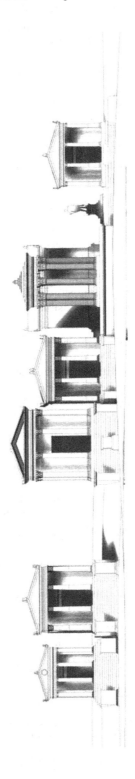

Fig. 10. The complex of the romans temples in the West End of the Forum. Eastern elevation. Graphic restoration. (Drawing by the author for his diploma's project)

2.2 Managing the West End of Roman Forum-Anastylosis Program for the Temples

In the context of this diploma thesis, the study focuses on the complex of temples across the West End's Terrace in order to maximize the readability of the monument and its importance, as it is a special feature of Roman Corinth. So the western boundary of the Roman market is defined not only by shops or administrative sites, but also by worship buildings. As a mild scenario, it is proposed that the archaeological site and the route among the monuments to be essentially unchanged. The interventions are of a rescue nature and are designed to help the visitor to inform the building site of the West End plateau. In contrast to the above, a more dynamic scenario consists of a different way of approaching the monument. It is well known that the route that was followed in Roman times is exactly the opposite of today, with the entrance to the city located to the north of the present archaeological site on the road (Cardo maximus) that leads the Roman Corinth to its harbor in Lechaion. A course that goes hand in hand with the ancient will dramatically alter the way the Forum and the temples' monuments approach the western boundary, gaining special weight within the space without requiring any special changes to the standing structure of the archaeological site.

Additionally, this proposal contains the possibility of the restoration both of the Temple H and the Tempietto of Babbius. The two buildings of West Terrace complex are the only ones that allow us to have a fuller picture of the size of the temples, as they preserve a sufficient number of blocks, allowing each monument to become an autonomous and self-contained entity with the possible addition of new material where it is required. The remains also determine the value of the restoration work, since they are mostly portions of the entablature, columns, capitals and not the alcove's body. Thus, although it is not easy to recover the whole of the temple, the nature of the building is understood by the visitor, since the part that is going to be restored defines the character and it's use, which in the particular case of the temple H is the extension and not the cella, or the wing in the case of the Tempietto.

As a result of this, the attempt of restoration with the remaining blocks is to bring emotional value to the visitor, allowing him to realize that in the place of the current ruins, there was once a temple complex, which reinforces the feeling of sanctity of the place and especially on a flat roof dominated by buildings of commercial and administrative nature. Additionally, it contributes to the scientific and historical value of the monument, because it allows the visitor to understand the entity of the site, the use and the important role of Architecture of Roman Temples. Also, the addition of new material reinforced the structural stability of the particular monument, since the area of Corinth from antiquity recorded intense seismic activity.

For the study of the Western Boundary complex, the relative restoration study was preceded by the systematic mapping of the standing condition of the plateau and the architectural members. The complexity of the West End building program, its size, the different levels of the floor of each temple in relation to the level of Terrace with the floor of the Forum, as well as the poor state of their structure, necessitated its study through different means each time. The basic method of surveying was made by topographic and photogrammetric methods, while photography and aerial photography took place. This was followed by the production of a three-dimensional digital model

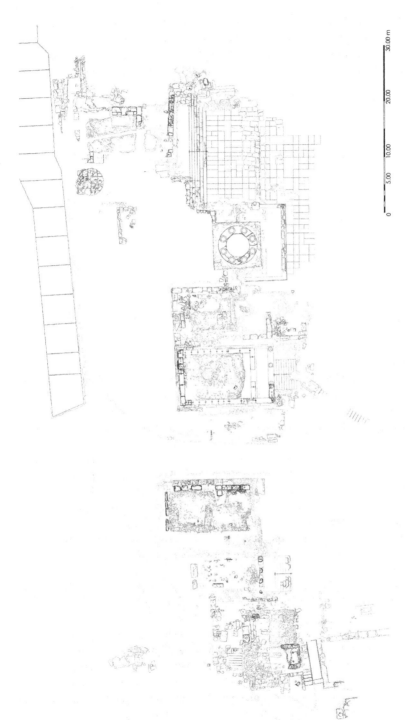

Fig. 11. The Romans temples in the West End of the Forum. Graphic restoration of the proposal (Drawing by the author for the diploma's project)

and the orthophotos, while at the same time the architectural members were recorded in order to gain a better understanding of the whole area and the way of organizing the archaeological site. The final design product, through a cad program, gave the form of the current situation, and it was the design infrastructure for the following proposal of restoration (Fig. 12).

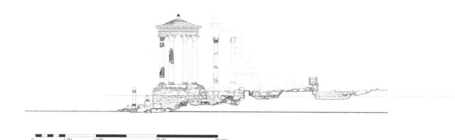

Fig. 12. Tempietto of Babbius and the temple H. Graphic restoration of the proposal. Elevation from the North side. (Drawing by the author for the diploma's project)

Temple H

Concerning temple H, a number of blocks have been identified in the southwest corner of the temple and in the west, so that its height and layout are recognizable. However, since the number of these blocks is limited, it is suggested to place artificial stones in the dimensions given by the graphic representation [20] which can easily be carved and imitated in the local limestone. Their dry conjunction will be made with tractors that will be held in pairs of two, giving the ability to create a sufficient bearing surface for the supernatural architectural parts. This solution has already been applied to the Propylon of Hadrian's Library, where it presents the same peculiarity about the base and the steps that lead to it [21]. Regarding the access of the temple, all capitals are located in situs, while parts of the columns are located to the east of the temple. From the way of placement, it is obvious that there's an attempt to put the broken members in place in order to give the final height of the layout. In the same space, the parts of the pediment of the temple, which give us the size of the temple, are placed in the same space, while members belonging to the alcove's door are also preserved. From the architraves, the corner is preserved on the southeastern side of the stand and one on the southern side belonging to the vault. Thus, in the end, the image of the porch and its continuation to the west side of the temple can be attributed.

Therefore, it is proposed to place the stones in their original position and height, while in the absence of architectural material, an artificial stone would be placed with a processing relative to that of the foot. Between the new material and the ancient material, a sheet of lead will be added in order to distinguish the material change. Finally, there will be holes in the columns, so that the parts will be shuffled, where titanium bars will be placed as a connection to each other, as well as the new material where necessary, which will be marble relative to the original (Fig. 13).

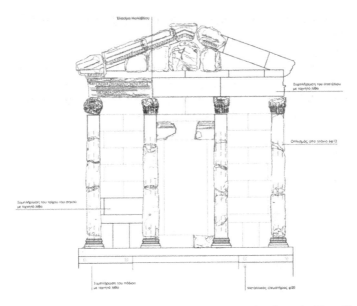

Fig. 13. Temple H. Graphic restoration of the proposal. Elevation from the East side. (Drawing by the author for his diploma's project)

Babbius's Tempietto

For the Babbius temple the works are located in two parts. The first concerns the structure of podium and the second one the temple itself. Regarding the podium from the graphic representation of Corinth I [22], it appears that it occupies more space than the one that is currently recorded in the archaeological site. The reason is that the stone on which it is placed is the core of the foot, in which the material is placed around it and it is covered with marble. Several members of the three rows of the wall are preserved, of the uprights and parts from the cornice. The number of these parts are certainly not enough, however, it gives us the height of the base, while it's area perimetrically from the eastern elevation can be determined by the trace located in that position and by the total length of the third step that approximates the proposed length of the restoration. Thus, the ancient material of the steps can be placed and supplemented by new material, possibly made of artificial stone or marble, thus giving the total length of its foot. In terms of height, however, the uprights are limited and their position has not been determined apart from the one which is inscribed and placed axially. They can form the molding's shape and height together with the cornice, while adding material to the cornice using the fasteners allows construction consistency and static capability.

Regarding the Temple, almost all the members of the temple are preserved in the field, as well as the maximum percentage of the stylobate, so that the diameter of the temple can be determined. Also the two bases of columns from the eight as well as the two capitals and sections of the cornice are saved. The part from which a large percentage of original material is missing is that of the columns. Thus, it is proposed to construct the necessary columns (the preferred material is marble) with which the ancient material will be incorporated and will be tied with the architrave in their original position. The construction will become more coherent since the seismicity of the area is an inhibiting factor for such constructions. On the roof, as part of it is preserved in the form of a flake, it is possible to construct the missing part in an abstract way, maintaining the relative perimeters defined by the scalloped portions, thus allowing the pine-like end to be placed. With respect to the construction of the roof, although the archive of ASCSA in its appendix for Corinth has versions of the graphic of the restoration, for the installation of the roof sections it is proposed that they are constructed in the way of the roof of the Horologion of Andronikos Kyrristos, in the Roman market of Athens [23], since it allows far better placement of members on the architrave (Figs. 14 and 15).

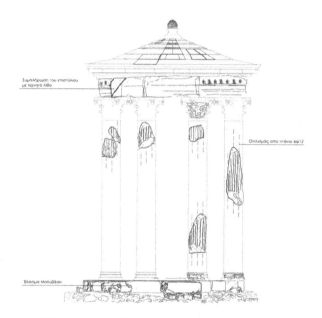

Fig. 14. Tempietto of Babbius. Graphic restoration of the proposal. Elevation from the East side. (Drawing by the author for his diploma's project)

Organization of the Construction Site

In addition to the restoration program of the two monuments, the organization of the constructions in the archaeological site, and particularly in the western boundary of the Roman Forum, is also required. The needs of the working process impose the organization of the construction site so as to make the possible interventions in the space, disturbing the smooth operation of the archaeological site to a minimum and the safe passage of the visitors. Thus, it is necessary during the period in which the work is to be done in Temple H, that the visitors pass through either the temple D or Temple F, which can be easily restored, since they do not require special massive constructions.

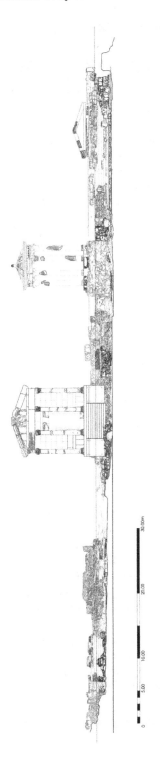

Fig. 15. The Roman temples. Graphic restoration of the proposal. Elevation from the East side. (Drawing by the author for his diploma's project)

At the same time, it is necessary to prepare the surrounding area to welcome the installation of an ambivalent elevated mobile bridge crane. This proposal offers the scaffolding adaptation to any ground conditions, provides the possibility of micro-processing of the blocks with millimeter accuracy and the ease of handling without requiring specialized personnel [24]. Finally, it is necessary to study the area where the maintenance and restoration of the architectural parts will be hosted. The limited area of the Terrace, as well as its double location in the main course of visitors, require that the site of maintenance and restoration work be placed in a different location from the temples. This suggests either the exploitation of existing workshops located in various places within the archaeological site or the placement in the south of the boundary near to the South Stoa of a relative space, since it is very close to the field of work. Alternatively, the area between Temple D and the Northwest Stoa is proposed, as it is downstream of the Terrace to a degraded level, making it inaccessible to the public (Fig. 16).

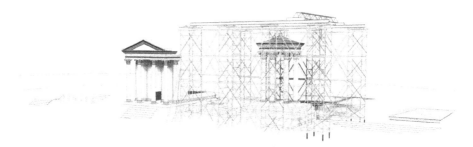

Fig. 16. Mobile bridge crane. The proposal of the construction at the archaeological site. East Elevation. (Drawing by the author for his diploma's project)

3 Conclusions

As part of the values of historic representation and the management of cultural heritage, in conclusion, the suggestion of a new management plan for the whole archaeological site of Corinth would organize visitors' circulation on a new basis giving them as a result a much different experience of the site. In terms of enhancing the understanding of the Roman forum, anastylosis strategies for the temples on the west terrace are important, because is the only boundary of the central area that is not recognizable as a whole. Two of the temples are able to be restored to their original height according to anastylosis principles, whereas the other temples' foundations are restored in order to give prominence to their plan outline. By implementing those strategies for the archaeological site of Corinth, it is estimated that visitors would understand the use of the Roman forum, the archaeological value of its monuments and especially the important role of its western boundary with the unique array of small temples.

References

1. Scranton, R.L.: Corinth. Monuments in the Lower Agora and North of the Archaic Temple, vol. I, Part III, p. 3. The American School of Classical Studies at Athens, Princeton, New Jersey (1951)

2. The present study is a summary of the postgraduate diploma thesis, supervised by Professor Emeritus Manolis Korres, and focuses on the study of the temples of West Terrace in the ancient Corinth Forum. I am in debt to the supervisor of my Diploma Thesis, Manolis Korre, to Dimitris Bartzi Architect and Candidate Doctor of the School of Architecture of the NTUA, to ASCSA school, to the former Director of the ASCSA's branch in Corinth, Guy Sanders and the architect of the school in Corinth James A. Herbst and to the Ephorate of Antiquities of Corinth

3. Παπαχατζής Δ. Νικ., Παυσανίου Ελλάδος Περιήγησις (Κορινθιακά-Λακωνικά), pp. 60–62. Εκδοτική Αθηνών, Αθήνα (1989)

4. Εφορεία Προϊστορικών και κλασσικών Αρχαιοτήτων Κορινθίας, Επιμέλειας Κίσσας, Κωνσταντίνος ΣΙ (Έφορος Προϊστορικών και Κλασσικών Αρχαιοτήτων Κορινθίας, Αρχαία Κορινθία. Από τους προϊστορικούς χρόνους έως το τέλος της αρχαιότητας, p. 54, Εκδόσεις του Φοίνικα, Αθήνα (2013)

5. Εφορεία Προϊστορικών και κλασσικών Αρχαιοτήτων Κορινθίας, op. cit., pp. 54–60

6. Εφορεία Προϊστορικών και κλασσικών Αρχαιοτήτων Κορινθίας, op. cit., p. 39

7. Scranton, R.L.: Corinth. Monuments in the Lower Agora and North of the Archaic Temple, vol. I, Part III, pp. 290–311. The American School of Classical Studies at Athens, Princeton, New Jersey (1951)

8. The temples now retain their name in Latin alphabet because it is not easy to identify them and the basic description of Pausanias (1st half of the 2nd century)

9. Scranton, R.L.: op. cit., pp. 52–57

10. Scranton, R.L.: op. cit., pp. 32–36

11. Regarding the dimensions of the basins according to Robert L. Scranton's publication, the eastern basin might have been about one meter from the front towards the back

12. Scranton suggests that water is transported by lead pipes from the aqueduct to the western edge of the central stores in Poseidon's Fountain

13. Παπαχατζής Δ. Νικ, op. cit., p. 64

14. Scranton, R.L.: op. cit., pp. 36–47

15. Scranton, R.L.: op. cit., pp. 36–39, 48–51

16. Scranton, R.L.: op. cit., pp. 17–32

17. The type of this temple is a special case since it is not often found in Greece and the use is intended either to honor a deity in accordance with the building program or a prominent person. An additional feature is the number of columns that the construction carries. Specifically, two types of temples are located, the first being the temples where the columns have a singular number, and the second refers to those where they have an even number. The utility is whether a strong element such as the statue of the deity is placed in the center of Tempietto, so the use of the single number is necessary to prevent the seamless goddess from going through the columns and to focus the visitor on the center of the circular temple. A typical example is the temple of Rome and Augustus on the Acropolis of Athens and the circular temple of Melikerti in Isthmia. On the contrary, monoptera circular temples with an even number of columns aim at the viewer's view of running the monument without focusing on the interior of the temple, but in the building itself and the complex in which it is placed. Examples of this type beyond the tempio of Babbius are the monopteros on the ancient market of Athens near the Stoa of Attalos and Nymphaeus in the city of Argos

18. Unfortunately for the restoration of the architectural members of the porosite that have been found it is not easy to repay because of their poor condition or the many fragments. So the published version of Robert L. Scranton's ASCSA issue, Corinth, Volume I, Part III, is a fantastic illustration, except the scale of the poem

19. Task Force for the Preparation of Management Studies for the Archaeological Site of Ancient Corinth, Plan for the Management of Cultural and Natural Heritage, in the Greater Archaeological Site of Ancient Corinth, Model of Analysis, December 2015

20. Αρχείο ASCSA, Corinth, Restored Elevation of Temple H, Buckley, J.A. (1985)

21. Ανθέμιον, ενημερωτικό Δελτίο της Ενώσεως Φίλων Ακροπόλεως, Τεύχος 11, pp. 5–17, Ιούλιος (2004)

22. Scranton, R.L.: op. cit., pp. 17–32, 308–309

23. Kienast, H.J.: Der Turm der Winde in Athens, Reichert (2014)

24. Μπάρτζης, Δημήτριος Γ., Αρχαίο Στάδιο Σικυώνας, Διπλωματική εργασία, pp. 89–95, Εθνικό Μετσόβιο Πολυτεχνείο, Σχολή Αρχιτεκτόνων Μηχανικών, ΔΠΜΣ «ροστασία Μνημείων», Κατεύθυνση Α': Συντήρηση και Αποκατάσταση Ιστορικών Κτηρίων και Συνόλων, Αθήνα (2015)

The Restoration of the Katholikon of Daphni Monastery: A Challenging Project of Holistic Transdisciplinary Approach, Novel Methodologies and Techniques and Digital Modelling

Androniki Miltiadou-Fezans[1(✉)] and Nikolaos Delinikolas[2]

[1] National Technical University of Athens, Athens, Greece
amiltiadou@arch.ntua.gr
[2] Hellenic Ministry of Culture and Sports, Athens, Greece
delinikolan@gmail.com

Abstract. The Byzantine monastery of Dafni, inscribed in the world heritage list of UNESCO (1990), is one of the most important monuments of middle Byzantine period, famous worldwide for the mural mosaics of its Katholikon. Being constructed in a seismic area, the monastery was affected throughout its history by many earthquakes that caused damages to the bearing structure and the mural mosaics. Due to severe damage provoked by the 1999 Athens earthquake, large scale structural restoration and conservation interventions were necessary to protect this world heritage monument. In this paper a concise synopsis of the overall structural restoration project of the Katholikon will be presented, together with a brief description of the investigations, research and novel techniques used and the 3D digital modelling of the monument. The variety and complexity of the problems and the importance of a holistic approach and of synchronized transdisciplinary efforts at various levels, in order to find the optimum solutions will be highlighted.

Keywords: Restoration · Pathology · Daphni Monastery · Digital modelling

1 Introduction

Daphni Monastery is one of the major middle - Byzantine monuments [1–4], situated in a forestry area at 11 km from Athens on the Sacred Way of the Classic Antiquity, from Athens to Eleusis (Fig. 1). The Katholikon (main Church) of the Monastery and its famous mural mosaics (Figs. 1a and b) have suffered severe damage during the 7th September 1999 earthquake that affected the region of Attica.

Immediately after the struck of the earthquake, a multidisciplinary working group [5] was formed by the Hellenic Ministry of Culture (HMC), with the assignment to do

---

A. Miltiadou-Fezans—Assistant Professor, National Technical University of Athens.
N. Delinikolas—Architect, Head in honour of the Section of Studies on Byzantine Monuments, Hellenic Ministry of Culture and Sports.

A. Moropoulou et al. (Eds.): TMM_CH 2018, CCIS 962, pp. 59–77, 2019.
https://doi.org/10.1007/978-3-030-12960-6_4

Fig. 1. N-E view of the church (left) and view of the dome (right), after the recent restoration works.

the necessary inspections, assess the nature and the significance of damage and prepare in collaboration with all competent services of the Ministry and a multidisciplinary Scientific Committee (composed by Professors: Ch. Bouras, T.P. Tassios, E. Mariolakos and N. Zias) set to this purpose a strategic plan for the protection, conservation and restoration of the monument, its mosaics included. The first inspections made evident the need for large-scale emergency measures to protect the monument from further deterioration and to provide safe working conditions for the personnel [6]. Besides, the extent of damage indicated that large-scale structural restoration and conservation interventions should be carried out, to reach the consolidation of all standing constructions.

Considering the importance of the monument, a close collaboration with Universities and other research centres was judged as absolutely necessary in order to investigate and record the state of the monument immediately after the earthquake in the best possible way, using innovative methodologies and digital modelling. Thus, in parallel to the design and application of the adequate urgent measures, a first series of research programs and investigations were undertaken. These programs were aiming at: (i) An accurate geometrical and architectural survey of the church, using novel survey methods and digital photogrammetry, in collaboration with the School of Rural and Surveying Engineering of the National Technical University of Athens (NTUA) [7–9], (ii) A preliminary survey of damages, assessment of mechanical properties of construction materials, monitoring of the width evolution of main cracks, and preparation of a first detailed numerical model to be used for preliminary structural analyses, (in collaboration with the Laboratory of Reinforced Concrete of N.T.U.A.) [10, 11], (iii) An investigation and assessment of the physicochemical characteristics of the construction and pointing mortars (in collaboration with the Laboratory of Construction Materials of the Aristotle University of Thessaloniki) [12], and (iv) A geophysical and geotechnical investigation of foundation and foundation soil (In collaboration with the Department of Geology of the University of Patras) [13–15].

The above research works were carried out in close collaboration with the responsible scientific personnel nominated by the competent authorities of the Hellenic Ministry of Culture (HMC), namely the Directorate for Restoration of Byzantine and

Post-Byzantine Monuments (DRBPM), the 1st Ephorate of Byzantine Antiquities (1st EBA) and the Directorate for the Conservation of Ancient and Modern Antiquities (DCAMA).

Based on a synthesis of the obtained results, adequately completed with detailed in situ observations and archive researches, the first structural restoration study was elaborated. Following the proposals of this study, approved by the Scientific Committee and the Central Archaeological Council, a step-by-step multidisciplinary approach was adopted, both concerning the design and the implementation of the restoration interventions, as proposed by ISCARSAH (International Scientific Committee on the Analysis and Restoration of Structures of Architectural heritage) [16]. Such an approach gives the possibility to start the execution of a series of interventions, and in the same time to perform the in situ and laboratory investigations that are necessary for the design of the next step. Furthermore, the assessment of the effect of the first step interventions to the structural behaviour of the monument can be carried out and be considered for the design of the works of the second step. The first phase of interventions comprises measures taken to repair and strengthen masonry elements. The strengthening interventions of the second phase aim to improve the overall behaviour of the entire structure.

In parallel with the works of the first phase, the following research and investigations were undertaken by the Directorate for Technical Research on Restoration (DTRR) of the HMC, in collaboration with Universities and Research Centres, to support the design of both phases of interventions: (i) Seismic monitoring of the structure for investigating its behaviour before, during and after interventions. In this framework a special study for the estimation of the seismic hazard was also carried out [17]; (ii) Design of adequate mortars and grouts [18–20] and development of a specific grouting application methodology adequate for masonry bearing mosaics and for the mosaics substrata [21]; (iii) Experimental investigation of the mechanical behaviour of brick enclosed masonry bearing mural mosaics, before and after grouting [22]; and (iv) Application of NDT's for investigating non-visible parts of the monument, monitoring the grouting movement during injection and controlling the grouting effect and efficiency after the completion of the injections [23–25]. A brief synthesis of all the main results of the afore-mentioned investigations is given in [26, 27].

The realization of the first phase of structural restoration interventions has been accomplished in 2007, together with the most of the aforementioned research, surveys and investigations. The design of the application project of the second phase interventions was elaborated by a multidisciplinary working group in 2012 [28]. In this framework, all aforementioned data and research results obtained were used for preparing and calibrating adequate numerical models and perform structural analyses and safety checks [29–31]. Moreover, all geometrical, structural and morphological data of visible and hidden parts were as well used for developing a 3D digital model of the entire monument, in which all decorative elements (mosaics, frescoes, marble cornices, etc.) were as well included. The works of the second phase interventions were carried out from 2011 to 2016. In this paper, the holistic transdisciplinary methodology and the "design in process" adopted will be presented together with the most important novel methodologies, techniques and digital modelling.

2   Brief Presentation of the Katholikon and Its Digital Survey

The Katholikon belongs to the octagonal type and preserves large part of the original mural mosaics of excellent art. It comprises the main church, the sanctuary, the narthex and four chapels, which complete its orthogonal plan. In the western part, only the perimeter walls of an exonarthex or portico and those of a spiral stairway tower leading to the upper floor have survived (Fig. 1). The central part of the main church is cross-shaped in plan, the hemispherical dome rising over its square core. The dome is 8,2 m in diameter and 16,4 m high, and rests on an almost cylindrical drum with 16 piers and 16 vaulted windows (Fig. 2). The dome and its drum are carried by eight pendentives and eight arches (four semi-circular and four embodied in the squinches of the corners), forming an octagon and achieving in this way the transition from circle to square. All the other parts of the monument are covered with byzantine groin vaults.

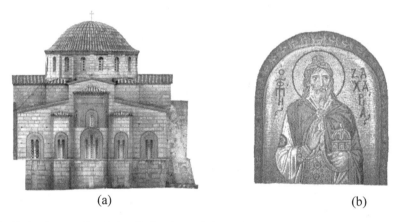

(a)   (b)

Fig. 2. Orthophotograph of the E façade and the orthophotograph and the corresponding development of a mural mosaic.

The exterior faces of the vertical perimeter walls are built according to the enclosed brick system (stones with bricks around them). Two different masonry construction types are followed, corresponding to their lower and upper parts [4]. Both types of the perimeter walls are constructed according to the three-leaf masonry type, with varying widths of their leaves [25]. Plain brick masonry was used for the construction of all the windows and doors (Fig. 2), as well as of the entire vaulted roof and all arches, apart from those of the west façade of the portico. In this part of the monument, curved stones were used by the Cistercians Monks, during the interventions carried out in the 13th century.

In order to proceed to the architectural and structural investigation and analyses and the design of interventions, a thorough geometrical and architectural survey of the monument at a general scale of 1:25 was decided. Furthermore, the recording of its mural mosaics at a very large scale (1:5) was considered necessary, in order to serve the need for a very detailed documentation and design of conservation interventions of the mosaics by the Conservators. Besides, the requirements included, among others, the raster development of those mosaics, which are constructed on developable, i.e. cylindrical or conical, surfaces. The aforementioned research program was carried out with the close collaboration of experienced scientific personnel of the DRBPM and 1st EBA of the HMC.

As reported by Georgopoulos et al. [7], the problems of recording such works of art at a scale of 1:5 were numerous, regarding the issue of the reference systems combination, the adequate way to perform the stereoscopic photography and the photogrammetric measurements performed on a digital photogrammetric workstation. For this task specialized approaches, procedures and software in order to accommodate for the data acquired during the surveying process have been developed. Detailed presentation of the novel methodologies undertaken are presented elsewhere [7–9]. In Fig. 2 the orthophotograph of the East façade of the monument and an example of a mural mosaic survey are presented.

3 Pathology-Qualitative Interpretation and Preliminary Numerical Verification

The Monastery is situated in a Neocene tectonic graben between the mountains Egaleo and Korydallos at the west side of Athens, 150 m away from the E-W trending marginal fault between the alpine Mesozoic limestone and the post-alpine deposits [32]. Located in a tectonically active area, many intensive earthquakes damaged Dafni Monastery throughout the centuries (from 11th c. to our era). Partial collapses and extended damage were provoked to the structure and the mural mosaics, which led to major interventions. Detailed presentation of the construction phases and the long history of the monument, including the extensive interventions undertaken (mainly during 13th–15th c. and the end of 19th and during the 20th c.) are reported in [4, 11, 33]. Figures 3, present the historical pathology and past interventions on the digital survey drawings presented above. The reconstructed parts of the monument are shown with solid colours. Yellow: 13th–15th c., orange: 1891, green: 1895, purple: 1897–1907, dark blue: 1955–60.

The 1981 earthquake (Alkyonides islands) caused numerous hair cracks to the building and severe damages to the mosaics. Thus, the pathology observed in the monument after the 1999 earthquake was only partly due to this severe earthquake.

After the 1999 Athens earthquake, an extensive network of shear and bending cracks appeared on the walls and piers of the monument, whereas numerous old cracks (due to previous earthquakes) increased in length and width (Fig. 4). Severe structural dislocation and outwards movement of the walls was recorded in the NE corner of the main church (~ 14 cm to the N and ~ 10 cm to the E). Significant out-of-plane displacement of the N and S arms of the cross (~ 16 cm and 21 cm respectively), and of the free standing west wall of the exonarthex were also recorded, due to further deterioration of previous deformations (~ 16 cm in the corners and ~ 25 cm in the middle). The damages were more extensive in the higher parts of the structure, especially in the sanctuary, the arms of the cross and all the arches below the dome area. As shown in Fig. 4, the NE and NW small arches just below the squinches presented severe dislocation near their crown, followed by out of plane deformations of the

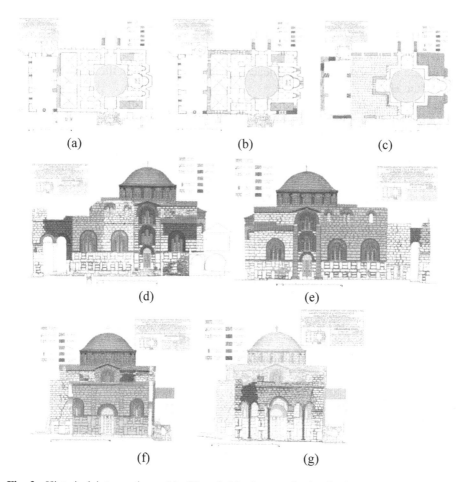

Fig. 3. Historical interventions: (a), (b) and (c) plans at the level of entrance doors, lower windows and squinches, respectively, (d), (e), (f) S, N and W façade of the church respectively and (g) W façade of the exonarthex. All uncoloured joints still preserve the old Byzantine pointing mortars (Color figure online)

squinches themselves. The structural condition of the dome (reconstructed at 1891 and damaged soon after its reconstruction at 1894) was assessed as extremely critical immediately after the earthquake. Horizontal cracks have appeared along the perimeter of the drum (both at its base and top). In the piers of the drum that are situated perpendicular to the East-West direction, horizontal cracks (due to out-of-plane bending) have opened at their top and bottom. In the piers that are situated parallel to the E-W axis, diagonal or bi-diagonal (shear) cracks have appeared. In the intermediate piers, mixed type of (less severe) cracks was observed.

This pathological image does seem to confirm seismological data regarding the predominant direction of the 1999 earthquake. Thanks to the external upper metal tie-rod, confining the hemispherical dome near springing level, the occurrence of severe cracking of the dome's shell was prevented.

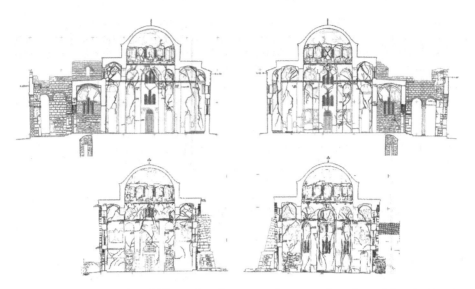

Fig. 4. Pathology after 1999's earthquake presented on central sections of the structure

One of the major steps of the assessment of an existing structure is the analytical reproduction of its pathological image. In the case of the Katholikon, preliminary linear parameter analyses were performed, as a means for selecting adequate emergency interventions and as a first attempt for the numerical verification of the pathological image of the monument. Thus, for the preliminary analytical study, using the computer code ACORD, the structure was modelled by shell elements (Fig. 5), whereas the mechanical properties of elements belonging to various parts of the structures were assumed on the basis of the available data for the construction materials. Linear elastic analyses were performed for various combinations of actions (self-weight alone or combined with seismic action). Both static and dynamic analyses were performed.

Theses analyses permitted to clearly distinguish the vulnerable areas of the structure and the obtained results were conform with the qualitative interpretation of the pathology of the monument.

(a)   (b)

Fig. 5. (a) The numerical model prepared for parameter analyses and (b) Deformed shape of the structure looking to the N (dynamic analysis, seismic action along the longitudinal axis).

The comparative examination of all available quantitative and qualitative data proved the inherent vulnerability of the structural system of the monument, in which (a) a stiff central cupola is resting (through the drum) on four major arches parallel to the two main axes of the church, as well as on four arches oblique in respect with the longitudinal and the transverse axis, (b) the vertical, as well as the horizontal component of the self-weight of the whole system of (intersecting) groin vaults arranged around the central dome, are transferred to rather flexible stone masonry piers, (c) there are not ties in the arches and vaults, or other elements to confine critical structural parts or link the various parts among themselves. Therefore, there is a tendency of the structural system to deform laterally in its upper region even under self-weight alone. Expectedly, this behaviour is deteriorating when a seismic event occurs, during which, according to the results of the preliminary dynamic analysis of the structure, out of phase movement of various parts of the monument occurs (Fig. 5b).

Due to the inherent vulnerability of the structural system of the Katholikon and the high seismicity of the area, proved also by its historic pathology, several parts of the structure have been deformed independently; multiple cracks have emerged, dividing the entire structure in separate parts, and thus, interacting with nonlinear behaviour. Such behaviour can explain the severe damages occurred to the drum of the dome, as well as to the arches and vaults supporting the dome. As aforementioned, this ascertainment was also confirmed by the preliminary numerical analysis (Fig. 5b) and reported in the first phase of the restoration study of the Katholikon [5].

On the other hand, such behaviour suggests that in addition to interventions that are necessary for the repair and strengthening of damaged masonry elements, adequate strengthening measures (in form of tie-rods, diaphragms over the extrados of vaulted roofs to connect the vertical walls, pillars and piers, bracing of the windows of the dome, confinements of drum and piers, etc.) should be taken. This conclusion is in accordance to the historical pathology of the monument and the fact that various parts that have been simply reconstructed in the past were cracked again (like the piers of the drum of the dome reconstructed in 1891 and cracked soon after, during the 1894 Atalanti earthquake) or presented out of plane deformation (as for example the south wall of the narthex, already reconstructed in 1896 and deformed again, as measured after the 1999 earthquake).

Consequently, besides the repair and strengthening of masonry elements, a series of interventions to improve the overall behaviour of the structure were also schematically proposed, by the first restoration study, for further elaboration [5, 27]. The aim was to alleviate the inherent vulnerability of the structure and, thus, to improve its future behaviour in some critical regions, avoiding the danger of further deterioration of deformations and eventual local collapses. The application of such measures was approved in principle by the Scientific Committee and the responsible authorities. Additionally, given the importance of the monument and the high values of its mural mosaics, the need for further research and investigations was also recognized.

4 Design in Process for Structural Interventions Optimization

Due to the high values of the monument and its vulnerability, the decision was taken to start the application in priority of a first series of interventions to achieve the best possible repair and strengthening of masonry elements, avoiding further deterioration of damage (which could lead to a total disruption of continuity and even local collapses) and in parallel to investigate thoroughly its structural behaviour under seismic actions. The aim was to design the minimum necessary strengthening interventions, which would be selected as optimum, considering all the values of the monument, and above all its mosaics. To this end additional data were necessary to perform all necessary numerical analyses and safety checks with calibrated models and reliable data, avoiding conservative assumptions (both for actions and resistances). Such assumptions would lead to extensive interventions that might not be needed and that would inevitably alter the architectural value of the monument.

Thus, a design in process was adopted, a series of research and investigations were initiated and an earthquake monitoring system was installed in the monument, to collect real data for the dynamic characteristics of the structure, before, during and after interventions. This step by step multidisciplinary approach, proposed as well by ISCARSAH recommendations for important monuments, led to the partition of the design of the project and the implementation of the works in two phases, giving the possibility for the necessary additional data to be collected and detailed structural analyses and safety checks to be carried out.

The study for the first phase of structural restoration interventions was completed in 2003 [4, 5]. The first phase of works (2004–2007) comprised all measures considered necessary for the repair and strengthening of masonry elements (stitching of cracks and deep re-pointing where necessary, systematic hydraulic lime grouting, local reconstructions, etc.), and some local morphological restoration interventions, to correct alterations provoked in the past.

Furthermore, detailed survey of the materials, geometry and pathology of many parts of the monument, that became accessible thanks to the realization of the works of the first phase, was carried out (e.x. internal faces of masonry walls and the intrados and extrados of vaulted roof) [27].

The design of the application project of the second phase restoration interventions of the Katholikon, including all necessary structural analyses, checks and calculations for the assessment of the efficiency and the dimensioning of optimum interventions, as well as detailed drawings, after in situ checks of their applicability (actual geometry, mosaics locations, etc.), was elaborated by a multidisciplinary working group in 2012 [28].

The second phase of works (2011–2016) comprised strengthening interventions designed and selected as optimum to improve the overall behaviour of the whole structure, such as ties and struts, diaphragmatic structures on the extrados of the vaults and in the exonarthex, bracing of the windows of the drum of the dome, confinement of structural elements, etc. The interventions of the second phase aim to reduce the vulnerability of the structure and the extent of the damage under future seismic events, especially in the upper level, where the mural mosaics are located. In this second phase are also included all the works for the morphological, aesthetic and functional restoration of the monument.

5 Numerical Analysis and 3D Digital Modelling for Structural Assessment and Architectural Representation

As it is reported in more details in other papers [29–31], detailed numerical simulation and structural analyses of the Katholikon was carried out by means of more sophisticated tools, considering the improvement of the structural behaviour after the implementation of grouting, on the basis of the results derived from the aforementioned experimental research on masonry behaviour, before and after grouting. Two separate finite element models have been developed using solid elements (Fig. 6).

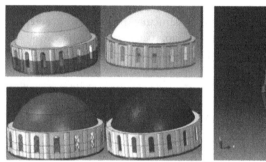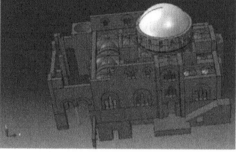

Fig. 6. The two numerical models for structural analyses, without and with interventions

The first model comprises the hemispherical dome with the drum, while the second simulates the entire structure of the monument. The numerical analyses were performed using the code ABAQUS (2011) and three-dimensional, tetrahedral finite elements. The separate drum-dome model was analysed with seismic actions applied as equivalent static loads, while linear time history dynamic analyses were performed for the entire structure model. To calibrate both models, the data recorded by the monitoring system, adequately analysed, has been exploited. Their capacity to numerically reproduce the historical and recent pathology, identifying the most vulnerable areas of the monument, was also checked. The necessity of applying strengthening measures to alleviate the structure, improve the inherent structural drawbacks and reduce its vulnerability, already identified by preliminary numerical analyses, was confirmed and further refined. Consequently, the calibrated models were used for the qualitative evaluation of the efficiency of alternative structural strengthening interventions (installation of ties, masonry confinement, diaphragms, etc.) or combinations of interventions, to improve the structural behaviour of the monument. For the design and dimensioning of various strengthening measures, the models were adequately modified and in some cases "cracks" were induced, to overcome shortcomings of elastic analysis. After the evaluation of the efficiency of each alternative intervention, further analyses were performed for selected combinations of measures.

Apart from developing adequate numerical models to perform structural analyses, a 3D architectural model was also developed, based on all geometrical and morphological data of visible and hidden parts, including decorative elements. The 3D structural and architectural numerical tools developed were complementary and necessary to design and dimension the various stainless steel or wooden elements and their connections with masonry, considering the real geometry. Thus, the applicability of proposed measures was ensured and their effect on the appearance of the monument was fully documented.

6 Brief Presentation of the Structural Restoration Interventions

6.1 Masonry Repair and Strengthening Interventions

A series of interventions were applied to improve in the best possible way the mechanical behaviour of masonry elements by (a) reinstating their continuity, (b) increasing their resistance (mainly to traction and shear), and (c) ensuring that no side effects might occur, due to possible durability matters. The masonry repair interventions comprised mainly the following works (Figs. 7, 8 and 9): (i) very careful removal of internal renderings and deteriorated pointing mortars applied during previous interventions, without harming the old ones, adjacent to or underneath them, (ii) removal of tiles and other covering and filling materials to reach the extrados of all vaulted roofs, (iii) stitching of the most severe cracks, using long stones, bricks, or thin titanium plates, (iv) few local reconstructions, necessary either for the repair of dislocated or collapsed parts or for the restoration of past morphological alterations, (v) deep re-pointing where necessary, (vi) preparation of masonry for grouting, (vii) implementation of grouting, (viii) removal of all injection tubes and filling of holes with mortar, (ix) in situ conservation of deteriorated old mortars, (x) execution of all necessary works to ensure the protection of the extrados from rainwater.

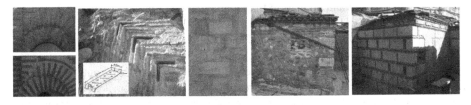

Fig. 7. First phase structural restoration interventions (1960's cement re-pointing mortars removal, stitching of cracks with titanium plates or stones and local reconstruction to correct past alterations).

(a)   (b)

Fig. 8. (a) North wall of the E arm of the cross bearing old byzantine mortars in its external façade and a heavily damaged mosaic in the internal one. (b) External old mortars of 1st and 2nd construction phase and internal precious mosaics, which had to be preserved in situ.

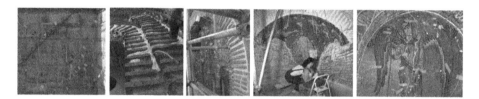

Fig. 9. Preparation for grouting of masonry bearing old mortars and of mural mosaics.

The application of adequate mortars and grouts compositions, constituted a key parameter to ensure a successful intervention, both for the masonry and the mosaics. Thus, extended laboratory research and in-situ pilot tests have been undertaken by the Directorate for Technical Research on Restoration (DTRR) of HMC to design the adequate mortars compositions, considering the characteristics of the existing materials, the requirements set by the restoration study and the worksite conditions. For most of the deep and surface re-pointing works lime-pozzolan based mortars were used. For the surface re-pointing of the façades old pieces of ceramics (from archaeological excavations) were grained in situ in the adequate granulometry and used in the mortar mixtures, to avoid modern bricks kilned in high temperatures. For local reconstructions, deep re-pointing of extremely damaged areas and crack-stitching using titanium plates, hydraulic lime based mortars were applied. Detailed information on existing mortars is given in Papagianni [12], while a synthesis regarding the old mortars and the basic compositions designed and applied to the monument are reported in Anagnostopoulou and Miltiadou-Fezans [18].

The in-situ application of grouting had to be implemented without removing the old pointing mortars and, of course, without removing the mural mosaics decorating its interior (Fig. 8). Consequently, due to the multitude of cracks, the grout for masonry was expected to penetrate not only to the masonry but also to the cracked substratum of the mosaics. Besides, the application of grouting was also proposed to improve the bond between mosaics substrata and masonry [34].

Thus, special attention had to be given to design adequate grouts, not only because grouting constituted the main of the works undertaken that could substantially improve the mechanical behaviour of masonry, but also since, it is an invisible, and thus, irreversible intervention, which affects the masonry elements and all the mosaics, frescoes and old mortars beard on them. To this end, a set of design requirements related to the properties of the grout (namely, injectability, strength and durability) was determined, with the purpose to reach a material suitable both for masonry and mosaics. Extended experimental investigations were carried out by the DTRR and NTUA to design adequate grout compositions, to assess the mechanical properties of masonry elements, before and after grouting, and finally to evaluate the effect of grouting on their mechanical properties [19, 20, 22]. Furthermore, as the proper design of a grout composition cannot ensure on its own the successful execution of the grouting intervention, an application methodology was developed [23] for the implementation of injections in a more rational and fully controlled way, using entrance and exit tubes of small diameter in mosaics and old mortars areas (Fig. 9), low pressures and detailed quality control of the grout and the application of the injections to ensure that the grout reached and filled the numerous cracks and fissures.

The application of grouts to masonry bearing mosaics being a delicate procedure, the decision was made to train the Personnel through trial applications. To this end, in addition to the personnel of DTRR and NTUA, the team of experienced Conservators of the DCAMM, working at Daphni Monastery, was also involved in the experiments carried out at the Laboratory of DTRR and LRC of NTUA, contributing to the construction of mosaic's replicas on the wallettes, to the installation of grouting tubes on the mosaics, and to the implementation of injections. The novel methodology was tested and further refined, during the injections of the specimens made for in-laboratory testing. Thus, the in-laboratory work served as a full-scale rehearsal of the entire grouting procedure on historic masonries bearing precious mosaics. Furthermore, this was the best way for training the involved personnel. In this way, collaboration and osmosis were developed within the multidisciplinary group.

The mutual understanding established led to a smooth and fruitful cooperation, between those working on masonry and those working on mosaics, during the in-situ implementation of grouting, necessitating meticulous and refined interventions. For instance, plastic transparent tubes (d = 4 − 10 mm) had to be installed on both faces of masonry and at adequate distances forming a grid (l = 0.5 to 1.0 m, both horizontally and vertically), together with similar, but finer, tubes (d = 1.0 − 3.3 mm) installed to the mosaics.

Moreover, various compositions of lime-pozzolan or hydraulic lime mortars had to be prepared and applied carefully to seal voids, fissures, cracks and detachments, while the tubes were inserted in various depths, with the purpose to reach the interfaces between the various layers of mosaics substrata, as well as interfaces between masonry leaves. Besides, all this information, along with the number allotted to each tube, had to be reported on drawings (Figs. 10 and 11). Throughout the whole project, entrances and corresponding exits of the grout were recorded on a time basis, together with any change of pressure at the entrance to the wall. All these data were noted on specially designed daily calendars, with reference to the area of grouting application. These manually collected data and those collected by the Grout Recorder (pressure, flow rate and grout volume consumed) were then combined, and the grout volume corresponding to the various groups of entrances and exits was determined, together with the total volume of the grout consumed during the corresponding day. All the recorded data were organized in tables and after appropriate processing the volume of the grout consumed into the masonry was extracted. Then the data collected were reported on the series of drawings presenting the positions and numbers of the grouting tubes. Thus, a good estimation of the grout movement and consumption in relation to the various regions of the structure was achieved. This procedure was followed during the whole project in the Katholikon, both for masonry and for mosaics. In Fig. 10 the corresponding drawings for the internal and external façade of the east wall of the monument are presented. One can easily recognize the areas with high grouting consumption, and those with a lower one. The tubes with no consumption are also noted. Thus, for this specific wall the volume of the grout consumption was estimated to reach 6.5% of the total volume of the wall. In the case of the west wall of the monument, which was reconstructed during past restoration interventions [4], the estimated grout consumption attained 2.5% of its total volume.

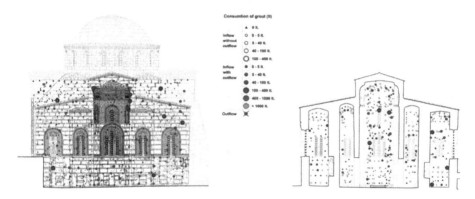

Fig. 10. Schematic presentation of grouting consumption on external and internal east façade.

Using this methodology, the movement of the grout inside masonry and behind the mosaics was continuously monitored (Fig. 11). Thus, both for masonry and mosaics, it was checked that the grout reached and filled the cracks and voids, ensuring the quality of the intervention. The seismic monitoring data and those obtained by sonic tomography [17, 24] carried out during and after the grouting, gave valuable information for the effect and the efficiency of the whole grouting intervention, and enhanced the validity of the methodology applied.

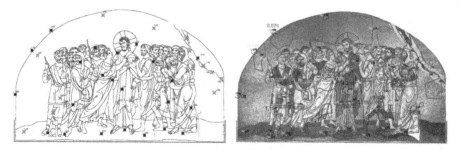

Fig. 11. Juda's Betrayal mosaic: survey of injection tubes and of grout volume consumed.

6.2 Strengthening Interventions Improving the Overall Behaviour of the Monument

In this paragraph, a concise synopsis, per area, of the strengthening interventions, selected as optimum to improve the overall behaviour of the structure, taking into account all the values of the monument and above all the protection of the mosaics, is given:

a. In the dome and its drum: Replacement of the existing confining steel ring-plate, situated at the upper level of the drum, by a stainless steel one. Installation of a second non-visible confining stainless steel ring in the extrados of the hemispherical dome, underneath the ceramic tiles. Insertion of stainless steel stiffening frames in the windows of the drum (Fig. 12b, c and d).

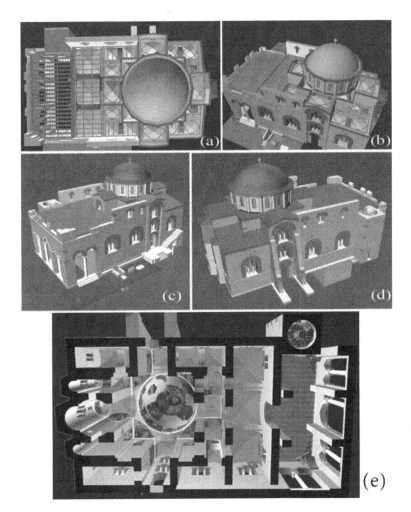

Fig. 12. The optimum strengthening measures, as presented in various images of the 3D model.

Fig. 13. External view of the restored monument from S and an internal view from S-E.

b. Over the extrados of the groin-vaults of main church: Installation of stainless steel "diaphragmatic" structures; their final geometry was adjusted to the empty space existed in each area (Fig. 12a and b).

c. Over the extrados of the groin vaults of the narthex and W chapels: Installation of a diaphragm composed of LVL wooden plates (kerto Q) supported by stainless-steel beams (for the vertical loads) and of stainless-steel beams at the perimeter, connecting the entire system with the walls (Fig. 12a and b).

d. In the exonarthex: Installation of a diaphragm composed of LVL wooden plates (kerto Q) supported by wooden beams (for vertical loads), and of stainless steel perimeter beams, connecting the entire system with the walls. Installation of stainless steel ties and wooden struts in the arches and replacement of the masonry piers, built by Stikas [2], with copies of marble ionic columns (Fig. 12a and b).

e. In the interior of the main church, the narthex and the crypt: Installation of stainless steel ties and struts in most of the arches. Confinement of west piers of the church and the piers of the crypt, with stainless steel plates (Fig. 12e).

f. In the tower of staircase: Confinement of the masonry walls in three levels, by insertion of titanium threaded tie-bars, bonded to masonry by means of hydraulic lime injections and anchored externally on titanium anchorage plates. Installation in the tower's interior of a round steel staircase (Fig. 12d and e).

As aforementioned, the application of the works of the second phase has been accomplished in 2016. In Fig. 13 an external view of the restored monument from S and an internal view from S-E is given.

7   Conclusions

The brief presentation of the restoration the Katholikon of Daphni Monastery attempted in this paper, highlights the variety and complexity of the problems and the importance of a holistic approach, of synchronized transdisciplinary efforts at various levels, of developing novel methodologies, innovative intervention techniques and digital modelling, in order to find the optimum solutions.

The design in process adopted for the restoration of the Katholikon proved to be very efficient, as it gave the possibility to carry out the necessary in-situ and in-laboratory investigations and research for the rational design of optimum structural interventions, using calibrated models and reliable data. Thus, with the implementation of the aforementioned optimum structural restoration interventions, the static and dynamic behaviour of the monument will be modified only at a level necessary for strengthening the structure, without altering radically its authentic existence and the integrity of its form and without essential impact to the today existing mural mosaics. The behaviour of the structure in future earthquakes will be improved, without excluding however, even for the level of seismic actions considered, the appearance of damage in case of a strong earthquake. Furthermore, most of the interventions were designed to be reversible so that to give the possibility to future generations of architects and engineers to attempt to re-intervene to the monument, probably by means of more advanced techniques.

In 2016, seventeen years after the 1999 Athens earthquake, the structural restoration of the Katholikon has been accomplished. It must be noticed however, that in such important monuments, preventive restoration and conservation interventions are always necessary and should be considered in the framework of a maintenance plan.

Last but not least, it is strongly believed that the experience and knowledge acquired by all levels of involved stuff in the framework of the structural restoration of this important monument of World Heritage constitute a solid base for research, investigations, studies and structural restoration interventions to other similar historical structures and monuments.

Acknowledgements. The Directorate for the Restoration of Byzantine and Post-byzantine Monuments (DRBPM) was responsible for the structural restoration studies and works. The Directorate for the Conservation of Ancient and Modern Monuments (DCAMM) was responsible for the mural mosaics conservation works. The 1st Ephorate of Byzantine Antiquities (1st EBA) was responsible for the overall surveillance of the works. The Directorate for Technical Research on Restoration (DTRR) was responsible for research and investigation works and in situ quality control of grouting application. All projects were co-funded by European Regional Development Fund (ERDF) and National Funds, in the framework of the 2nd and 3rd Community Support Framework and the "National Strategic Reference Framework-NSRF" Operational Program.

References

1. Millet, G.: Le Monastère de Daphni. Histoire, Architecture, Mosaïques, Monuments de l'Art Byzantine I, Paris (1899)
2. Stikas, E.: Restoration and consolidation of the exonarthex of Dafni's Katholikon. DChAE Ser. **4**(3), 1962–1963 (1962)
3. Bouras, Ch.: The Dafni monastic complex reconsidered. AETOS Studies in honour of Cyril Mango, B. G. Teubner Stuttgart und Leipzig (1998)
4. Delinikolas, N., Miltiadou-Fezans, A., Chorafa, E., Zaroyianni, E.: Study on restoration of the Katholikon of Dafni Monastery: Phase A. Internal report, HMC/DRBPM/1st EBA (2003). (in Greek)

5. Delinikolas, N., Miltiadou-Fezans, A., Tsofopoulou, E., Minos, N., Chrissopoulos, D.: Initial Inspections, Assessment and Proposals for Protecting the Monastery of Daphni, Internal report. HMC/1st EBA/DRBPM/ DCAMM (1999). (in Greek)
6. Miltiadou-Fezans, A., Tassios, T.P., Delinicolas, N., Chorafa, E., Zarogianni, E., Chandrinos, I.: Earthquake structural problems and urgent measures undertaken to support the Katholikon of Daphni Monastery in Athens, Greece. In: Proceedings of the 8th International Conference on STREMAH 2003, Halkidiki, Greece, pp. 533–542 (2003)
7. Georgopoulos, A., Daskalopoulos, Th., Makris, G.N., Ioannidis, Ch.: A trip from object to image and back. Int. Arch. Photogramm. Remote Sens. Spat. Inf. Sci. **34**(5/W12), 141–144 (2003)
8. Georgopoulos, A., Ioannidis, Ch., Karkanis, M., Iliopoulou, Ch.: Large scale monument data base design. In: CIPA XIX International Symposium, Antalya (2003)
9. Georgopoulos, A., Ioannidis, C., Daskalopoulos, A., Demiridi, E.: 3D reconstruction and rendering for a virtual visit. In: International Congress of ISPRS, Istanbul, July 2004
10. Vintzileou, E., Miltiadou-Fezans, A., Palieraki, V., Delinikolas, N.: The use of radar techniques and endoscopy in investigating old masonry: the case of Dafni Monastery. In: Proceedings of the 4th International Seminar on Structural Analysis of Historical Constructions, SAHC, Padova, pp. 351–359 (2004)
11. Miltiadou-Fezans, A., Vintzileou, E., Delinikolas, N., Zaroyianni, E., Chorafa, E.: Pathology of Dafni Monastery: survey, monitoring of cracks, interpretation and numerical verification. In: Proceedings of the 4th International Seminar on Structural Analysis of Historical Constructions, SAHC, Padova, vol. 2, pp. 1285–1294 (2004)
12. Papagianni, I.: Investigation of physicochemical characteristics of the historic mortars of the Katholikon of Daphni Monastery. Research Report, AUTH and HMC/DRBPM (2002). (in Greek)
13. Papamarinopoulos, S.: Geophysical investigations of the foundations of the Katholikon of Daphni Monastery. Internal report DRBPM (2001). (in Greek)
14. Polymenakos, L., Papamarinopoulos, S., Miltiadou, A., Charkiolakis, N.: Investigation of the foundations of a Byzantine church by three-dimensional seismic tomography. J. Appl. Geophys. **57**, 81–93 (2005)
15. O.M.T. Geotechnical investigation of the foundation soil of Daphni Monastery, Internal report, DRBPM (2004). (in Greek)
16. International Scientific Committee on the Analysis and Restoration of Structures of Architectural heritage (ISCARSAH). The Principles were adopted by ICOMOS General Assembly in Victoria Falls (2003)
17. Mouzakis, Ch., Miltiadou-Fezans, A., Touliatos, P., Delinikolas, N., Dourakopoulos, J.: Earthquake based condition monitoring of the Katholikon of Dafni Monastery. In: 6th International Conference on SAHC, Bath, UK, pp. 621–628 (2008)
18. Anagnostopoulou, S., Miltiadou-Fezans, A.: Design of restoration mortars of the Katholikon of Dafni Monastery. Internal report, DTRR, Hellenic Ministry of Culture (2007). (in Greek)
19. Miltiadou-Fezans, A., Kalagri, A., Delinikolas, N.: Design of hydraulic grout and application methodology for stone masonry structures bearing mosaics and mural paintings: the case of the Katholikon of Dafni Monastery. In: Proceedings of the International Symposium on Studies on Historical Heritage, Antalya, Turkey, 16–21 September 2007
20. Kalagri, A., Miltiadou-Fezans, A., Vintzileou, E.: Design and evaluation of hydraulic lime grouts for the strengthening of stone masonry historic structures. Mat. Str. **43**, 1135–1146 (2010)
21. Miltiadou-Fezans, A., et al.: Methodology for in situ application of hydraulic grouts on historic masonry structures. The case of the Katholikon of Dafni Monastery. In: Proceedings of the 6th International Conference on SACH, Bath, UK (2008)

22. Vintzileou, E., Miltiadou-Fezans, A.: Mechanical properties of three-leaf stone masonry grouted with ternary or hydraulic lime based grouts. Eng. Struct. **30**(8), 2265–2276 (2008)
23. Côte, Ph., Dérobert, X., Miltiadou-Fezans, A., Delinikolas, N.: Mosaic-grouting monitoring by ground-penetrating radar. In: Proceedings of the 4th International Seminar on Structural Analysis of Historical Constructions, Padova, pp. 401–406 (2004)
24. Côte, Ph., et al.: Application of non-destructive techniques at the Katholikon of Daphni Monastery for mapping the mosaics substrata and grouting monitoring. In: Proceedings of the 6th International Conference on SACH, Bath, U.K., pp. 1149–1156 (2008)
25. Palieraki, V., Vintzileou, E., Miltiadou-Fezans, A., Delinikolas, N.: The use of radar techniques and boroscopy in investigating old masonry: the case of Dafni Monastery. Int. J. Arch. Herit. Conserv. Anal. Restor. **2**(2), 155–186 (2008)
26. Miltiadou-Fezans, A.: Multidisciplinary approach for the structural Restoration of the Katholikon of Daphni Monastery in Attica Greece. In: Proceedings of the 6th International Conference on SAHC, Bath, UK, pp. 71–87 (2008)
27. Miltiadou-Fezans, A.: Design in process, multidisciplinarity and synergy: key issues of the structural restoration of Daphni Monastery. In: Proceedings of the 10th International Conference on SAHC, Leuven, Belgium, pp. 23–34 (2016)
28. Miltiadou-Fezans, A., et al.: Application design for the restoration of the Katholikon of Daphni Monastery, Phase B, HMC, DRBPM (2012). (in Greek)
29. Miltiadou-Fezans, A., Delinikolas, N., Dourakopoulos, J., Giannopoulos, P., Vintzileou, E., Mouzakis, Ch.: Interventions to complete the restoration of the Katholikon of Daphni Monastery. In: Proceedings of the 4th Greek Conference on Restoration, ETEPAM, Thessaloniki, Greece, pp. 602–622 (2015)
30. Miltiadou-Fezans, A., Delinikolas, N., Dourakopoulos, J., Giannopoulos, P., Vintzileou, E., Mouzakis, Ch.: Numerical analysis of the structural behaviour of the Katholikon of Daphni Monastery. In: Proceedings of the 10th International Conference on SAHC, Leuven, Belgium, pp. 1559–1566 (2016)
31. Miltiadou-Fezans, A., Vintzileou, E., Mouzakis, Ch., Dourakopoulos, J., Giannopoulos, P., Delinikolas, N.: Structural analyses of the Katholikon of Daphni Monastery with alternative interventions improving its overall behaviour. In: Proceedings of the 16th E.C.E.E., 18–21 June, Thessaloniki, Greece (2018)
32. Mariolakos, I., Fountoulis, I., Andreadakis, E.: Engineering geological problems caused by human interference on monuments that influence their seismic behavior. In: Proceedings of the Compatible Materials Recommendations for the Preservation of European Cultural Heritage, PACT 59, Ac. Civ. Eng. of Greece, pp. 53–63 (2000)
33. Delinikolas, N., Miltiadou-Fezans, A.: The construction phases and the historical pathology of the Katholikon of Dafni monastery. In: The Proceedings of the 1st Conference on Restoration, ETEPAM, Thessaloniki (2006). (in Greek)
34. Chryssopoulos, D., Anamaterou, L., Georganis, F.: Documentation study for the mosaics of the Katholikon of Dafni monastery, after the 1999 earthquake, HMC/DCAMM (2003). (in Greek)

Restoration and Consolidation of the Byzantine Church of Palaia Episkopi in Tegea, Arcadia

Konstantina Siountri[1,2,3(✉)]

[1] Hellenic Ministry of Culture, Athens, Greece
[2] Department of Informatics, University of Piraeus, Piraeus, Greece
ksiountri@unipi.gr
[3] Cultural Technology and Communication Department,
University of the Aegean, Mytilene, Greece
ksiountri@aegean.gr

Abstract. The church of the Dormition of the Virgin in the area of Palaia Episkopi in Tegea of Arcadia is dated to the second half of the 10th century. The central dome collapsed during the decade of 1850s resulting to its gradual damage. In 1884 the Tegean Association (*Tegeatikos Syndesmos*) undertook the restoration of the Byzantine church and invited Ziller to execute the designs of the reconstruction works. The project was completed in 1888 and the church officially opened in 1889. Between 1936 and 1939 the well-known artist Aginoras Asteriades completed the internal decoration of the church with a series of wall-paintings. The following years a numerous problems related to its construction put the building and its murals at risk.

The study "Restoration and Consolidation of the Byzantine Church of Palaia Episkopi in Tegea, Arcadia" was presented originally as a thesis at the MSc "Protection of Monuments", Track A, National Technical University of Athens (NTUA) with Supervisor Pr. Spyridon Raftopoulos. Then it was approved by Central Archaeological Committee (*KAS*) of Hellenic Ministry of Culture in 2006 and was implemented by the Directorate of Anastylosis of Byzantine and Postbyzantine Monuments (*DAVMM*) and the 26th Ephorate of Byzantine Antiquities during 2007–2009. It aimed to eliminate the causes of damage and tried to ensure the continuous protection of the building as well as the promotion of the historical, archaeological and aesthetic qualities of the monument.

Keywords: Protection of cultural heritage · Palimpsest · Social engagement

K. Siountri—Postgraduate Programme (M.Sc.) "Digital Culture, Smart Cities, IoT and Advanced Digital Technologies", Track "Digital Culture".

© Springer Nature Switzerland AG 2019
A. Moropoulou et al. (Eds.): TMM_CH 2018, CCIS 962, pp. 78–87, 2019.
https://doi.org/10.1007/978-3-030-12960-6_5

1   Introduction

The byzantine church of the Dormition of the Virgin in Palaia Episkopi was constructed over a roman theatre[1], erected by Antiochus IV Epiphanes (175 - 164 BC), using material (spolia) from it and the nearby early-Christian basilicas. The dome that collapsed in the middle of the 19th century and the upper parts of the masonry were reconstructed under Ziller's instructions (1884–88). The murals are made by Aginoras Asteriades (1936–1939), a painter, hagiographer and engraver of the "Generation of 1930s". Therefore, it constitutes a monument where three periods of the Greek civilization co-exist, i.e. Antiquity, Byzantium and the contemporary period.

The project "Restoration and Consolidation of the Byzantine Church of Palaia Episkopi in Tegea, Arcadia", originally presented as thesis at the MSc "Protection of Monuments", Track A, National Technical University of Athens (NTUA) with Supervisor Pr. Spyridon Raftopoulos, aimed to give technical solutions to a numerous problems related to its construction that put the building and its murals at risk.

Furthermore, the project documented the construction phases of the monument and contributed to the bibliography by bringing to light unpublished historical information related to its original construction, the date of the fall of the dome, information and designs of its reconstruction and decoration made by Ziller, A. Asteriadis and E. Stikas. The whole study after its completion was donated to Tegean Association and to the Hellenic Ministry of Culture.

During 2007–2009 the Directorate of Anastylosis of Byzantine and Postbyzantine Monuments (DAVMM) and the 26th Ephorate of Byzantine Antiquities implemented the above mentioned restoration project with EU funds, after its approval by the Central Archaeological Committee (KAS) in 2005.

The restoration and consolidation works have finished almost a decade before. But still new historical evidence is coming out. The same applies to the problems of the monument. For that reason the Tegean Association has the intention to move to digitization of all the documents concerning the church, the past interventions and the history of it. This database will function as an integrated monitoring system of the monument for its future researchers and administrators.

2   Historical Outline

Tegea is one of the most important archeological sites of Greece. The region has been continuously inhabited since the Neolithic era and always played an important role among the ancient cities of the Peloponnese, both in terms of wealth and population. According to Pausanias, Tegea, which became the capital of the homonymous region,

---

[1] The theater is referred to be erected after the generous sponsorship of the Syrian-Greek ruler Antiochus IV Epiphanes (146–330 AD), according to Tito Livio. The first who noticed the remains of it was L. Ross, (Reisen in Peloponnes , Berlin 1841). The French Archaeological School followed with René Vallois at 1912 (BCH 50, 1926).

was built by King Aleos. The name is etymologically related to the τέγος (tegos); the covered, fenced city, as its plain is surrounded by mountain ranges, while on the west is Lake Taka.

Pausanias[2] refers frequently to the importance and the glory of Tegea, to the inscriptions and to the big number of its remarkable monuments and buildings, like the Temple of Athena Alea. Indeed, the excavations of the 5th EPKA of the 1990s brought to light a part of the Agora, roman buildings, a large three-aisled early Christian basilica, a part of a bath complex, decorated spaces with mosaic floors and a small middle-Byzantine temple.

Throughout the period of Christianity in Tegea, there are two main periods that left us with remarkable monuments of art: the Early Christian one, until Justinian (565 AD) and the Middle Byzantine (10th–12th centuries). From the first period we distinguish the three-aisled basilica of Thyrsos and from the second one the church of the Palaia Episkopi.

Since the 5th century, Tegea is a Bishop's base. We are informed about *Ofelimos*, the only known Bishop of Tegea from the acts of the Fourth Ecumenical Council of 451 AD and from an inscription found in 1910 at the church of St. Ioannis Provantinos. In 530 AD the region is still named as Tegea in the list of the Provinces of the Byzantine Empire of Synecdemos[3], at the beginning of the reign of Justinian, with a geographical position between Argos and Lacedaimon. Also, at the end of the 10th century, Tegea is referred at the life description of the monk Nikon the Repent.

Among these two time references, 530 AD and the end of the 10th century, the time of the Slavic invasions and settlements (567–805 AD), the name of Tegea is gradually replaced by the name Amikle and Nykli, (with frequent references into the Chronicle of Morea[4]). Anastasios Orlandos[5], in the 12th volume of the Archive of the Byzantine Monuments of Greece the origin of the Nykli name, quotes various versions of scholars such as A. Bon, A. Chatzis and Ger. Kapsalis.

During the Byzantine era it seems that the town of Amyklion occupied only the area around the Byzantine Church of Palaia Episkopi. After the extermination of the Slavs at the beginning of the 9th century and the Byzantine Empire's domination in the Peloponnese, the settlement of Amyklion evolved into a prosperous city. At that point the city was fortified with a strong wall, as due to the flatness of the ground, its residents wanted to protect it from a future attack. Indeed, in 1208, Geoffroi de Villehardouin[6] conquered the city.

---

[2] Nikolaos Papachatzis, Pausanias' Peregrination of Greece - D - Details of Achaia & Arcadia - volume 4, Ekdotiki Athinon, Athens 1980.

[3] Le Synecdème Du Hieroclès - E. Honigmann (Corpus bruxellense historiae byzantinae Forma Imperii Romani I), Bruxelles 1939 pp. 1–48. - look A. Bon, Le Pèloponnèse bysantin, Paris 1951, p. 23.

[4] Lyrics 1715, 1752, 2027, 3332, 4878, 5147, 6711 etc. edited by Peter P. Kalonaros.

[5] A. Orlandos, - Bulletin of Byzantine and Christian Monuments of Greece Volume IB 1973, pp. 141–162.

[6] Look "Chronicle of Morea".

In 1283, during the re-establishment of the hieratic thrones under the emperor Andronikos Palaiologos and Patriarch Joseph, Grigorios of Monemvasia re-established the Diocese of Amyklion and consecrated Nikephoros Diakin[7] as bishop. Since then, the bishopric of Amyclion was moving around other Metropolises i.e. Patras, Lacedaimonos, etc. for many centuries.

In 1294 Nykli returns back to the Byzantines. However, its inhabitants decided to move to the closest high mountain and established their new city Muchli that was fortified with a triple strong wall. Thus, the place name Nykli disappears and is replaced by the name "*Palaia Episkopi*".

In 1458, the Peloponnese is occupied by Mohammed II. In 1688, the wider region of Tegea, after 230 years of continuous Turkish domination, came under the authority of the Venetians. One of the first actions of the new masters was to make inventories in order to better understand their acquisition. Within this framework, the Venetians ask the leaders of the church of Morea to register their ecclesiastical property, both monasteries and parish churches.

At that time, Tegea belonged to the bishopric of Amyclion at Neochori. In 1700, through the census[8] of the Tegea parish temples, we confirm the operation of the church of Palaia Episkopi. Also, we obtain the information that the castle was demolished.

In 1715, the Turks conquer again Arcadia, until its final liberation. This is where the first period of prosperity of the temple ends and its gradual devastation begins, as the native population has moved to Muchli. In the post-revolutionary period, under the patronage of the Tegean Association, the local community will reconstruct and reoperate the temple of Dormition of the Virgin, bringing the second revival of the monument that lasts until today.

3 Typology

The church belongs to the type of the three-aisled cross-in-square church with a narthex and five domes. There are different opinions about the date of its construction. Anastasios Orlandos, after observing details of the building, puts the date back to the mid-10th century; Charalampos Bouras believes that this church belongs to the group of temples at the end of the 10th and the first half of the 11th century, while George Sotiriou placed it in the late 11th or early 12th century.

In the framework of this research, an article of 1857[9] has been found that is reporting that on the south and west side of the church of St. Dimitrios in the nearby village Stadion, the builders are using material from the templon of the ruined church of Palaia Episkopi. The spolia that we find today on St. Dimitrios are decorated with

---

[7] N. Veis, Byzantis A', 1909, p. 68, 112.

[8] State Archives of Venice, Grimani Archive, folder 54, subfolder 157, information provided from the Historian Vasilis Siakotos for the needs of this research.

[9] Veltiosis no. 297, 12-9-1857. The information is provided from the Historian Vasilis Siakotos for the needs of this research.

twin columns that support arches under which plant decorations are being developed. This decoration theme seems to be applied after the 11th century, but we must not forget that the altarscreen is placed in the church after its construction.

4 Post-revolutionary Period

During the years of the second Turkish occupation the temple was neglected. Nikolaos Alexopoulos[10] writes in his book that the villagers used the church as a stable. However, since the first years after the Liberation, the people of Tegea established a festival around the monument every Easter Friday, at the feast of "*Zoodohos Pigi*"[11].

The travelers who visited the site of Tegea, such as Nicholas Biddle[12] in 1806, Edgar Quinet[13] in 1828–1830, A. Blouet[14] in 1833 and Kyriakos Pitakis[15] in 1837, saw the church being in ruins but also they reported the extraction of material from it and the nearby monuments, e.g. the temple of Athena Alea, by the local people.

In 1857, a local newspaper[16] refers to the fall of the dome and places it around 1830. However, in 1842, J.A. Buchon[17] visited Tegea and was the last to mention the existence of the central dome of the temple. The two above dates (1842 and 1857) combined with two strong seismic vibrations in the central area of the Peloponnese in the years 1842 and 1846 lead to the conclusion that part of the superstructure had to collapse in the second half of the 1850s. After the fall, the gradual destruction begun and the church did not operate again.

In the copper engraving, published by N. Moraitis[18], we see a picture representing the church before its reconstruction. Although the cloisonné masonry is not right, we get the information that the central dome and the arcs that support it have collapsed. The four small domes, the three semicircular apses of the sanctuary and the lunette of the southern transept remain at their position.

After the revolution, the estates around the church eventually came under the property of the Tegean Association, either through purchase or through donations. It is a total area of approximately 80 acres including the medieval walls.

---

[10] N. Alexopoulos, History of the medieval cities of Peloponnese, 1951, p. 89.

[11] The feast was called the day of "Kallos", because during the dance the future brides were selected.

[12] R.A. McNeal, (ed.), Biddle in Greece: The journals and letters of 1806, Pennsylvania State University Press 1993, p. 177 [17/06/1806].

[13] Edgar Quinet, Greece in 1830 and its relations to antiquity, Translation by Lila Ginaka, Tolidi Pulications, Athens 1988, p. 132.

[14] Expéndition Scieintifique de Morée II, Paris 1833, p. 84.

[15] Archaiological Newspaper, issue b, November 1837.

[16] Veltiosis no. 297, 12-9-1857. The information is provided from the Historian Vasilis Siakotos for the needs of this research.

[17] J.A. Buchon, La Grèce continentale et la Morée, Paris 1843, p. 420.

[18] Artist Unanimous, N.Moraitis, History of Tegea, 1932.

In 1884, the newly created Tegean Association undertook the repair of the Byzantine church and invites architect Ziller[19], who executed the designs for the reconstruction and restoration of the Christian temple.

Until today the existing bibliography (including my references in my previous research) attributed the reconstruction of the Dormition of Virgin in Palaia Episkopi to Ernst Ziller, the famous Saxon architect and designer of royal and municipal buildings in Athens, Patras and other Greek cities.

However, the designs (two facades) that came recently to the knowledge of the Tegean Association through a private collector are signed in 1885 by the architect Paul Ziller[20], the brother of Ernst, who worked as an assistant to his brother until 1878. During 1878–1890 Paul Ziller runs his own architectural studio, so it is very likely that he is the designer of the restoration of the church of Palaia Episkopi.

The facades are representing the church before and after its reconstruction. The first one has many similarities (perspective and information) with that one that four decades later N. Moraitis published.

The temple ends on 20 November 1888 and is inaugurated on 14 April 1889. Nevertheless, the effort of maintenance, repair and decoration continues. In 1917, twenty nine years after the restoration of the monument, the President of the Tegean Association commissioned the architect Emmanuel Kriezis[21] to examine the causes of moisture that appeared on the interior surfaces of the temple in order to propose a solution. Unfortunately, there is no evidence that the proposals of Em. Kriezis had been implemented on the monument. During 1917–1922, the interior of the temple was painted with lime[22] and during 1923–1924 the four small domes are covered with coper. Also, in 1924 the external joints are covered with cement in order to prevent moisture. In 1927 the old wooden doors of the temple were removed and new metal[23] ones were placed. In 1932 the church was internally lime plastered[24].

This last action opens the way for the famous painter Agenoras Asteriadis, to whom the Tegean Association commissioned the execution of the internal iconography and decoration of the monument (1936–1939). Along with Asteriadis' work, in 1939 the architect Eustathios Stikas designed the marble templon and the despotic throne.

In 1954, the Service of Anastylosis implemented the construction of the internal marble floor. Finally, during the period 1987–1989, works on the roof were carried out in order to diminish the humidity of the temple. These works failed to deal with the problem that over one century was causing damages to the monument.

---

[19] Proceedings of the Administration. Council of Tegean Association 7-5-1889.

[20] https://fr.wikipedia.org/wiki/Paul_Ziller_(architecte) (retrieval 30-09-2018).

[21] Archive of Directorate of Anastylosis of Byzantine and Post-Byzantine Monuments.

[22] N. Moraitis, History of Tegea, 1932, p. 240.

[23] N. Moraitis, History of Tegea, 1932, p. 720.

[24] N. Moraitis, History of Tegea, 1932, p. 732.

5   Description of the Contemporary Church

The exterior dimensions of the church are 14.20 m × 19.20 m. Inside, the church is divided into two uneven spaces, the narthex and the main church.

The central entrance to the narthex is in the middle of the west side of the temple, while two smaller arched openings lead to the north and south sides of it. The narthex communicates with main temple through a central arched opening and two smaller side doors.

One can also enter the main temple by two arched single-sided doors at the north and south side and a later north arched opening, which leads to prothesis. Inside the main church, the sanctuary's space has three semicircular apses and is separated from the rest of the interior by a marble altarscreen. The northeast and the southeast corner partitions are used as parabemata, which communicate with the central bema through two arched openings.

The central dome has an internal diameter of 5.65 m and a height of 15.11 m. The other four smaller domes are raised at the four corners of the monument measuring 2.50 m in diameter and about 9.50 m in height.

The exterior walls of the monument are about 1.00 m thick and combine large stone and marble semi-finished courses, probably taken from the ancient theater, with dentil courses, red bricks and spolia from neighboring monuments (marble cornices and sculptures from ancient gravestones or early Christian basilicas).

The masonry at the apses of the sanctuary and at the higher zone of the small domes, due to their curved shape, ware made of small stones, as well as small square brick pieces placed in successive horizontal zones. These materials were bonded with plenty of plaster.

The temple originally was originally decorated with chequerboard ornament of strips of diagonal square tiles (opus reticulatum). Today, the above decoration of the eastern façade has been removed or locally covered by plaster.

As we have already mentioned, from the byzantine church, the masonry is preserved up to the height of the springing of the semicircular barrel-vaults, at the four small domes and part of the lunette of the southern cross-arm. The marble toruses, at the base of the tympanums, were removed from the base of the outer semicircular retaining wall of the ancient theater[25].

The masonry of the tympanums (all except the south one, which is the only original from Byzantine phase) is constructed with cloisonné masonry, which is one of the Ziller's intervention characteristics (dome etc).

It is as if (Paul) Ziller already knew the principles that derive from the Venice Charter (1964) at article 9 and suggest that any restoration "*must stop at the point where conjecture begins, and in this case moreover any extra work which is indispensable must be distinct from the architectural composition and must bear a contemporary stamp*".

---

[25] René Vallois (BCH 50, 1926).

6 Pathology

At the end of the 20th century although the monument did not seem to have severe structural problems, a number of shear cracks appeared at multiple levels, with top-down and east-to-west direction, especially at the semicircular barrel-vaults, at the openings and the western gable wall. A deviation of 5 cm from the vertical at the north-east and south-east corner was also measured.

The problems of the rising and descending damp were serious, leading to the destruction and to the total loss of the wall paintings. The cracks at the cooper cover and the removal of the tiles at the roof, along with the cement that had covered all the joints during of the intervention in 1924 caused gradual damages that threatened all the internal decoration.

The aesthetic problems deriving from the electricity cables that existed all around the church and the organic pollutants and cement that had covered the surfaces degraded also the value of the monument.

7 The Restoration and Consolidation Project

During 2006–2009, the Hellenic Ministry of Culture executed the project of "Restoration and Consolidation of the Byzantine Church of Palaia Episkopi in Tegea, Arcadia", which was funded by European Union. The restoration and consolidation works aimed:

(a) to solve the structural problems that have occurred during the last decades due to seismic actions, e.g. repairing cracks in masonry with partial application of grouting and placing prestressed beam of stainless steel,

(b) to solve the problem of rising and descending damp through the consolidation of the foundations, the restoration of the vaults, the reconstruction of the roofs, and the removal of the cement of the joints with the simultaneous application of new compatible ones,

(c) to enhance to the monument by removing the biological pollutants from the external surfaces, replacing the electrical installations and the sanctuary floor etc.

The implementation of a collateral study by Directorate of Conservation of Ancient and Modern Monuments, which was also approved by the Central Archaeological Committee, allowed specialists carry out conservation works (2011–2014) on the murals, which have been in a very bad condition. The rising and descending damp phenomena, which would eventually lead to the destruction of the paintings, were suspended, however the destabilization of the internal microclimate had led during the restoration and consolidation works to the rapid deterioration of the exceptional work of Agenora Asteriadis.

8 Future Projects

During this research a lot of historical evidence was collected or discovered. The archive of the Tegean Association was proven valuable for the documentation of the church of Palaia Episkopi. The original designs of Aginoras Asteriadis and Efstratios Stikas are an example of these important documents.

For that reason, Tegean Association has the intention to move to the digitization of its archive creating a database of the relative bibliography, proceedings and designs to the monument. This act will allow the management of the information and will provide a monitoring system for the future administrative boards. The design of this system with all the necessary data and metadata will be realized with the help of University of Piraeus.

9 Conclusions

The case of the church of Palaia Episkopi in Tegea proves that such great monuments are not built for only one generation. It is inevitable to freeze time and to stop the deterioration. It is also not right to believe that the documentation of cultural heritage stops to the existing bibliography. The example of Paul Ziller, as presented in this paper for the first time, proves the above.

For that reason, we believe that the use of advanced digital technologies will allow us to safeguard our cultural resources, create new knowledge using prior (hidden) knowledge and develop a reflective society.

References

Alexopoulos, N.: History of the medieval cities of Peloponnese, Synodinos – Lekkas, Athens (1951)

Bérard, V.: Tégée et la Tégéatide BCH 18 (1893)

Blouet, A.: Expéndition Scientifique de Morée II, Paris (1833)

Bon, A.: Le Péloponnèse bysantin jusque' en 1204, Paris (1951)

Bouras, Ch.: Byzantine and Post-Byzantine Architecture in Greece, Melissa, Athens (2001)

Buchon, J.A.: La Grèce continentale et la Morée, Paris (1843)

Quinet, E.: Greece in 1830 and its relations to antiquity, Translation by Lila Ginaka, Tolidis, Athens (1988)

Leake, W.M.: Travels in the Morea, London (1830)

Leake, W.M.: Peloponnesiaka – A supplement to travels in the Morea, London (1846)

McNeal, R.A. (ed.): Nicholas Biddle in Greece: The Journals and Letters of 1806, Pennsylvania State University Press (1993)

Moraitis, N.: History of Tegea (1932)

Orlandos, A.: Bulletin of Byzantine and Christian Monuments of Greece, vol. IB, pp. 141–162 (1973)

Papazachos, V., Papazachou, K.: The earthquakes in Greece, Ziti, Thessaloniki (1989)

Papachatzis, N.: Pausanias' Peregrination of Greece - D - Details of Achaia & Arcadia, vol. 4, Ekdotiki Athinon, Athens (1980)

Pittakis, K.: Archaeological Newspaper, issue b, November 1837

Siountri, K.: "The Holy Temple of the Dormition of the Theotokos in Palaia Episkopi of Tegea", The Archaology in Peloponnesse, vol. 1 Proceedings of the Conference in Memory of Archaeologist Panagiotis Velissarios "Panagiotis Velissarios, Roman and Byzantine Peloponnese", Archaeological Institute of Peloponnesian Studies, Megalopolis (2012)

Sotiriou, G.: Christian and Byzantine Archeology, vol. A (1942)

René, V.: Le Théâtre de Tégée, BCH 50 (1926)

Veis, N.: Byzantis A' (1909)

Archives of (a) Tegean Association (b) Directorate of Anastylosis of Byzantine and Post-Byzantine Monuments, Hellenic Ministry of Culture

Decryption, Through the Collaboration of Historical Accounts and Technology, of the Hidden Phases of Construction of the Church of the Holy Monastery of Kykko in Cyprus. A Preliminary Report

Nasso Chrysochou[✉]

Frederick University, Nicosia, Cyprus
{nassoch,art.cn}@frederick.ac.cy

Abstract. The Church of the Holy Monastery of Kykko has undergone many reconstructions and constructional changes since its Imperial Byzantine establishment at the end of the 11th and beginning of the 12th centuries. The monastery, housing the famous Icon of the Holy Virgin, is said to have been established by Constantinopolitan funding from the Emperor Alexios Comnenos. Historical records have informed us that the monastic complex burnt down in the 14th century and was rebuilt in 1366 by the Frankish King, Peter Lusignan. It was destroyed by fire a second time and was rebuilt a year later in 1542. In 1751 a third fire partly destroyed the monastic complex and most likely became the pretext for the expansion of the church complex, defying strict Ottoman rules against any such action. The church underwent a large-scale embellishment in the 1970's and acquired a new concrete roof which hid the existing one, preserving many ancient secrets in the in-between space. The monument was architecturally documented and researched for the first time in 1998 by the author, but the roof was then inaccessible for study. A recent preliminary research of the old roof structure, which had been hidden and stripped of its more recent coverings, confirmed the historical records referring to the many reconstructions and also revealed many constructional phases of the famous edifice. Through three-dimensional modeling of the church, the newer phases are stripped away, thus revealing possibly the original church and its preserved typological integrity, hidden inside the many reconstructions carried on during the centuries. Step by step, the newer phases are visually added on, to allow a new understanding of this much venerated but so far little studied monument.

Keywords: Kykko · Historical phases · Reconstruction · Three-dimensional modeling

© Springer Nature Switzerland AG 2019
A. Moropoulou et al. (Eds.): TMM_CH 2018, CCIS 962, pp. 88–103, 2019.
https://doi.org/10.1007/978-3-030-12960-6_6

1   Historical Information Regarding the Architecture of the Church of the Monastery of Kykko

The Monastery of Kykkos is located in the province of Nicosia, in Marathasa, at an altitude of 1,200 m. According to tradition, it was founded at the end of the 11th or the beginning of the 12th centuries under the reign of the Byzantine emperor Alexios Komninos, who was responsible for the bequeathal of the Holy Icon of the Virgin from Constantinople and the establishment and economic support of the monastery [1–3] (see Fig. 1).

Fig. 1. The church of the Monastery of Kykko from the south.

The original church was destroyed by fire in 1365, a period during which Cyprus was under Frankish occupation, but the following year it was rebuilt by the French King of Cyprus, Peter Lusignan, in honor of his wife Eleonora from Aragonia, on whose feudal land the monastery was built. The church, which is referred to as being wooden although this more likely meant that it had a wooden roof, was destroyed by fire for a second time in 1541 and was rebuilt during Venetian rule in 1542, by the then Abbot Symeon [4, 5]. Some information regarding the disaster is given in an archival report of the Venetian state in 1542. It refers to the buildings of the Kykko monastery, describing them as abandoned after the fire of 1541. It appears, however, that the monastery had not been totally destroyed. It maintained its importance during the Venetian occupation, since it is recorded to have received financial support from the Venetian government [6, 7]. At this time a scriptorium and manuscript binding probably operated there [8].

Traveler Richard Pocoche, who visited Cyprus about two centuries later in 1730, reported that the monastery received many visitors from many parts of the world, including Russia [9]. Among these is the well-known monk Vasil Grigorovich-Barsky, who visited many of the known monasteries of the island, for most of which he left descriptions and sketches.

Barsky visited Kykkos twice, in 1727 during his second visit to the island and in 1736 during his fourth visit to the island. In his diary, he has bequeathed us with vital information regarding both the church as well as the monastic complex of Kykko, which he described as stone-built with a tile-covered wooden roof and a dome with windows. During his second visit to the monastery, the famous Russian monk had an opportunity to stay there for 23 days and to complete his description in more detail. The church, according to his diary, was in the northern part of the monastery and not in the center (as was traditional with Orthodox monasteries). It was equal in width, length and height and similar to those of the Agio Oros in Greece. It had five doors, three on the west side,[1] and one on each of the south and north sides. The propylaea had one entrance. The church had vaulted ceilings (Barsky specifically stated that it had "vaulted roofs on all four sides"), with the dome visible only from the inside. Externally, the vaulted roof was covered by a wooden, tiled roofing structure, similar to the Catholicon of St John Lampadistis in Kalopanayiotis. He also notes that the Church of Kykkos is the most beautiful church he has described after that of St. Mammas (Morphou) [10].

The first reference by Barsky to a dome with windows is confusing as to the form of the Catholicon's roofing. One possible scenario is that, between his two visits (1727–1736), the church vaults were covered with a new wooden roof, which also covered the original dome as a measure of weather protection, as was the case in other wooden roofed churches of Troodos, a feature unique to Cyprus and particular to the region [11–14].[2]

Small renovations and additions to the monastic complex were carried out throughout the Ottoman domination. In 1734, the great cistern was constructed by the then Abbot Sophronios in the southwest of the church, lining up with the southwest corner of the 3rd bay of the church from the east or, as we shall soon see, the south-western corner of the then church [15] (see Fig. 2).

Surprisingly, traveler Alexander Drummond, who visited the Monastery in 1742, only six years after the visit and description by Barsky, notes, contrary to the description of the latter, that the church has no entrances from the west [18].

In 1751 a third big fire broke out, which destroyed the cells of the monks. Seraphim Pisidios in the republication and completion of the book of Ephraim Athenaios reveals the destruction in 1751, as well as the rebuilding of the Monastery in 1755 under the guidance of the then Abbot Parthenios [2]. The works lasted for five years.[3] The church was rebuilt first as what was described as "three aisled, vaulted" but it seems that part of the roof collapsed during the works and had to be rebuilt.

Significant information on the destruction of the church after the fire of 1751 is found in Ottoman documents discovered in the monastery and subsequently published. As stated in a May 1754 letter, a court decision was issued by the Ottoman administration, authorizing the construction of the church, provided that the new building

---

[1] Here I am correcting what I believe is a mistake in the translation by Grishin. He mistakenly refers to five doors on the west side instead of three plus two on the other two sides. Five doors on a west elevation of any church are almost improbable.

[2] Lampadistis monastery in Kalopanayiotis has an extensive wooden roof that covers the whole system of vaults and domes underneath.

[3] An inscription with the year 1760 and the name of the master builder «Ιωάννη» (John) is carved on the round window of the refectory.

Fig. 2. Plan of the Monastery of Kykko (drawing by Nasso Chrysochou, Eleni Petropoulou, Artemis Pseftodiakos, Maria Philokyprou).

would be rebuilt on the old foundations, without elevating its original roof or any additional decorations. This document also provides valuable information on the dimensions of the ruined church, though the extent of the destruction is unknown. According to the *firman*, the church, prior to its reconstruction, had a length of 16 Ottoman yards and a width of 11 yards, which today stands at 12.12×8.33 m [16]. The year 1755, carved on the Bishop's throne, testifies that the wood-carved iconostasi was constructed around the middle of the 18th century, immediately after the rebuilding of the church.

Several items regarding the architecture of the Catholicon of Kykko after its rebuilding in 1755, are also illustrated in a 1778 etching, drawn by Michael Apostolos and etched by Innocentedri and Pietro Scattaglia, which was printed in Venice for the Kykko Monastery [17]. In this synthesis, the church is depicted from the north with a high-pitched gable roof with a dome but without the present northern extension. Their also appear only two bays, after the dome instead of the four existing today.

Conservation and small-scale works were conducted at regular intervals in the church, such as the opening of the furthest west door on the south wall of the church in 1795 [6]. Alas, the Holy Monastery was doomed to suffer more destruction by fire eighteen years later.

The fire of 1813 seems not to have damaged the church and its paintings. According to George Sotiriou, who visited the Monastery in 1931, an inscription with the year 1500 survived in the eastern apse of the sanctuary inside the round window. In the semi-dome of the apse is a depiction of the Platytera and other religious scenes dated to the end of the 18th - beginning of the 19th centuries by the Cretan painter Ioannis Kornaros, executed during the reign of the Abbot Meletios [11, 19].

In a sketch executed in 1862 by the traveller to Kykko Edmond Duthoid, the vaulted church appears to be covered by a wooden roof [20]. In photos of the 1960's the western part of the roof is raised above the rest, almost hiding the dome from that direction.

The latest and most important change was carried out on the church in 1973–74, when the external walls of the church were raised and it was covered by a concrete, cross-pitched, tiled roof. The contemporary roof covered the old vaults and the base of the dome, while its tympanum was elevated so as to be visible above the new roof and a false semispherical dome was cast over it. The inclined concrete slabs were supported on a series of beams and pillars that were placed directly onto the old roof structure, above its columns and its walls. Subsequently, a void was formed between the two roofs, that was never fully examined until now (see Fig. 3).

Fig. 3. The space between the two roofs of the Catholicon of Kykko Monastery.

2   The Church Today

The church today has local masonry walls, the stone interspersed with pieces of broken ceramics. Limestone was used in the cornerstones and as decorative frames around the openings. It has the form of a three-aisled basilica with a dome and internal dimensions of 24.9 × 12.72 m, while internally it is divided in 6 bays by 8 round columns and two square piers (including the area of the apse). The central aisle is covered with a vault, supported on arcades.

Fig. 4. The central aisle of the church.

The side aisles are covered with cross vaults, which are interrupted by arches and allow for the formation of tall windows on the two side facades. The dome is not in the center of the church, but immediately in front of the bay of the apse, supported by the two square piers and two columns (see Fig. 4). The tympanum of the dome is cylindrical internally, and eight sided externally. The northern arm of the cross was extended by 7 m and to its west was created a large space used as a baptistery. Further west, a cell with a cross-vault roof appears at one point to have communicated through a large opening with the westernmost part of the church.

Fig. 5. Plan of the church today, showing additions and their constructional joints through the centuries (drawing by Nasso Chrysochou, Eleni Petropoulou, Artemis Pseftodiakos, Maria Philokyprou).

The two orthogonal piers east of the dome and the 18th century iconostasis form the boundaries of the sanctuary, its floor raised 0.17 m higher than that of the main space. The sanctuary has only one central apse, circular internally and semi-hexagonal externally. The corner apartments have a square shape and do not terminate in semi-circular apses. All three sanctuaries are roofed with vaults in an east-west direction. The northern apartment (prosthesis) has an external entrance, with three semicircular steps. In the southern part of the sanctuary (diaconico), there was a stair that is today blocked off, leading to a small underground crypt, while four steps lead to a small room above it. The underground room might have served as an ossuary but has now been sealed. Similar spaces were found in the monasteries of Panagia Kriviotissa [21], Stavrovouni [22] and Apsinthiotissa near Sygrari [23]. The stair leading upstairs was probably a secret passage to the Abbot's quarters [24].[4] This also has recently been sealed (see Fig. 5).

The entrances to the Catholicon today are four, two on the south side, one on the north façade with a Propyleon and one opening directly into the Sanctuary, again on the north side. Propylae also existed in the Byzantine monastery of Machairas as

---

[4] These were documented in the year 1998 by the team of architects: Nasso Chrysochou, Maria Philokyprou, Eleni Petropoulou, Artemis Pseftodiakos. Since then they have been closed off.

mentioned in the founder Nile's Typicon [25]. The high windows of the church are large and symmetrical, two in the south and two in the north. Symmetrically-installed windows exist in both the western and the eastern facades. Three small windows exist in all three parts of the sanctuary.

3 New Evidence Discovered Under the New Roof

Below the modern concrete roof, the older one is of an inscribed cruciform shape to the east, the arms of the cross forming a steeply inclined, gable roof. The eastern arm, part of the northern arm (the extension is vaulted) and the southern arm of the cross, are covered in a most unusual fashion with flat limestone slabs, which terminate in a decorative cornice. The western arm only partly preserves its pitched form as it was also extended and appears to have been reconstructed with a vaulted roof when the church was extended westwards. The base of the dome, which is most likely the oldest part, appears to be built entirely of bricks, while the square base formed over the pendentives, is also covered with the flat calcareous slabs (see Fig. 6). These are of a 0.40×0.40 m^2 and have been discovered to have lead grouting instead of the usual lime and were most likely covered with lead sheeting, which helps to explain the 1896 reference to the "lead tiles of Kykko" [26].

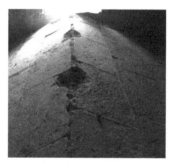

Fig. 6. The eastern arm of the cross of the roof of the church covered with calcareous slabs and grouted with lead.

Observing the old roof west of the dome, constructional joints appear at two places on the vault of the central aisle (see Fig. 5). These are at the end of the first and the third from the west, bays. The cross-vaults seeing from inside the church covering the side aisles on the roof appear as vaults constructed perpendicular to the central one. They also appear to be of a different constructional material and method of building (see Fig. 7).

Fig. 7. The side aisles are covered with vaults perpendicular to the central one.

As can be seen from photographs of the middle of the 20th century, the old roof of the church was covered with Byzantine type tiles (with flat and semicircular tiles), much like those found in many Byzantine monastic churches.[5] This type of tile was used during the Byzantine years until the Latin era and beyond. In archival photos of the beginning to mid 20[th]century, there appears to be lead sheeting in certain areas underneath the tiles. With the construction of the new reinforced concrete roof, the older tiles were removed [4].

4   Reconstructing the Phases Through Historical Information, in situ Study and 3D Architectural Modeling

The form of the original church of the 12th century is unknown. It most likely had a wooden roof but all evidence disappeared with the destructive fire of 1365. It was rebuilt the following the year but it was destroyed again in the fire of 1541 and rebuilt the following year. Since there was no recorded destruction or rebuilding of the church between 1542 and the visit of Barksy in 1735, we can safely assume that the church that he so elaborately described, is the Venetian period (1489–1571) edifice. Through his description and the aid of three dimensional architectural modeling, we can build a very clear picture of the Venetian period church which, according to Barsky, was very similar to that of the Lampadistis monastery, vaulted ("on all four sides"), of similar width, length and height, with a dome, but covered with a second wooden roof as was customary in the Troodos area [28][6]. Moreover, his account of three doors in the west cannot support the previous researchers' claims that it was a single-aisle church [4]. The in-situ evidence in the roof shows a cross-inscribed roof and a constructional joint at the end of the 3rd bay from the east. This is also reinforced by the most important discovery, the fact that the dimensions given to us by the Ottoman delegation and which refer to the 16th century church, if not an even older edifice, match the length of

---

[5] The tiles appear in old archival photos, but we cannot exclude the possibility that these are the tiles from an earlier roof.

[6] Wooden roofed churches of the Troodos region and of the same period are Panagia Podithou Galata (1513 and 1514) and the church of Panagia Kakopetria (1520).

the 3 eastern bays of the present church. This dimension of 12.12 m also matches exactly the internal height of the dome, as described by Barky.

From all the above evidence, we can surmise that the church that existed in 1542 and possibly also in 1366 was of a cross-inscribed plan type, domed building, covered with a wooden roof. The exact angle of its slope and its direction are unknowns, but examining the monastery of St Ioannis Lampadistis in Kalopanayiotis, which according to Barky had a similar roof, as well as existing conditions and relationships of the church to the monastic complex, we can make an assumption on the direction of its slope. As the church was positioned on the north side of the monastic cloister with its south elevation facing it, we can assume like in the example of the Catholicon of Lampadistis, that the roof had a west-east slope to direct the falling snow away from the courtyard. Like in Lampadistis, a lower roof possibly protected the south elevation and the door that connected the southern door of the church with the cloister and the rest of the monastic complex (see Fig. 8).

Fig. 8. A three dimensional model depicted from the northwest, of the possible shape of the wooden roof covering the Catholicon of Kykko (drawing by Andrea Michaelidou). A similar roof, covers the monastic church of St Ioannis Lampadistis in Kalopanagiotis (picture on right).

Similar monastic churches were those of Agio Oros, St Nicolas of the Roof in Kakopetria which dates to the 11th century, St Ioannis Lampadistes in Kalopanayiotis of the same era, mentioned above, and the cross inscribed Church of the Holy Cross in Lefkara dated to the 14th century if not earlier, and whose extensions and alterations bear an uncanny resemblance to the reconstructions implemented on the Kykko church [12].[7] The cross inscribed plan was popular in the island from the 12th to the 14th century and therefore points to the direction that the 16th century building was possibly a mere restoration of the 1366 church rather than a new building with a different

---

[7] This monument has similarly undergone many restorations and extensions but has started like the Kykko Catholicon as a cross-inscribed plan which based on surviving frescoes, has a terminus ante quem in the 14th century. What remains of the old plan, as in the case of Kykko, is the cross with the dome, while the west arm of the cross was removed in 1867 (according to an inscription over the western door) in order to extend the church westwards with cross vaults in the main and side aisles. The west most bay was added later in 1910, as was the case of the Kykko Catholicon.

typology. Furthermore, the dated epigraphy of 1500 inside the apse also gives us a terminus ante quem for at least the eastern part of the church to the beginning of the 16th century or earlier. As the width of the apse has a direct correlation to the width of the central aisle, this is additional information to be used in the reconstruction of the historic phases (Fig. 9).

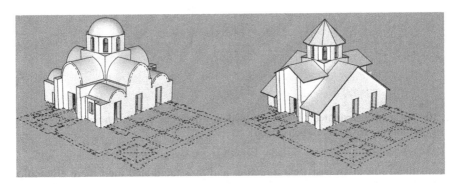

Fig. 9. An architectural three dimensional modeling depicting the church as it might have stood in the 16th century, if not also earlier. The first model shows the church as it was first constructed and the second the church covered with the gable roof (drawings by Andrea Michaelidou).

What though of the width of 8.33 m given to us in the Ottoman document? This does not match the 12.8 m of today's church width and neither does it support Barky's description of a church of equal width and length. Could the church have been widened? The 1754 dimension given in the Ottoman document for the width of the then edifice can support neither the theory of a single-aisled church (a dome of approximately 8 m diameter is not probable for a Cypriot church) nor that of a cross inscribed type with such a difference between length and width.[8] Therefore we must conclude that some mistake was made in the documentation of the measurement of the width, which must have been closer to 12 m as is the case with the present day, three eastern bays of the current church. It is of interest to note that the Ottoman dimensions of 12.12 × 8.33 m match exactly the dimensions of the church, in width and length of the first two bays starting from the east.

A second phase that can be ascertained is the extension of the church westwards. The destruction of the western arm of the cross coincides with the historical information provided by Seraphim Pisidios, who informs us that in 1755 the church was "rebuilt as a three aisled vaulted basilica" [4] (see Fig. 10). It is obvious that the cross in square church was not totally rebuilt but only elongated by two bays, as is confirmed by the in-situ roof inspection showing a second constructional joint at that exact point. The extension of the vaults of the existing cross-inscribed building would have been

---

[8] In a cross inscribed church, the proportions of the middle aisle to the side aisles is usually that of 1:2 as is the present church. Given that the apse is dated to the 16th c then the middle aisle must follow a similar width of approximately 4.7 m as is the present one, giving us an approximate width of 2.35 m for the side aisles. Their actual width at the piers is about 3 m which is close enough.

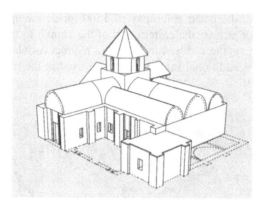

Fig. 10. An architectural three dimensional modeling depicting the church with the 1755, west extension (drawing by Andrea Michaelidou).

easy, as these run in an east-west direction and thus facilitated their mere elongation. During these works, it seems that possibly the central vault of the western arm of the cross failed, partly or totally destroying the dome. Thus the western supports of the dome were rebuilt as round columns rather than pillars, to match the rest of the extension. Drummond's 1742 report of nonexistent doors to the west could be explained as an early attempt by the monks to stabilize an old and weakened church building (two to four hundred years old) by filling in the western doors. The removal of the western wall to facilitate the extension of the church possibly caused the above described collapse.

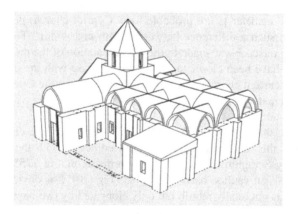

Fig. 11. An architectural three dimensional modeling depicting the church with the final western extension and the change of the side vaults to cross vaults (drawing by Andrea Michaelidou).

The church was extended further west by another bay, giving it its present form, possibly after the destruction of 1813 even though there is no historical information confirming this, other than the physical evidence of the constructional joint. The westernmost bay was added to be used most likely as a narthex, even though today it is an integral part of the church. The cell on the northwest corner of the church, possibly part of a now destroyed wing, was most likely used as an entrance to the narthex from the north.

The extension of the north arm of the cross possibly took place in the 19th or 20th century, even though there is no recorded information on this. The latter extension is already present in a photo by Theodoulos Toufexis in the year 1907 [18] and thus could have been built under the reign of the Abbot Sophronios (1862–1890), who was responsible for many conservation works in the monastery.

Unfortunately nothing is documented after the destruction of 1813 but works on the church were carried on continuously in the 19th and 20th centuries, down to the present day. Probably at the beginning of the 20th century, the side vaulted aisles were changed to cross vaults as is shown from the different construction of their roof. This was very common in Cypriot churches of the time, as an introduction of a neo-Gothic morphological feature made prevalent by colonial British architects [30] (see Fig. 11).

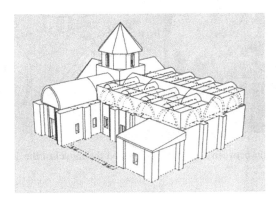

Fig. 12. An architectural three dimensional modeling depicting the church with the side aisle vaults covered by perpendicular to the central one, vaults (drawing by Andrea Michaelidou).

The in-situ documentation of the roof shows vaults over the side aisles, perpendicular to the central one. These were created most likely over the cross vaults, to allow for the water to drain off the roof prior to it being covered by a second wooden one, as cross vaults tend to produce a flat roofing surface inappropriate for the wet and snow producing winters of the Troodos mountains (see Fig. 12).

The last major change to the edifice was the introduction of the reinforced concrete roof, which covered and kept secret all the above discoveries. Its possible impact on the monument will be the next item to be investigated (see Fig. 13).

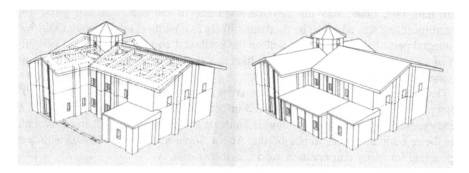

Fig. 13. An architectural three dimensional modeling depicting the church with the modern concrete roof and the latest addition of the northern baptistery (drawing by Andrea Michaelidou).

Fig. 14. A comparison between the 1778 etching and the model of the 16th century church.

5   Conclusion

After this first attempt at reconstructing the architectural phases of this so important monument, a lot of questions arose, awaiting answers through further research. The architectural three dimensional reconstruction modeling aided the visual understanding of the building but was also a tool to experiment with the unknown forms of the building, such as the wooden superimposed roof that Barsky describes and whose evidence has been completely lost to us.

Questions remain as to the unexplained defiance of the Ottoman authorities' orders regarding the maintenance of the exact church size and form. These appear to have been disregarded in 1755 and the church extended. Were the Ottoman authorities more lax than was thought? Obviously, it seems that what were thought to be strict rules could be bent or bypassed, as was also the case with Cretan churches under Ottoman rule.

Another question that remains open is the dating of the high-pitched gable roofing covered with stone slabs and lead grouting. As the western arm has been partly destroyed in order to be extended, it reveals to us that these possibly pre-existed the western elongation of 1755. This type of construction is unique to this Cypriot monument and gives us information on new constructional techniques so far not known to have been used on the island. Gabled roof churches were popular in the 15th–16th centuries and were found in Orthodox edifices mainly in the Famagusta region, possibly an influence from the Latin architecture imported to the island by the Francs and Venetians [28]. Such gables were covered by a hydraulic lime plaster in order to water proof them. The use of the porous stone on the Kykko roof, in a mountainous region with extreme winter conditions, can only be justified if the roof surface is protected, in this case by lead sheeting. Was this then a new innovation used during the rebuilding of the church in 1542? Also, does this justify the depiction of the church with high pitched gables in the 1778 etching by Michael Apostolos? This was previously assumed to have been an exaggerated artistic whim, used by the artist to flavor his work with a western nuance [17]. Indeed this might be the case with the depiction of the height of the buildings. However a comparison between the print etching and the three dimensional model allows us to think that the artist was indeed depicting a fairly realistic view as concerning the roof forms (see Fig. 14).

Lastly, the periods during which the church was covered with a wooden roof are also undefined. Wooden roofs feature in the early descriptions of the first churches of the monastery. Whether these were wooden roofs or merely wood-structure coverings of vaulted roofing, remains unclear as is their constructional techniques and morphology. Even Barky's descriptions create some confusion as to the form of the wooden roofing of the church[9]. If this was an integral part of the protection of the church and its frescoes, then when was it introduced and if it was applied after the lead covering, what was the reason? In pictures of the 20th century the roof of the church appears to have been covered by tiles of various types, attached no doubt to the steep inclinations below by a wooden frame. However, no evidence of this roofing survives, minus a few holes on the stone covering the pediments.

Obviously these new discoveries are just the beginning for a further scientific investigation on this so important and yet little studied monument of Cypriot Byzantine and post Byzantine ecclesiastical architecture.

References

1. Μαχαιράς, Λ.: Χρονικόν Κύπρου, Dawkins, R.M. (ed.) Oxford (1932)
2. Αθηναίος, Ε.: Η περιγραφή της Σεβάσμιας και Βασιλικής Μονής του Κύκκου, Ενετίησιν (1751)
3. Κυπριανός, Αρχ.: Ιστορία Χρονολογική της νήσου Κύπρου, Ενετίησιν (1788)
4. Dionysios Kykkotis (Abbot of Kykko): The Holy monastery of Kykko founded with a cross, Nicosia (1979)
5. Gunnis, R.: Historic Cyprus, Nicosia (1936)

---

[9] In his first account he mentions the dome windows and in the second he describes the roof and dome as covered by a wooden roof.

102   N. Chrysochou

6. Grivaud, G.: Η Ιερά Μονή Κύκκου και τα εισοδήματά της στα 1553. Επετηρίς Κέντρου Μελετών Ιεράς Μονής Κύκκου 1, 75–93 (1990)
7. Αριστείδου, Α.: Η Ορθόδοξη Εκκλησία της Κύπρου κατά την περίοδο της Βενετοκρατίας. Θησαυρίσματα 23, 200–201 (1993)
8. Constantinides, C.N., Browning R.: Dated Greek manuscripts from Cyprus to the year 1570, Nicosia 1993
9. Guillemard, F.H.H.: Transcript of four diaries written by, Cyprus 1887–88. (Typed copy in the library of C.A.A.R.I, Nicosia)
10. Grishin, A.: A pilgrim's account of Cyprus. Bars'kyj's travels in Cyprus. In: Wallace, P.W., Orphanides A. (eds.) Sources for the History of Cyprus, vol. III, New York 1996
11. Σωτηρίου, Γ.Α.: Τα Βυζαντινά Μνημεία της Κύπρου. Εν Αθήναις (1935)
12. Παπαγεωργίου, Α.: Οι ξυλόστεγοι ναοί της Κύπρου. Τόμος αναμνηστικός επί τη 50ετηρίδι του περιοδικού "Απόστολος Βαρνάβας"(1918–1968), pp. 361–556, Λευκωσία (1975)
13. Κοκκινόφτας, Κ.: Τα μοναστήρια της Ιεράς Μητροπόλεως Μόρφου. In: Πολιτιστικό Ίδρυμα Τραπέζης Κύπρου - Ιερά Μητρόπολις Μόρφου (εκδ.), Ιερά Μητρόπολις Μόρφου. 2000 χρόνια Τέχνης και Αγιότητος, pp. 179–192, Λευκωσία (2000)
14. Παπαγεωργίου, Α.: Η βυζαντινή αρχιτεκτονική (4ος-12ος αι In: Πολιτιστικό Ίδρυμα Τραπέζης Κύπρου - Ιερά Μητρόπολις Μόρφου (εκδ.), Ιερά Μητρόπολις Μόρφου. 2000 χρόνια Τέχνης και Αγιότητος, pp. 63–70, Λευκωσία (2000)
15. Κυριαζή, Ν.Γ.: Λαογραφικά. Κυπριακά Χρονικά (13), pp. 228–237 (1937)
16. Θεοχαρίδης, Ι.: Οθωμανικά έγγραφα 1572–1839. Κέντρο Μελετών Ιεράς Μονής Κύκκου (1) XXXIX, Β' (1993)
17. Περδίκης,Σ.: Η περιγραφή της Ιεράς Μονής Κύκκου σε μία Χαλκογραφία του 1778, Επετηρίς Κέντρου Μελετών Ιεράς Μονής Κύκκου1, pp. 31–50 (1990)
18. Σοφρονίου, Σ.: Ξένοι επικέπτες στην Μονή Κύκκου. Ιερά Μονή Κύκκου, In: Α. Τσελικά, Σ.Περδίκης (επιμ.) Εικών ασπέρου φωτός. Πολιτιστικό Ίδρυμα Ιεράς Μονής Κύκκου, pp. 220–233, Αθήνα (2010)
19. Άσπρα-Βαρδάκη, Μ.: Χείρ ην ζωγράφου Κορνάρου Ιωάννου Κρητός εν τη Μονή της Κύκκου. Πρακτικά Συνεδρίου: 'Η Ιερά Μονή Κύκκου στη βυζαντινή και μεταβυζαντινή αρχαιολογία και τέχνη' (Λευκωσία 14.5.1998–16.5.1998), pp. 344–347, Λευκωσία (2001)
20. Σεβέρη, Ρ.: Ταξιδιώτες ζωγράφοι στην Κύπρο 1700–1960. Π. Νικολάου(trans.), Ίδρυμα Κώστα και Ρίτα Σεβέρη (2003)
21. Παπαγεωργίου, Α.: Κρινιώτισσας Παναγίας Μοναστήρι. Μεγάλη Κυπριακή Εγκυκλοπαίδεια, vol. 14, pp. 297–298, Φιλοκυπρος Λευκωσία (1991)
22. Η Ιερά Μονή Σταυροβουνίου: Ιστορία-Αρχιτεκτονική-κειμήλια, Ιερά Μονή Σταυροβουνίου (εκδ.), Λευκωσία (1998)
23. Παπαγεωργίου, Α.: Η παλαιοχριστιανική και βυζαντινή αρχαιολογία και τέχνη εν Κύπρω κατά το 1964, (ανατύπωσις εκ του περιοδικού Απ. Βαρνάβας), Λευκωσία, 10 of 8–20 (1965)
24. Χρυσοχού, Ν.: Η εξέλιξη της αρχιτεκτονικής της Ιεράς Μονής Κύκκου. Ιερά Μονή Κύκκου, In: Τσελικά, Α., Περδίκης, Σ. (eds.) Εικών ασπέρου φωτός. Πολιτιστικό Ίδρυμα Ιεράς Μονής Κύκκου, pp. 48–61, Αθήνα (2010)
25. Μέναρδος, Σ.: Η εν Κύπρω Ιερά Μονή της Παναγίας του Μαχαιρά. Ιστορικό διάγραμμα, εν Πειραίει (1929)
26. Κοκκινόφτας, Κ.: Ιερώνυμος Μυριανθέας (1838–1898). Ο λησμονημένος δάσκαλος του έθνους, Λευκωσία (1990)

27. Pelekanos, M.: The role of a post-byzantine timber roof structure in the seismic behavior of a masonry building—the case of a unique type of timber-roofed basilicas in Cyprus (15th–19th century). In: Cruz, H., Saporiti Machado, J., Campos Costa, A., Xavier Candeias, P., Ruggieri, N., Manuel Catarino, J. (eds.) Historical Earthquake-Resistant Timber Framing in the Mediterranean Area. LNCE, vol. 1, pp. 17–31. Springer, Cham (2016). https://doi.org/10.1007/978-3-319-39492-3_2

28. Χρυσοχού, Ν.: Ορθόδοξη εκκλησιαστική αρχιτεκτονική στην Κύπρο κατά την περίοδο της Ενετοκρατίας, Μελέται και Υπομνήματα, vol. VIII, pp. 118–125, Ίδρυμα Αρχιεπικόπου Μακαρίου, Λευκωσία (2018)

29. Stylianou, A.: Αι περιηγήσεις του μοναχού Βασίλειου Βάρσκυ, εν Κύπρω. Κυπριακαί Σπουδαί, vol. 21, pp. 62–68, Nicosia (1957)

30. Ντέλλας, Ν.:Οι σταυροθολιακές εκκλησιες της δωδεκανήσου (1750–1924). Διαδακτορική διατιβή, Παραρτημα Β, Κατάλογος σταυροθολιακών εκκλησιών Κύπρου, Αθηνα (2011)

The Oslo Opera House – Condition Analysis and Proposal for Cleaning, Protection and Maintenance of Exterior Marble

Pagona-Noni Maravelaki[1(✉)], Lucia Toniolo[2], Francesca Gheraldi[2], Chrysi Kapridaki[1], and Ioannis Arabatzis[3]

[1] Technical University of Crete, 73100 Chania, Greece
pmaravelaki@isc.tuc.gr
[2] Politecnico di Milano, 20133 Milan, Italy
[3] NanoPhos SA, Science and Technology Park of Lavrio, 19500 Lavrio, Greece

Abstract. The photo-oxidative degradation of the polymers applied in the past onto the marble slabs of the Oslo Opera House (OOH) led to the yellowing of the coatings and/or to the formation of insoluble fractions of polymer. Despite the high reduction of the $b*$ parameter (yellowness) observed after the mechanical abrasion, this cleaning treatment cannot be recommended for further application, due to the unacceptable removal of structural material during the abrasion. Cleaning of historic masonry has progressed significantly in recent years, thanks to methods that rely on nanoscience. Therefore, in this research project instead of using organic solvents exhibiting toxicity and poor performance, advanced nanostructured fluids, such as micelles solutions (MC) and microemulsions (ME), embedded in nanogels with industrial solvents were tested aiming at the efficient removal of the hydrophobic coatings from Carrara marble slabs. These nanogels are amphiphilic-based formulations using a system of water, oil and surfactant. However, lack of reproducibility characterized the application of these advanced nanogels, due to partly removal of the unevenly distributed polymer products. Therefore a three-step cleaning methodology was adopted including: the introduction of a chelating agent capable of dissolving the hard film created by the applied coatings, thus allowing any subsequent MC and ME solutions applied in the second step, to swell the polymeric chains and to facilitate their removal, which was further completed by suitable oxidants. Optical and scanning electron microscopy coupled with EDX, along with colorimetric measurements proved that the three-step cleaning tests applied to yellowed samples can successfully remove the colour discoloration.

Keywords: Oslo Opera House · Yellowing identification ·
Cleaning with nanogels

1 Introduction

1.1 Problem Identification

The Norwegian National Opera and Ballet House (OOH) is a contemporary architecture designed by Snøhetta Architects and built in Oslo in the period 2000–2008 [1].

© Springer Nature Switzerland AG 2019
A. Moropoulou et al. (Eds.): TMM_CH 2018, CCIS 962, pp. 104–116, 2019.
https://doi.org/10.1007/978-3-030-12960-6_7

The OOH was characterized by cladding mainly Carrara marble tiles and represents a connection of the city with the fjord landscape. The construction of the Opera House is the largest single culture-political initiative in contemporary Norway, and is therefore of outstanding importance for the country and its population.

In 2007–2008 the external marble surfaces of OOH underwent protective treatment aimed to facilitating subsequent regular cleaning operations. The treatment involved impregnation of all the surfaces with a fluorinated acryl copolymer, Faceal Oleo HD (PSS Interservice AG, Switzerland), and application of PSS 20 (PSS Interservice AG, Switzerland), a polysaccharide-based product for anti-graffiti protection of surfaces [2]. The functionality of Faceal Oleo HD for horizontal and vertical surfaces was guaranteed for 3 and 10 years, respectively, while the PSS 20 guaranteed for 3 years for all surfaces [2].

Despite cleaning operations that were regularly performed on the surfaces, a yellow colour discoloration of the horizontal surfaces appeared since 2007, inducing considerable aesthetic alteration to the aspect of the building.

Given the poor performance of Faceal Oleo HD [2], alternative products were tested in the course of 2011, after which the product showing the best performance was applied extensively in 2012 for a second impregnation. Official reports and documents of previous analytical investigations pointed out to the recovery of a high amount of silicon on surface, rather than the chemical nature of the relevant product.

Subsequent investigations aimed at clarifying the source and mechanism of the yellowing were also conducted but did not yield significant result. The early hypothesis that the fluorinated acryl polymer might be involved in the discolouration of marble through some degradation product was questioned on the grounds that the removal of the polymer was not able to eliminate the yellowing [2]. The new protective treatment did not prove to be effective in preventing the discolouration of marble, suggesting that the cause of the problem should be addressed differently.

To complicate things further, efforts made so far to remove the yellowing can hardly be considered satisfactory in the absence of an explanation and prevention of the phenomenon. Indeed, even though some cleaning methods have been proven able to bring the marble back to its pristine state [2], the high costs of the operation cannot be justified without gaining insights into the cause of discolouration and the strategy to effectively tackling the problem.

1.2   Damage Classification and Actions to Be Undertaken After the Building Inspection

In the framework of the Project "The Oslo Opera House – Condition analysis and proposal for cleaning, protection and maintenance of exterior marble" the research team of Politecnico di Milano and Technical University of Crete, inspected the OOH in 2015 and identified the sampling areas taking into account: (a) the different decay pattern morphology; (b) the different orientation and exposure of tiles to the weathering conditions; (c) the macroscopic differentiation in the color intensity of the observed yellowing; (d) the different surface finishing of the tiles (polished or rough); (e) the necessity to obtain the minimum but sufficient number of samples.

In a first attempt to approach the condition survey according to the Standard Protocol EN 16096:2012 [3], we suggest that Symptoms are varying from CC0 (no symptoms) to CC2 (moderately strong symptoms), the Urgency risk classification ranges from UC0 (long term) to UC1 (intermediate term) and the possible measures range from RC1 (maintenance/preventive conservation) to RC2 (moderate repair and/or further investigation). Several characterization techniques were applied to investigate five marble samples of exposed tiles coming from OOH. The key points of our investigation are:

– to gain a deep insight into the causes of the discolouration (chromatic alteration);
– to assess the conservation condition of Carrara marble, with particular regard to the severe environmental conditions;
– to set-up tailor-made cleaning methodology of the marble tiles (Fig 1).

Fig. 1. Yellowing observed in the cladded façade with Carrara marble tiles of the Opera Oslo House.

2 Materials and Methods

2.1 Marble Samples and Applied Coatings

The Carrara marble tiles from OOH were demounted and collected from different areas of the building (P1, P2, P3A, P3B and P6), as reported in Fig. 2. In particular, from P1–P6 tiles several different small blocks were cut and sent to the laboratories in Italy and Greece.

The tiles coming from the building are characterized by different surface finishings: P1 tile has a "heavy bush hammered" finishing, P2 and P3B tiles have a "rough bush hammered" finishing and P3A and P6 tiles have a "sawn" finishing (Fig. 2) [2]. Some specimens, which were cut from the internal part of a Carrara marble tile "not exposed" (as referred from Statsbygg) on the roof of the building but conserved in the backyard at OOH, were considered as reference sample for the unaltered material. Moreover, reference products of the two surface coatings used for the protection of the marble,

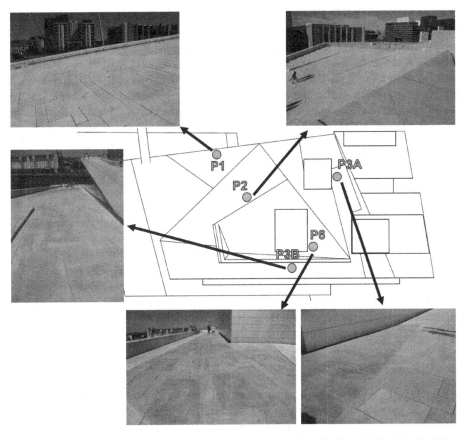

Fig. 2. Original localization of the marble tiles in the areas P1, P2, P3 and P6 of the building.

such as Faceal Oleo HD and PSS 20, PSS Interservice AG, Switzerland, were also investigated for comparison purposes.

2.2   Instrumental Methods

VIS spectrophotometric measurements were carried out on the surface of each tile with a Konica Minolta CM-600D instrument with a D65 illuminant at 8°, wavelength range between 360 nm and 740 nm. Measurements were elaborated according to the CIE L*a*b* standard colour system, which describes a colour with three different parameters, L*, a*, and b*, measuring respectively brightness, red/green and yellow/blue colour intensities. 180 measurements were performed on each area (20 × 20 cm) of the exposed marble tiles, whereas 100 measurements were carried out on the samples taken from the unexposed marble tile. The average results of L* a* b* were used to calculate the colour difference ΔE^* ($\Delta E^* = [(L_2^* - L_1^*)^2 + (a_2^* - a_1^*)^2 + (b_2^* - b_1^*)^2]^{1/2}$) between the exposed and the unexposed marble tiles.

From each tile, different micro-samples were collected from the surface. They were observed with a Leica M205C stereomicroscope equipped with a Leica DFC 290 video camera. Then, the morphology of the stone specimens was analyzed by Environmental Scanning Electron Microscopy (ESEM) and EDX analyses (Zeiss EVO 50 EP ESEM, equipped with an Oxford INCA 200 - Pentafet LZ4 spectrometer). The exposed surface, fresh fractured surface and the polished cross-section of the samples were investigated. Moreover, samples from the mortars collected from P2 and P6 marble tile were analyzed with ESEM-EDX.

In order to identify the protective treatments applied to the surface of both the unexposed and exposed marble tiles, a small amount of carbonatic powder was collected from the surface by a scalpel and the liquid extracts with appropriate solvents were deposited on KBr window then analyzed by Fourier-Transform Infrared Spectroscopy (FTIR), using a Nicolet 6700 spectrophotometer equipped with a DTGS detector (acquired between 4000 and 400 cm^{-1} with 128 acquisitions and 4 cm^{-1} resolution), after solvent evaporation. For comparison, the two surface coatings used for the protection of the marble tiles (Faceal Oleo HD and PSS 20) were analyzed by FTIR in the same conditions.

Moreover, from each tile, the deposits on the surface were collected with a pin and analyzed by micro-FTIR with Nicolet Continuum FTIR microscope equipped with a MCT detector (acquired between 4000 and 600 cm^{-1} with 128 acquisitions and 4 cm^{-1} resolution) using a micro compression diamond cell accessory.

2.3   Cleaning Methodology

To remove the previously applied treatments in OOH a cleaning methodology using appropriate pure solvents, nanofluids and nanogels in different cleaning supports, such as pulp poultices, cotton fabrics, gauze, agar was attempted. At the end of the tests, the poultices were removed and the surface was gently rinsed with hot water to remove possible residues. Table 1 lists the categories of the cleaning agents tested for each tile.

Table 1. Overview of the tailored-made cleaning formulations.

Cleaning type	Cleaning code
Pure and mixture of solvents	SLV
Nanogels of advanced micelles solutions	NGMC
Nanogels of microemulsions	NGEM
Nanogels of micelles solutions and microemulsions	NGME
Nanogels of microemulsions and micelles with Industrial solvents	NGEMS
Oxidizing agents	OxA
Chelating agents	CA

The nanogels and cleaning agents listed in Table 1 were specifically formulated for the cleaning of OOH. These formulations consist of an anionic surfactant, water and solvents (a non-solvent and a good solvent for the polymer, respectively). The presence of the surfactant guarantees a decrease of the interfacial tension between the stone and the swollen polymer droplets and, therefore, the stabilization of these irregular structures. The shape of the water droplet test evidenced the polymer removal and the hydrophilic behaviour of the surface. Factors such as the application time of the cleaning agent and the ageing of the polymers influenced the removal of the applied materials. It is of great importance to have in mind that an aged polymer cannot be fully removed from the surface, probably because of the stone's porosity. Carrara marble is a low porosity stone but it is still a porous material and protective treatments penetrate inside the pores for a thickness of hundreds microns till 1 mm. In particular, detachment and enlargement of crystalline grains' boundaries can be observed on the examined surface and this has surely increased the surface porosity of the marble tile, favouring the penetration of polymeric materials.

Cleaning tests, coupled with the direct observation of the morphological and aesthetical changes of the polymer film by means of optical and electronic microscopy and colorimetric measurements, have shown that the nanogels applied are not always efficient in polymer removal.

We suggest that this behavior, which is related to the superimposing of the different protective layers, made the system extremely difficult to be dissolved in a one-step advanced cleaning formulation. This was confirmed in lab experiments which were characterized by scarce reproducibility, due to the poor adsorption of the nanogel ingredients on the external layers of the polymeric coating. These structural differences can be responsible for the different cleaning efficacies and for the different kinetic profiles observed. However, the synergistic action between water, organic solvents and surfactant is of fundamental importance to obtain excellent cleaning performances. In fact, water causes the structural reorganization of the external polymeric layers, while the organic solvent swells the entire film increasing the mobility of the polymeric chains. At the same time, the surfactant interacts with the outer polymeric layers and decreases the interfacial energy at the marble/polymer interface, promoting the detachment of the polymer from the marble substrate.

At this point a smart selection on the key-steps recommended for cleaning was the introduction of a chelating agent capable of dissolving the PSS 20 polysaccharide layer, which created a hard film, not allowing any solvent to swell the polymeric chains. Therefore, after the sufficient removal of polysaccharide, as proven by the increase of water absorption, the nanogel application followed. The third key-step was the application of an oxidizing agent to facilitate the removal of polymeric residues decomposed from the previous treatments.

3   Results and Discussion

3.1   Aesthetic, Morphological and Compositional Parameters of Yellowing

Figure 3 shows the color parameters as obtained by colorimetric comparison between the yellowed and reference Carrara tiles. The color parameter b*, related to the yellowing, along with the total reflectance showed the highest diversification in the P6, P2 and P3B Carrara marble tiles comparing to the P1, P3A and reference marble.

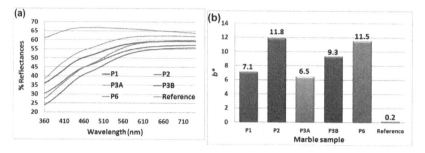

Fig. 3. (a) Reflectance spectra and (b) $b*$ values of the yellowed marble originating from all the levels (P1, P2, P3A, P3B and P6) in comparison with the reference Carrara marble.

The samples collected from the surface of P1–P6 tiles are characterized by a significant inter- and intra-granular decohesion, where the crystals show corroded gemination planes and detached grains near the surface. This important degradation phenomenon of the crystal structure is well described also in the document "ICOMOS-ISCS: Illustrated glossary on stone deterioration patterns" [4]. It can be related to changes in temperature due to severe microclimatic condition of the building, with freezing-thawing cycles, to surface erosion phenomena caused by the weather precipitations (rain and snow) and to salt crystallization cycles caused by NaCl (halite) marine spray and therefore cyclic dissolution and re-crystallization inside the surface porosity of the stone. In Fig. 4, the electron microscopy images of the surface of sample collected from P6 tile are reported. This sample is characterized by an intense yellowing with $b* = +11.5$, and an inter-granular decohesion. In particular, the surface of some grains is characterized by the presence of some "drops" of an organic material (Fig. 4, darker areas) with high signals of Si and C, according to the elemental analysis, and ascribed to the presence of a siloxane treatment by means of solvent extraction and FTIR spectroscopy [5].

In addition to the solvent extractions, the deposits observed on the surface of the tiles were collected and analyzed by FTIR spectroscopy equipped with an optical microscope to analyze sample at the micron scale. The spectrum obtained from one of these micro-samples is reported in Fig. 5, and compared to the spectrum of the anti-graffiti coating PSS 20, based on vegetal polysaccharides. A significant overlay of the absorption bands of the two spectra can be observed, in particular, the absorption bands at 3349 (OH stretching), 2926 (CH stretching), 1651 (OH bending), 1566 (C = O stretching in COO-), 1410 and 1365 (CH bending) and the main peak at 1012 cm^{-1} related to C-C and C-O stretching of the pyranose ring of polysaccharides [6]. This result clearly indicates that these yellow-brown deposits are residuals of PSS 20 treatment.

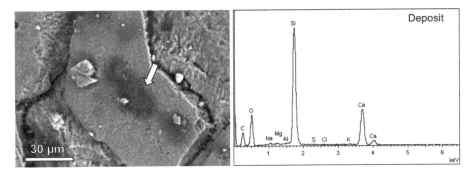

Fig. 4. Electron microscopic image of a crystal of P6 sample with accumulation of an organic treatment and the corresponding EDX spectrum.

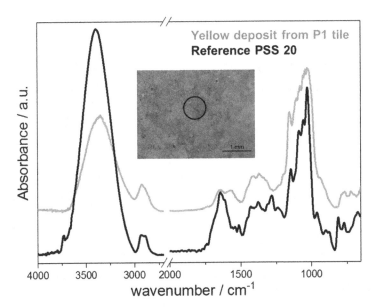

Fig. 5. Microscopic image of yellow-brown surface deposit on P1 tile with the FTIR spectrum of the deposit (grey) and the spectrum of the vegetable polysaccharide-based treatment (PSS 20) (black). (Color figure online)

3.2 Cleaning Results

Research has been carried out in order to remove the polymeric Faceal Oleo HD and PSS 20 that were used as protective and antigraffiti, respectively, and have been recognized as responsible for the discoloration due to their photochemical instability. Our research studies on cleaning were mainly focused on marbles originating from P2 and P6 levels which exhibit the most intense staining comparing to the other levels.

The most commonly used method to remove deteriorated polymers from treated surfaces is the use of solvents usually organic, such as acetone, xylene, toluene, ethyl acetate, alcohols or mixtures of them. Unfortunately, these solvents are not only toxic and harmful to workers' health, but they might induce further migration of the dissolved polymer inside the stone. Moreover, when solvents are used for the removal of the applied polymer, swelling of the polymer can be observed on the surfaces, thus demanding further mechanical abrasion in order the polymer removal to be completed. However, through this abrasion, especially in porous materials, parts of the material underneath could be disrupted. As expected, none of the colour parameters were improved and in almost all cases the spreading of the polymers inside the bulk of the marble resulted in a decrease of lightness and an increase of the yellowing parameter b*, as illustrated in Fig. 6.

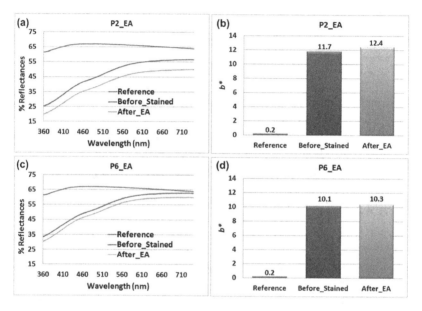

Fig. 6. (a) and (c) Reflectance spectra of the surfaces and (b) and (d) b* parameter values of P2 and P6 marbles before and after the immersion in ethyl acetate.

The core of our research was to assess the efficiency of advanced cleaning systems, such as nanogels with microemulsions and/or micellar solutions [7, 8]. Therefore, the performance of the designed solutions applied with a nanogel support was also assessed in all of the samples. However, the changes induced from the applied treatments in the colour parameters and the reflectance spectra indicated insignificant improvement of the colour parameters with lack of reproducibility.

The previously described attempts to remove the yellowing resulted in a poor performance due to durability of the applied coatings and the superimposing of different layers rendering their detachment extremely difficult. It was deemed important to combine the potentiality and efficiency of advanced formulations recently proposed, such as microemulsions and micellar solutions with well-established organic industrial solvents (NGEMS). Following this direction several promising results were obtained and the relevant formulations were designed in close collaboration with NanoPhos. However, no reproducibility characterized the application of these advanced cleaning systems and in some cases the state of the treated surface was clearly worsened (Fig. 7b).

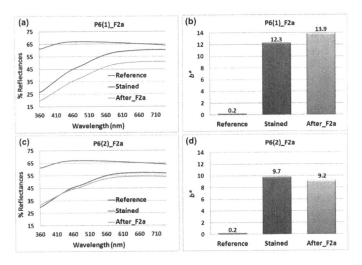

Fig. 7. P6 marble tile treated with NGEMS (F2a agent): (a and c) Reflectance spectra of the surfaces and (b and d) b* parameter values before and after cleaning.

The inefficiency of the advanced nanogel system NGEMS reinforced with the most commonly used industrial solvents into removing the yellowing, addressed research activities into adopting a scenario with a separate key-step before advancing with the application of the nanogels. This determinant cleaning key-step comprised of an application of a chelating agent capable of dissolving the polysaccharide film and capturing the Ca^{+2} ions in order to accomplish the polymeric film detachment [9]. Then the advanced nanogel system was applied and the coating removal was accomplished by a final application of formulations with oxidants.

At this point no further information could be provided about the chemical composition and specific ingredients of the three-step cleaning methodology due to constraints with the industrial partner and the relative patent application.

Table 2 lists the changes induced after the three-step cleaning in the b* colour parameter and the total colour difference ΔE. The three - step cleaning methodology applied to P1–P3B marble tiles induced more than 50% change in the $b*$ parameter related to yellowness, while luminosity increased too. The P6 tile exhibited a 42% reduction of the yellowness and the highest total colour difference towards yellowed and reference sample. No morphological changes were observed after the three-step cleaning, as illustrated in the electron microscopy image of P6 tile in Fig. 8 comparing

Table 2. Change of colorimetric parameters after cleaning

Samples	% Change of b*[a]	ΔE*[a(yellowed)]	ΔE*[b(reference)]
P1	−53.3	4.7	3.6
P2	−54.3	7.5	7.4
P3A	−64.9	3.9	2.7
P3B	−54.9	5.3	5.6
P6	−42.4	8.0	8.8

[a]referred to yellowed, [b]referred to Reference

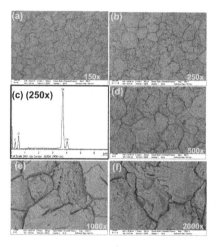

Fig. 8. SEM-EDX images of the cleaned P6 at different magnifications: (a) 150x, (b) 250x, (c) EDX spectrum (250x), (d) 500x, (e) 1000x and (f) 2000x.

with the corresponding image of the yellowed sample in Fig. 4, and in the macroscopic images of Fig. 9 for the P3B. The cleaned P6 surface exhibited a grain detachment and etching along with a visible boundaries decohesion, but not to a greater extent than that observed for the yellowed counterparts.

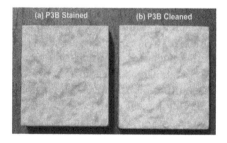

Fig. 9. Macroscopic images of the yellowed and cleaned surface for the P3B tile. (Color figure online)

4   Conclusions

The photo-oxidative degradation of the polymers applied onto OOH marble tiles, such as Faceal Oleo and PSS20 led to the yellowing of the coatings and to the formation of insoluble fractions of polymer, which makes the treatment not compatible and impossible to be completely removed. The removal of these polymeric coatings was therefore one of the main challenges in this scientific project, without suggesting mechanical abrasion treatment often associated with unacceptable removal of structural material.

Advanced cleaning strategies using nanogels with microemulsions and micelles exhibited low performance and lack of reproducibility, due to the partial removal of the polymers. A smart selection on the key-steps recommended for cleaning was the introduction of a chelating agent capable of dissolving the hard coating not allowing any solvent to swell the polymeric chains. The polysaccharide and the silicon based layers removal was checked and proved by the absence of Si in the electron microscopy study. Then, as a second key-step the nanogel application followed. The third key-step was the application of an oxidizing agent assisting in the removal of polymeric residues decomposed from the previous treatments.

Optical and scanning electron microscopy coupled with EDX and colorimetric measurements proved that the three-step cleaning tests applied to yellowed samples can successfully remove the residues of the polymer degradation and maintain the surface aesthetical characteristics.

Acknowledgements. This research was carried out in the framework of the competitive Project "The Oslo Opera House – Condition analysis and proposal for cleaning, protection and maintenance of exterior marble" funded by Statsbygg, Oslo (Norway). The Authors gratefully thank Dr. Adriana Eidsvik and all the Statsbygg staff for the constant support and advice during the implementation of the Project.

References

1. http://snohetta.com/projects
2. Nilsen, B.: Maintenance operations management (MOM) information for the project: the oslo opera house – condition analysis and proposal for cleaning, protection and maintenance of exterior marble. s.l.: Betong Consult as, 04.11.15. doc 201501489 Operaen i Oslo - Utendørs marmor, rev. 02
3. 1: Standard Protocol EN 16096:2012, Conservation of cultural property - Condition survey and report of built cultural heritage
4. ICOMOS-ISCS: Illustrated glossary on stone deterioration patterns, ICOMOS. International Scientific Committee for Stone, Paris (2008)
5. Gherardi, F., Kapridaki, C., Roveri, M., Gulotta, D., Maravelaki, P.N., Toniolo, L.: The deterioration of Apuan white marble in contemporary architectural context. J. Cult. Herit. (2018, in press)
6. Oh, S.Y., Yoo, D.I., Shin, Y., Seo, G.: FTIR analysis of cellulose treated with sodium hydroxide and carbon dioxide. Carbohydr. Res. **340**, 417–428 (2005)

7. Baglioni, P., et al.: Micelle, microemulsions, and gels for the conservation of cultural heritage. Adv. Colloid Interface Sci. **205**, 361–371 (2014)
8. Grassi, S., Favaro, M., Tomasin, P., Dei, L.: Nanocontainer aqueous systems for removing polymeric materials from marble surfaces: a new and promising tool in cultural heritage conservation. J. Cult. Herit. **10**(3), 347–355 (2009)
9. Deng, J., Shi, Z., Li, X., Liu, H.: Soluble Polysaccharides Isolation and Characterization from Rabbiteye Blueberry (Vaccinium ashei) Fruits. BioResources **8**(1), 405–419 (2013)

Modern Architecture and Cultural Heritage

Agnes Couvelas[✉]

3 Dyrrachiou Street, 15669 Papagou, Athens, Greece

Abstract. The problems of modern architecture as they pertain to the physical and social environment are acute and interconnected. Architecture has to contribute to an integrated collective effort for solutions.

Beyond traditional practice, architects need to adopt ecological design and be trained accordingly. Further to bioclimatic research and practice, teachings should also include the history of cultural heritage and its influence on design, intellectual development and ethical issues.

Architecture today is globalized; the same materials are used all over the world to reproduce a limited array of forms, while the principles of eco-design are often reduced to commercialized applications.

Believing that design is improved by taking into account experiences from different localities, a concept of Inter-locality is introduced as a synthesis of the design principles of different peoples in different localities facing similar problems. In my view, inter-locality is a vital alternative to the worldwide imposition of stereotyped forms that lead to an architecture without local character, without alternative expressions. As a major step, the establishment of a novel inventory of bioclimatic solutions as applied to traditional settlements of interest to architectural heritage and to individual anonymous architecture structures therein is proposed.

In my opinion, the study of architecture should adhere to the synergy between bioclimatic design and cultural heritage through inter-locality. None of the above can be achieved, though, without multidisciplinary collaboration starting at the borders of the University.

Keywords: Modern architecture · Cultural heritage · Sustainable design

1 Locality an Route to Inter-locality

Prominent architects around the world have vitally participated over the years in an active dialog between present and temporal localities.

The concept of **locality** can be defined as the quality that renders a product of architecture part of its surroundings (locus) both natural and cultural. Locality derives dynamically from the locus and includes the existing social relations at a particular time. Consequently, it is a specific time-space phenomenon.

Every locality is a potential form of universality. Its main characteristics are: dialog between artifact and nature, human scale, response to substantial human needs, rational and true usage of materials in structures, clarity by restriction in simple and purposeful forms, familiarity, and endowment with a universal spirit.

© Springer Nature Switzerland AG 2019
A. Moropoulou et al. (Eds.): TMM_CH 2018, CCIS 962, pp. 117–128, 2019.
https://doi.org/10.1007/978-3-030-12960-6_8

Architecture, "globalized" in many ways, should be guided, instead, by the principles of **Inter-locality** [1], a synthesis of new design principles that incorporates experiences from different temporal local identities. This term is introduced to emphasize the need to consider localities of different civilizations, as well as local values when creating a new design. Contrary to abstract universality, Inter-locality penetrates localities and derives from their qualities.

Being multicultural, it retains and enriches local identity and, contrary to what one may think, creates diversity, a quality that today, more than ever before, we need to preserve and enhance. Inter-locality has an open character and may lead to the appearance of new localities or enrich a work with universal qualities. It can thus minimize the damage and open prospects for the future.

2   Investigating Inter-locality

2.1   Anonymous Examples from the Past

The principles of inter-locality were used ante litteram in anonymous structures worldwide. Two of the innumerable examples of a universal locality appear in Figs. 1 and 2, where one can see it applied anachronistically, before the term was coined. Their qualities derive from local and are connected both to each place and all places. They are pan-cultural, running across cultures, not imposed or imported, though.

Fig. 1. Badgir (wind tower), Jazd, Iran (photo 1995) and pigeon coop, Olympia, Greece (photo 1968) in juxtaposition

Fig. 2. Traditional house in Santorini, Greece (photo 2014) and Bam, Iran (photo 1995) in juxtaposition

2.2 Present Examples from Our Work

Antiquities are characterized by locality, being inseparable from the place where they were found. But, at the same time, they contain inter-locality because of their global radiance and messages they carry to humanity. Since they derive from different historical periods, they also express temporal Inter-locality. Moreover, inter-local is the way we treat them by obeying to widely accepted museographic principles and rules.

The In Situ Museum of Naxos. In designing the Museum, we attempted to develop a new interpretation of the genius loci by expressing the identity of the historic landscape.

Fig. 3. The in situ Museum of Naxos. Agnes Couvelas, Architect (1994–1999)

Bricks were used as a modern version of an ancient material, mud-brick, which was widely used in the excavated area. As an industrial material, it both refers to and marks out the exhibits creating a new locality and constituting a means of symbolism of present time (see Fig. 3).

The museum empowers spiritually the local community, whereas the public square above it and the surroundings fulfill the principles of ecomuseology. Although design took place before these principles were widely deployed, the complex can now well be interpreted as a territorial heritage museum.

Therefore, it can be characterized as an emerging ecomuseum effectively capturing strong local distinctiveness. Thus, the rich material and immaterial heritage aids community sustainability [2].

The Marathon Tumulus Archaeological Park. Around the Marathon tumulus, an archaeological park is arranged to illuminate the historical battleground. Our intervention maintains the character of the plain (see Fig. 4), thus enhancing the Tumulus, this earthwork-like burial monument of a conical, strictly geometrical form whose shape repeats the maternal shape of the mountain (see Fig. 6). The landscaping is indigenous, similar to the one applied in ancient times. Accordingly, we used earthly materials but demanded precise marking to illustrate that the texture is the work of man and not of nature. The visit traces the periphery of the tumulus. A path defined by its material, crushed rock, intersects with a perceived, horizontal plane of planted flora like a fissure on the ground of a drained marsh, thus alluding to the ancient site [3].

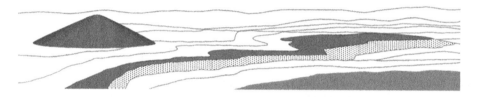

Fig. 4. Planters in different zones on the plain, parallel to the seashore (2004)

The structures testify the identity of the natural and cultural locality and express it in contemporary terms in connection with its historical tradition.

The strong horizontality refers to the marsh covering the ancient battle site and stresses the presence of the monument. The Tumulus emerges from the organized natural texture even more emphatically. The undulating pathway around the monument looks like a crack on the ground, resembling the dried marsh covering the ancient battlefield (see Fig. 5).

3   Education and Research

Teaching in architecture should include the history of cultural heritage and its influence on design, intellectual development and ethical issues. The approach should avoid the sterile "return to the past" that in architecture is often expressed as copying forms of the past in today's works [4]. It should focus, instead, on a universal consideration of the links between ecology, available means and way of life, traditional heritage and

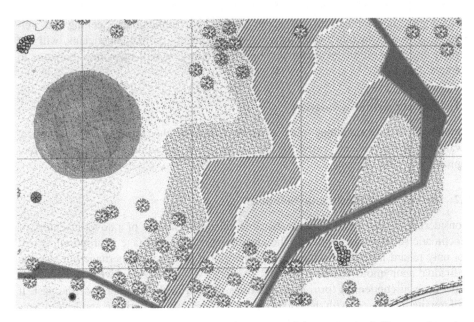

Fig. 5. The undulating pathway followed by visitors around the monument is like a crack on the ground, resembling the dried marsh covering the ancient battlefield.

Fig. 6. The Tumulus repeats the maternal shape of the mountain.

technological development in the search of inventive applications. Seeking and welcoming advice from other specialists, as e.g. engineers or mathematicians, will not restrict design intuition for architects; on the contrary, it will open new horizons to creative design [5].

3.1 Environmental Awareness

Schools in architecture may play a decisive role in the formation of ecological awareness of the students, provided that all, teachers and students, have already recognized this vital necessity.

3.2 Inventory of Bioclimatic Solutions

Considering the fact that local is sustainable, the establishment of a novel inventory of bioclimatic solutions as applied to the global architectural heritage is proposed. It will not only regard traditional settlements of the global architectural heritage but also the individual anonymous buildings therein. The latter would provide a wealth of practical experience in condensed form from solutions applied in the structures of the past. It will reunite the present with the wider meaning of the cultural heritage and lead to a re-evaluation of qualities that are almost forgotten.

The creation of such an inventory will require renewed research efforts in a vital field that is under gradual disappearance. The utilization of state-of-the-art technology in the documentation and visualization of tangible heritage will effectively produce accurate models for further inspection in three dimensions.

Rigorous investigation will lead to a codified body of knowledge in the form of a historic resource inventory. This investigation focused on ecology-centered assessments will guide anyone interested in gathering more data on sustainable solutions of the past constituting a reference guide for an evidence-based design.

Moreover, blending high-tech design and ancient sustainable applications into an integral corpus, will provide those interested in passive energy design with a reservoir of ideas and demonstrate that good design was and is sustainable design.

Jaalis. These historical perforated stone or latticed screens, common in Islamic Architecture, were filtering solar radiation and enhancing the wind passing through them. Both ornamental and fully functional - their properties, widely recognized now- are applied in analogous modern structures around the world. Further research on the initial applications will certainly offer invaluable information (see Fig. 7).

3.3 Shape and Topology Optimization

The recent methods of "topological optimization" should also follow the quantitative criteria of energy saving and quality of living conditions.

Due to the significant energy problems worldwide, bioclimatic approach is a topic of increasing interest in architectural design. The basic idea behind bioclimatic design is to incorporate passive (natural) systems into the buildings, which utilize environmental sources in order to cover housing needs, such as thermal or lighting comfort.

Jaali pattern	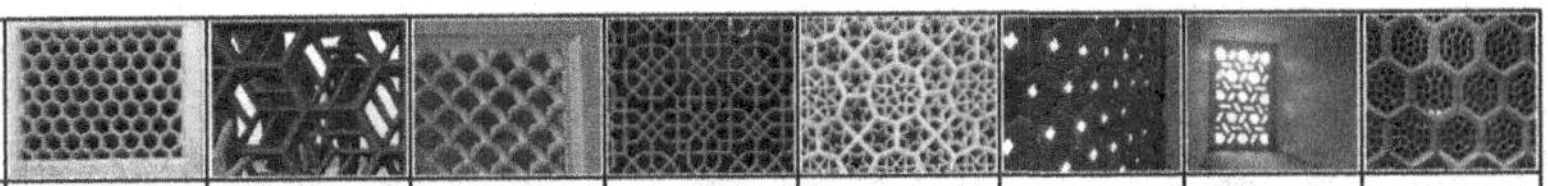							
Location	Taj Mahal	Taj Mahal	Fatehpur Sikri	Fatehpur Sikri	Fatehpur Sikri	Hawa Mahal	Hawa Mahal	Agra Fort
Use	Wall-Window	Railing	Railing	Window	Wall-Window	Wall-Window	Window	Window
Module	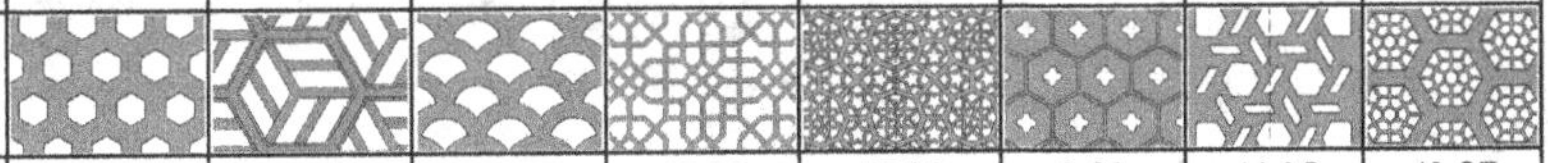							
% Void	34,12	47,31	40,52	67,12	57,38	8,91	46,15	41,07
% Baluster	65,88	52,69	59,48	32,88	42,62	91,09	53,85	58,93
Depth (cm)	4	5	5	5	5	6	5	5

Fig. 7. Jaalis. Investigation we performed in Rajasthan, India (January 2017)

In most cases, such solutions derive mainly from smart, practical ideas, rather than scientific rigor. Thus, they may probably be further improved by employing optimization methods. The efficiency of the proposed methodologies will be illustrated via three real-life examples.

4 Form Optimization Examples Related to Our Cultural Heritage

Currently, in collaboration with academic engineers with expertise in computational methods, we are analyzing the functional behavior of some of our projects [6].

This is the topic of a study under the European Research Program Horizon 2020. Shape and performance optimization techniques are applied.

4.1 "Sails", Shipping Company Building Improved Through Performance Optimization

In this project, each of the facades was specifically retrofitted to address its solar orientation (see Figs. 8 and 11). Externally mounted custom-made shades protect the interiors from overheating and glare, their shape alluding to the sails that once dominated the nearby port of Piraeus. The effectiveness of this external shading system is currently studied through simplified ray methods for modeling the light (see Fig. 9). Future research will encompass a parametric analysis and assessment of the performance of the proposed adaptive façade system through geometric configurations of the external shading structure and louvers [7].

Sailing boats crossing the Aegean today make the sail a living record of our cultural heritage and symbol of our maritime tradition (see Fig. 10). Similar research will be applied in the "sails" of this project to attain use of the prevailing wind through computational fluid mechanics (CFD), a further step based on optimization programs (see Fig. 11).

Fig. 8. SAILS headquarters, Piraeus, Greece (photo 2014), Agnes Couvelas, Architect, and private house in Kyoto, Japan (photo 1995) in juxtaposition

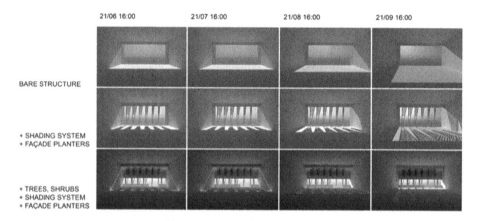

Fig. 9. SAILS headquarters, Piraeus, Greece (2012) - Geometric configuration diagrams (2018)

Fig. 10. Yachts are a living part of our cultural heritage. Modern sails in Santorini and SAILS shading devices that trap the fresher north-eastern summer breeze.

4.2 Handling the Wind at the House of the Winds

In designing this house in Santorini [8], handling the wind was an early concern. The solution was to exploit the prevailing wind was thus exploited as a passive means of

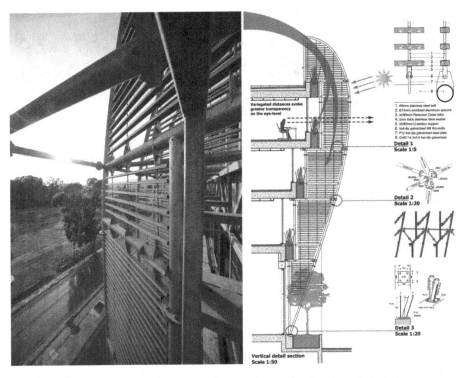

Fig. 11. SAILS shading devices - West façade detail and vertical section

protection against its own force. Funnels ending in slit-like outlets at the window-reveals were "burrowed" into the wall mass. Air entering the funnels accelerates and as it exits from the slit it functions as a protective curtain to the opening. The air, i.e., that exits the funnel forms an invisible curtain in front of the nearby window or door and deflects the oncoming wind that would otherwise invade the room [9].

Funnels allude to the earliest rock-hewn chambers, structures created by the wind as a consequence of the effect of natural forces on the rock itself. Their quality of form, dugout from the exterior towards the interior, was dictated by Aeolian energy and natural sandblasting and underlines potent messages associated with the place (see Figs. 12 and 13).

The shape and size of the funnels were initially determined in an empirical way without modeling or performance studies. The optimization of these parameters is currently the topic of a study under the European research program OptArch H2020 (see Fig. 14). The study, entitled "Improvement of bioclimatic design through optimization of performance," aims to optimize and explore the practical application of such techniques.

Fig. 12. The House of the Winds in Santorini, Agnes Couvelas, Architect (1994) - North façade detail and rock-hewn cave on the island [8, 9]

Fig. 13. The House of the Winds in Santorini - East façade detail showing wind deflection

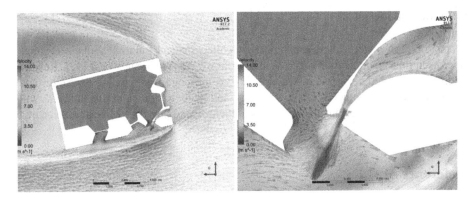

Fig. 14. The House of the Winds in Santorini - Upper storey plan diagrams [10]

4.3   Low Walls Against Sand Accumulation and Wind

Our proposal mainly consists of snakelike walls situated at right angles to the prevailing winds, their shape alluding to the wave formations on the sandy beach. The walls mitigate sand-blasting by deflecting the lower layers of local wind hitting against the walls. Their organic shape blends uses applicable to both local and global conditions.

The immediate additive effect of such low "ramparts" is to address local conditions by enclosing areas of cultural importance, such as archaeological findings. Differing in appearance and blending with the landscape, they will not cause to the visitor perceptual confusion with the relics.

What we search here is to optimize the wall shape by applying techniques related to fluid mechanics (see Fig. 15).

Fig. 15. Sand Containment Wall Sections & diagram, Agnes Couvelas, Architect (2013) [11]

5 Conclusion

In my opinion synergy between modern architecture and cultural heritage requires interdisciplinary cooperation and a leading role by the Academic world [12]. Related approaches, under a continuous broadening of the cognitive base of education and simultaneous development of new topology and shape optimization techniques, will lead to innovative buildings that enrich local identity and improve the performance of a built structure.

References

1. Couvelas, A.: Locality - interlocality: tools for architectural creation. Official J. Camra tal-Periti (45), 23 (2008)
2. Couvelas, A.: Metropolis square at Naxos Island: creation and functional evolution (1999–2012). In: 1st International Conference Proceedings on Ecomuseums, Community Museums and Living Communities, ECOMUSEUMS 2012, pp. 71–78. Green Lines Institute for Sustainable Development, Portugal (2012)
3. MONUMENTA: The Marathon Tumulus and its environs, Agnes Couvelas-Panagiotatou. https://www.monumenta.org/article.php?IssueID=3&lang=en&CategoryID=6&ArticleID=132. Accessed 18 July 2018
4. Rapoport, A.: A framework for studying vernacular design. J. Archit. Plan. Res. **16**(1), 52 (1999)
5. Dapogny, C., Faure, A., Michailidis, G., Allaire, G., Couvelas, A., Estevez, R.: Geometric constraints for shape and topology optimization in architectural design. Comput. Mech. **59** (6), 933–965 (2017). Springer
6. Agnes Couvelas is scientific responsible for Work Package 5: Bioclimatic building design improved through performance optimization, OptArch project of the European H2020-MSCA-RISE-2015 program
7. Couvelas, A., Matheou, M., Phocas, M.C.: Analysis and development of an adaptive façade system integrated on a multi-storey office building. In: The Tenth International Conference on Engineering Computational Technology, Barcelona, Spain (2018)
8. Couvelas, A.: House at Santorini. Domus Des. Sustain. **789**, 32–36 (1997)
9. Couvelas, A.: The House of the Winds in Santorini. Shape IKE, Athens (2016)
10. Prof. Yiannis Andreopoulos and his assistant Joan Gomez, PhD student from CUNY have kindly offered to study one of these funnels and test the simulation models to be developed
11. Michailidis, G.: Civil Engineering, PhD in applied mathematics (shape optimization), Ecole Polytechnique (X), Paris, "H. Navier & G. Stokes, some results on the examples of A. Couvelas" (2017, unpublished)
12. Couvelas, A: Synergy of sustainable design and cultural heritage. In: International Conference on Structural Engineering Education without Borders IV, Conference Proceedings, pp. 518–527. ACHE, Madrid (2018)

Cross-Discipline Earthquake Protection and Structural Assessment of Monuments

Post-seismic Restoration Project of Basilica Churches in Kefallonia Island

Themistoklis Vlahoulis, Apostolia Oikonomopoulou[(✉)], Niki Salemi, and Mariliza Giarleli

Restoration of Byzantine and Post-Byzantine Monuments,
Ministry of Culture of Greece, Athens, Greece
{dabmm, aoikonomopoulou}@culture.gr

Abstract. Engineers working on the preservation of cultural heritage, confronted to the ubiquitous demand for the restoration of historic and monumental buildings, need transdisciplinary and efficient methods for the study of ancient, yet highly complicated, structures. The severe 2014 Kefallonia earthquake, in combination with the vulnerable, in terms of geometry and materials, structural system of Basilica churches resulted to severe damage in two significant post-byzantine churches: the Basilica of Saint Marina in Soullaroi and the Basilica of the Virgin in Roggoi. The restoration study presented in this paper, performed by the architects and the engineers of the Directorate for the Restoration of Byzantine and Post-Byzantine Monuments of the Ministry of Culture of Greece, was carried out by following several successive research stages in order to result in a detailed intervention project for these Basilicas. The project was based on the digital representation and modeling, the architectural-morphological analysis, the historical survey of these monuments, as well as of similar buildings of the Eptanesa Island Complex and on the study of the construction technique of the two Basilicas. The monitoring of the damage patterns in comparison with the results of structural analysis and response under seismic loading, led to the adoption of intervention measures in order to repair and strengthen the masonry and the wooden structural elements of the Basilicas. For the restoration of the studied Basilica Churches of Kefallonia, which are considered as two of the most representative post-byzantine religious monuments in Greece with occidental architectural elements, techniques and materials for the consolidation of the structural systems have been proposed, in accordance with the limitations imposed from the monumental character of the churches for sustainability and reversibility, while fully respecting and safeguarding their authentic architectural and structural features.

Keywords: Restoration · Post-Byzantine monuments ·
Basilica masonry churches

1 Introduction

Historic masonry religious monuments are significant parts of the world cultural heritage and exhibit a large diversity of geometries and building materials. Particularly, in the Eptanesa Island Complex a unique school of post-byzantine religious architecture,

© Springer Nature Switzerland AG 2019
A. Moropoulou et al. (Eds.): TMM_CH 2018, CCIS 962, pp. 131–142, 2019.
https://doi.org/10.1007/978-3-030-12960-6_9

which emulates occidental architectural features, has been developed [9]. Two significant religious monuments, perfect examples of the Eptanesian architectural school, which do not exhibit severe structural and morphological alterations, the Basilica of Saint Marina in Soullaroi and the Basilica of the Virgin in Roggoi, have been studied in the Directorate for the Restoration of Byzantine and Post-Byzantine Monuments (Ministry of Culture of Greece) so as to create solid restoration projects to be executed after the important damage that occurred during the strong seismic activity in the island of Kefallonia from 26th of January till the 3rd of February 2014 [2]. The analysis of the stability of such rigid structural systems that suffer important loading during earthquakes and the proposition of suitable restoration techniques constitute a major task for engineers working on the preservation of cultural heritage. The conception of the restoration projects presented herein demanded multidisciplinary collaboration of engineers, was adapted to the intrinsic mechanical behavior of the building materials of old structures and aim to prolong the life and the durability of these two important monuments of Kefallonia.

2 Post-earthquake Emergency Interventions and Investigations in Two Damaged Basilicas of Kefallonia

2.1 Urgent Safety Measures

Two months after the main seismic activity in the island of Kefallonia in 2014, provisional works aiming at stabilizing the structural systems of the two Basilicas were, already, carried out. As significant aftershocks followed the main seismic activity, the temporary installation of scaffolding, shuttering, propping, confinement etc. was crucial so as to stop the evolution of damage to the already weakened structural systems of the Basilicas and provide sufficient time for all the necessary stages for the perception of a complete restoration study [5].

2.2 Digital Representation and Modeling

The monumental character, as well as the collapse state of the structural systems of the Basilicas, demanded a representation in detail. For these reasons, the laser scanning technique was implemented for both the interior and the exterior of each monument, using a Faro Focus 3D laser scanner with an integrated HDR-camera. Points were scanned at 5 mm intervals with ±2 mm precision and color rendering.

The results were processed in CAD software, so as to produce 1:50 architectural drawings (plans, sections, architectural details). The digital restitution of the Basilicas that included the damage patterns of their structural systems served as the basis for the design of the architectural restoration study [8].

2.3 Geotechnical Investigation

A geotechnical study was performed, with two drillings, one at each basilica, in the surrounding area near the monuments, to a maximum depth of 20 m. Soil conditions for each monument were found to be rather satisfactory. An upper narrow layer of backfill was found, over a layer of hard clay with low permeability for the Basilica of Saint Marina in Soullaroi and over a layer of calcareous sandstone for the Basilica of the Virgin in Roggoi, whereas the water level was not found. The bearing capacity of the soil was determined as, also, the necessary for the analysis soil parameters (modulus of subgrade reaction for the modeling of the soil-structure interaction, the corresponding soil category to the Eurocode 8, etc.).

3 Architectural Analysis

3.1 Basilica of Saint Marina in Soullaroi

The cemetery church of Saint Marina in Soullaroi is a listed monument (Government Gazette 152/B/9-4-63). It is dated at the end of the 17th century, according to inscriptions on the north part of the monument. It is a typical example of Eptanesian religious architecture, which emulates Baroque features. The church belongs to the usual type of single-nave wooden-roofed Basilicas. The overall dimensions of the ground plan are 22.20 m × 9.35 m, excluding the protruding semi-hexagonal apse (Fig. 1). The church is internally divided into three parts: the rectangular nave, the sanctuary, which is raised by three steps and the vestibule and women standing area, which is also raised by three steps in relation to the nave.

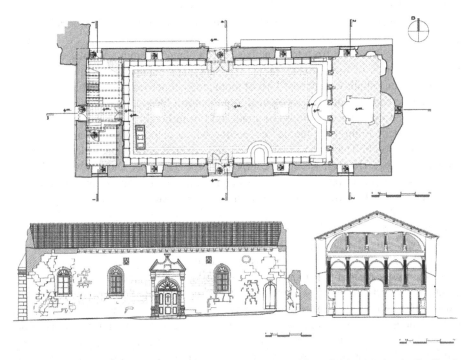

Fig. 1. Plan, north facade and transverse section of the Basilica of Saint Marina in Soullaroi.

All exterior facades are plastered, except from the north facade and built in ashlar masonry of 0.75 m thickness. Doors open on the three sides of the church. Each one of them has different morphological features. Following the norm of Eptanesian architecture, the facade is one of the long sides and not on the west side, which is the case with most Byzantine churches. The west side is not preferred, since, on the one hand, it is narrow and, on the other hand, the placement of the women's area on the upper story of this part restricts the height of the ground story. Consequently, such space restrictions would not allow the development of a monumental decoration on the west side of the church. The placement of the facade on the north or the south longer side of the Basilica has the obvious disadvantage of providing access to the interior through the short axis of the building, which does not allow the full appreciation of its spatial configuration that would be facilitated by the entrance through the west side and along the long axis of the nave. However, it is possible that the Baroque influence on Eptanesian architecture may have rendered the development of an imposingly monumental plastic ornamentation more important than circulation patterns or the experience of the building as a whole. Such ornamentation could only be placed on the long sides of the Basilica.

Thus, the main entrance to the Basilica of Saint Marina is on the north side. It is framed by a limestone moulding and rather baroque-styled double jambs with ante capitals and bases. The south side door is crowned by a limestone triangular pediment, which is simpler than the one above the north door. It is supported by a horizontal cornice which is in its turn based on relief ribbed jambs with ionic ante capitals and high bases. The third door on the west has a simpler arched lintel supported by jambs with ante capitals. The west door leads directly into the vestibule. A wooden trellis screen separates it from the nave.

The nave is separated from the sanctuary by an elaborate high gold plated wooden carved iconostasis of high artistic value. The sanctuary features one main niche and three smaller and shallower niches. Two of them are symmetrically placed in relation to the main apse and the third apse is opened into the north facade. The altar is stone-built and is covered by a wooden ciborium about 2.20 m in height.

The wooden ceiling or "ourania" covers the whole nave, curves towards the long walls of the church. The roof is dual-pitched and covered with French tiles resting on wooden sheeting. The roof structure comprises three triangular and twelve A-shaped trusses placed at 1.5 m intervals. An interesting feature of the interior of the church is the series of small vertical relief wooden piers that stress the intervals between openings on the long internal sides of the church. The piers support architraves at intervals of 2 m to 2.5 m, where upon rests the cornice of the wooden "ourania" (ceiling), which is also suspended by the A-shaped trusses.

The Basilica features important plastic decoration apart additionally to its door frames themselves. Saint Marina is on the arch key above the north door. Each side of the same door retains relief coats of arms, featuring a knight's armour, a tower and flowers. The south side features small votive ships carved on the wall plaster. The frame of the window of the east side is distinctively decorated with a relief Mandylion or "Image of Edessa", surrounded by an arched limestone frame. Above it, there is the relief figure of an angel. The triangular end of the jamb on the apse dome retains the following relief inscription ΓΧΠ 1786.

Based on the above description, it is possible to argue that the Basilica of Saint Marina in Soullaroi is a typical example of the dominant type of church in Kefallonia, whose ecclesiastical architecture constitutes a large and representative example of the Eptanesian school architecture. The ratio of the narrow and long sides of the Saint Marina Basilica which falls within the standard ratio of 1:1.15-2.5, the ceiling and the bell tower are indicative features of this school. Eptanesian churches are well-known for their emulation of Occidental influences, such as Baroque architecture. Such influences are particularly evident in the rich plastic decoration of the Saint Marina Basilica, including triangular or convex and concave pediments, the cornices, the half-columns, the frames of the openings and the keys of the arches. The importance of the decoration is evident in the emplacement of its main entrance on the north long side.

3.2 Basilica of the Virgin in Roggoi

The parish church of the Virgin in Roggoi is a listed monument (Government Gazette 621/B/11-11-1968). It is dated to the 17th century. It is also a single nave wooden roofed Basilica with Baroque Eptanesian features, such as the ones mentioned above for the Basilica of Saint Marina in Soullaroi. The dimensions of the rectangular ground plan are 15.30 m × 7.40 m (Fig. 2).

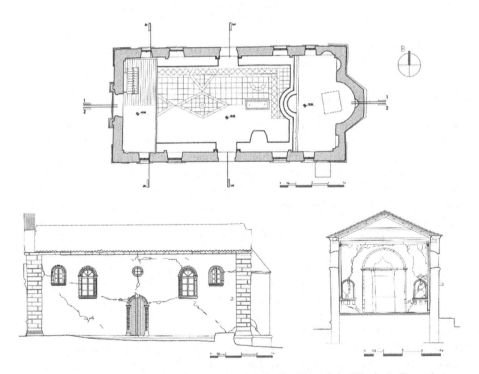

Fig. 2. Plan, south and transverse section of the Basilica of the Virgin in Roggoi.

The interior of the church is accessed through three doors, symmetrically placed along the longitudinal and vertical axes of the building. Contrary to the Saint Marina Basilica and the Eptanesian norm, the main facade of the Virgin in Roggoi is the west narrow side. The door of this facade is surrounded by a limestone semi-circular frame with prominent Baroque features, such as double jambs, ante capitals and bases and a monumental multicurved pediment. The north and south facade have symmetrically placed doorways with gothic style occuli above them. Arched windows open on either side of each doorway.

The masonry of the church is plastered stone ashlar of 0.70 m thickness. Well-cut limestone cornerstones are visible on all corners of the building. The same type of stone has been used for the arched openings of the north and south side. A cornice crowns the masonry, while a semi-cylindrical cordon runs around the monument at the threshold level. The roof is composed of 16 trusses at 0.90 m intervals. Each truss has two rafters, one tie beam and double vertical beams in the middle of the rafters. The wooden sheeting rests on the rafters. The latter are placed on modern concrete beams which are placed above the north and south wall masonry and are connected with metal beams. The interior of the wooden roof is accessed via a small opening in the ceiling of the women's standing area. The nave is covered by a wooden flat ceiling ("ourania"). It has simple geometric motifs. Similar to the Saint Marina Basilica, small wooden piers are placed on either side of the long interiors at intervals of about 2 m to 2.5 m.

4   Structural Analysis and Description of the Damage Patterns

4.1   Structural and Geometrical Characteristics of the Two Basilicas

For the majority of masonry structures, the problem of stability is mostly related to the geometry of the structure rather than the resistance of the building materials. Their geometry-related stability is an immediate consequence of the resistance criterion of structural masonry. As a matter of fact, masonry, viewed as a homogenized material, exhibits a very small tensile strength and a very high compressive strength, usually much larger compared to the stress states commonly imposed within a masonry structure [3].

The studied Basilicas, exhibit similar structural systems and are made of similar building materials. In particular, they constitute simple orthogonal structures of bearing two-leaf masonry, without wooden reinforcing inside masonry, featuring well-built corners in ashlar masonry. They are covered by wooden dual-pitched roofs and their wooden ceilings are suspended by the roof and supported by a secondary system of wooden columns and lateral beams. Their openings are reinforced with limestone frames. The west part of each Basilica is strengthened by the structural elements of the vestibule. Nevertheless, they present some different geometrical and proportional features (Table 1) which are indicative for the kind and the size of damage.

The Basilica of Saint Marina in Soullaroi is significantly longer and larger and covers almost twice the area of the Basilica of the Virgin in Roggoi. In terms of volume, the Saint Marina Basilica is also more spacious and in terms of percentage of

Table 1. Geometrical and proportional features of the two Basilicas.

	Saint Marina Basilica	Basilica of the Virgin
Longitudinal length [m]	22.29	15.46
Transverse length [m]	9.43	7.43
Area [m^2]	224.82	119.44
Transverse length [m]/ Longitudinal length [m]	1 : 2.36	1 : 2.08
Wall height [m]	6.22	5.82
Roof height [m]	8.26	7.49
Volume [m^3]	1398.38	695.14
Thickness of Wall section[m]	0.75	0.60
Percentage of openings in the N facade	10.00%	7.20%
Percentage of openings in the W facade	7.33%	7.86%

openings, it is more vulnerable in the long sides than the Basilica of the Virgin, which, on the contrary, is more vulnerable in the west facade. These conclusions are coherent with the damage patterns observed in the Basilicas, presented in the following section.

4.2   Damage Patterns of the Basilicas

The two monuments exhibit severe damage with similar patterns. The most important damage is partial collapse and, also, cracks along the entire thickness of the walls, which are developed in the east facade and on the long sides, near the areas where they cross with the east facade.

Fig. 3. Damage and safety measures of the Saint Marina Basilica and the Basilica of the Virgin.

In particular, the following damage has been observed (Fig. 3):

- Diagonal cracks across both sides of the niche of the sanctuary, as well as a horizontal crack in the base of the pediments are developed.
- The upper part of the pediments in the narrow sides has collapsed, as well as the north-east corner of the Saint Marina Basilica.

- On the long sides near the corner with the east facade, large-almost vertical- cracks are developed. Particularly, in the Basilica of Saint Marina in Soullaroi there is detachment of the south-east corner.
- The south facade of the Saint Marina Basilica exhibits a small incline from the vertical.
- In general, there are many cracks starting from windows and doors.
- The roof of the Saint Marina Basilica exhibits extensive damage as deflections of wooden elements and section reductions and, also insufficient support of the wooden skeleton (lack of the horizontal wooden beam that transfers uniformly loads to walls)
- Deflections are observed in the wooden columns that support the ceiling and at the iconostasis.

A significant conclusion about the damage observed in the studied Basilicas concerns the damage patterns and the geometrical proportions of the monuments. As the Saint Marina Basilica is proportionally larger than the Basilica of the Virgin, it presents more severe cracks (the detachment of the SE corner and the collapse of the upper part of NE corner). On the other hand, in the west facade, the Basilica of the Virgin exhibits more severe damage than the Saint Marina Basilica, due to the important percentage of opening and the tall multicurved pediment.

From investigation of the archives of the Directorate for the Restoration of Byzantine and Post-Byzantine Monument, it seems that the Basilicas, before the execution of restoration works that took place in the '80 and '90, presented damage patterns similar with the ones that the monuments present, nowadays, after the seismic activity of 2014. It is, also, of interest to point out, that both Basilicas exhibited a relatively sufficient behavior, but with significant damage, during the main earthquake in January 2014. The second main earthquake in February 2014, almost of equal intensity with the first main earthquake, found the structural systems altered and weaken, with severe cracks that have cancelled the continuous and relatively uniform behavior of the masonry walls and the wooden roof and, consequently, led the structures to partial collapse (Fig. 4).

Fig. 4. Damage of the Saint Marina Basilica after the first and the second main earthquake in 2014.

5   Structural Analysis and Results

5.1   Modeling and Analysis

The structural system of each Basilica was modeled with the following features. Masonry walls were modeled with shell elements and wooden beams and steel bars with frame elements. A larger strip of masonry was considered for the shell elements of the foundation and, in order to take into consideration the soil-structure interaction, properly calibrated springs where used at the foundation level [1].

Concerning the types of analysis the following has been performed: (a) modal response spectrum analysis and (b) lateral force analysis. The structural systems were considered as linear elastic.

For the response spectrum analysis, the Eurocode 8 type I spectrum was used with the following characteristics: (a) reference ground acceleration for seismic zone III, according to seismic zone map of Greece, $a_{gR} = 0.36$ g, (b) importance factor for monumental architecture $\gamma_I = 1.40$, (c) characteristic periods for soil type B, $T_B = 0.15$ s, $T_C = 0.50$ s, $T_D = 0.20$ s, (d) soil factor $S = 1.2$ and (e) behavior factor for unreinforced masonry structure with no ductility $q = 1.0$.

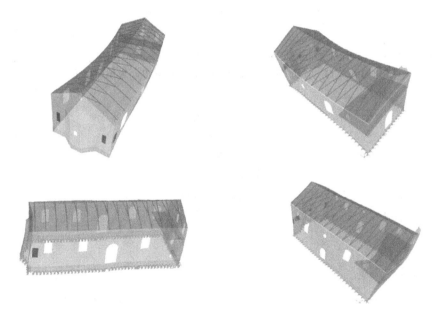

Fig. 5. Forms of the first and the second dominant eigen-modes of the Basilica of Saint Marina and of the Basilica of the Virgin with consideration of calibrated springs on the foundation level.

Both the first and the second dominant eigen-periods ($T_1 = 0.30$ s, $T_2 = 0.15$ s and $T_1 = 0.23$, $T_2 = 0.16$ s for the Saint Marina Basilica and for the Basilica of the Virgin, respectively) correspond to the horizontal branch of the spectrum, so the highest value of the acceleration was considered for the two directions of the lateral force analysis. As presented in Fig. 5, the first and the second dominant eigen-modes are translational with respect to the North-South axis (x-x) and the East-West axis (y-y).

5.2   Analysis Results and Corresponding Crack Patterns

In the following figures, the stress contours obtained from structural analysis are presented with reference to the cracks that were developed in each structure (tensile stresses are positive). Cracks are expected to open in the areas of the highest concentration of tensile stresses.

As presented above, for in-plane seismic action the development of the diagonal crack across the sides of the sanctuary of the Saint Marina Basilica is identified from comparison with major principal stresses (Fig. 6). Similarly, in the west facade of the Basilica of the Virgin, for out-of-plane seismic action, the horizontal crack under the circular window is identified from comparison with major principal stresses (Fig. 7).

The adopted linear elastic approach was proved to be a sufficient method for the assessment of the behavior of the Basilicas, as the results of the linear analysis are in agreement with the observed damage for the two masonry structural systems. These results are, mostly, coherent with the damage observed after the first main shock of the 2014 earthquake sequence, as already stated in the discussion of the damage patterns. After this first strong seismic shock, an ultimate state of equilibrium has been established for the structural systems of the Basilicas, which managed to avoid collapse. However, parts of their systems, as the pediments, became insufficient to withstand a second seismic shock of similar intensity and, consequently, were led to collapse [6].

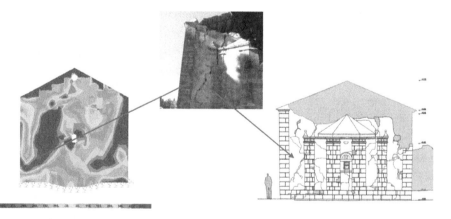

Fig. 6. East facade of the Basilica of Saint Marina: major principal stresses for in-plane seismic action and corresponding damage patterns.

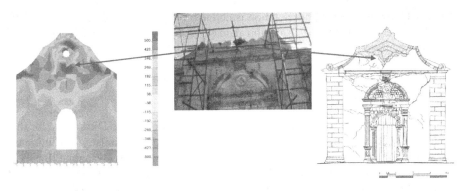

Fig. 7. West facade of the Basilica of the Virgin: major principal stresses for out-plane seismic action and corresponding damage patterns.

6   Restoration Scheme

In view of restoring the bearing capacity and strengthening the vulnerable to seismic action structural systems of the Basilicas, while respecting the architectural and structural characteristics of the monuments, the following interventions were proposed:

1. Consolidation of masonries through grout injection, repointing (design of repair material based on white cement in low concentration) and strengthening of the walls with fibre-reinforced plastering, including light stainless-steel grids, on the exterior facades [7].
2. Reconstruction of the collapsed areas and effective connection with the existing masonries through short stainless steel anchors.
3. Connection of cracked marble elements with stainless steel clamps and dowels.
4. Installation of long anchors alternately in the walls of the south-east and the north-east corners.
5. Installation of a frame on the level of the roof, consisting of slender reinforced concrete beams placed within the masonry longitudinal walls, which are connected through stainless steel beams anchored in the interior facade of masonry at the narrow sides.
6. Repair and strengthening of the roofs with additional wooden beams in the trusses, stainless steel ties and supports of the trusses.
7. Stiffening with installation of plywood panels over the inclined beams of the trusses and over the floor of the vestibule.
8. Stiffening at the foundation level with an external reinforced concrete beam and four masonry connecting beams for the Saint Marina Basilica.

The aim of the intervention measures is to strengthen the vulnerable to lateral loading areas of the structures that were, already, identified by the damage patterns after the recent and the previous earthquake sequences [4]. The assessment of the studied structural systems has been performed according to the Eurocodes 1, 5, 6 and 8. The structural systems of the Basilicas, strengthen with the above interventions, for which the highest reference ground acceleration is defined by the current Greek seismic code for Kefallonia, are anticipated to exhibit damage to a future earthquake, but significantly less compared to their initial state prior to the interventions.

7 Conclusions

In this article we presented the holistic approached followed in the Directorate for the Restoration of Byzantine and Post-Byzantine Monuments of the Ministry of Culture of Greece for the conception of restoration studies for two post-byzantine Basilica Churches in Kefallonia after the earthquakes in 2014. The immediate safety measures that were applied to the monuments allowed their survival to the aftershocks of the two strong earthquakes and allowed all the necessary investigations to take place concerning the geotechnical profile, the architectural representation and the assessment of their structural behavior. The restoration measures were selected with care for safeguarding the original architectural and structural systems of the two Basilicas. The knowledge acquired from the study of the Basilica of Saint Marina in Soullaroi and the Basilica of the Virgin in Roggoi served for the appraisal and verification of the restoration studies for other similar Basilicas in Kefallonia.

References

1. Gazetas, G.: Foundation vibrations. In: Fang, H.Y. (eds.) Foundation Engineering Handbook, 2nd edn, Van Nostrand Rheinhold (1991)
2. GEER - EERI - ATC: Earthquake Reconnaissance 26th January/2nd February 2014 Cephalonia, Greece Events, Version 1 (2014)
3. Heyman, J.: The Stone Skeleton: Structural Engineering of Masonry Architecture. Cambridge University Press, Cambridge (1995)
4. Lourenço, P., Mendes, N., Ramos, L.F., Oliveira, D.V.: Analysis of masonry structures without box behavior. Int. J. Archit. Herit. **5**, 369–382 (2011)
5. Modena, C., Valluzi, M.R., da Porto, F., Casarin, F.: Structural aspects of the conservation of historic masonry constructions in seismic areas: remedial measures and emergency actions. Int. J. Archit. Herit. **5**, 539–558 (2011)
6. Oikonomopoulou, A.: Numerical approach for the study of the behavior of ancient masonry structures. Ph.D. thesis, School of Architecture Paris-la Villette (2009). (in French)
7. RILEM: State of the Art Report of RILEM Technical Committee TC 203-RHM: Repair Mortars for Historic Masonry, RILEM publications SARL (2016)
8. Salemi, N.: Preservation and restoration of the residential quarters of the Kaisariani Monastery. MSc thesis, Postgraduate Program in Protection of Monuments: Conservation and Restoration of Historic Buildings and Sites, School of Architecture NTUA, Greece (2001). (in Greek)
9. Zivas, A.D.: The architecture of Zakynthos from 16th until 19th century. Technical Chamber of Greece, Athens (2002). (in Greek)

Investigation of the Structural Response of Masonry Structures

Georgios A. Drosopoulos[1], Jan Phakwago[1], Maria E. Stavroulaki[2], and Georgios E. Stavroulakis[3(✉)]

[1] Discipline of Civil Engineering, Structural Engineering and Computational Mechanics Group (SECM), University of KwaZulu-Natal, Durban, South Africa
drosopoulosg@ukzn.ac.za
[2] School of Architecture, Applied Mechanics Laboratory, Technical University of Crete, Chania, Greece
mstavr@mred.tuc.gr
[3] School of Production Engineering and Management, Institute of Computational Mechanics and Optimization, Technical University of Crete, Chania, Greece
gestavr@dpem.tuc.gr

Abstract. Methods used for the structural evaluation of masonry structures and monuments are presented in this article. Limit analysis is initially presented as a tool for the study of the collapse mechanism and limit load of masonry structures. Unilateral contact-friction laws are introduced between the stones, to describe damage due to opening-sliding. Related applications to masonry arch bridges as well as comparison with experimental investigations, are presented. Non-linear finite element analysis, using similar concepts taken from non-smooth mechanics, is also presented. Eventually, numerical homogenization methods are adopted for the evaluation of masonry walls. A microscopic sample (Representative Volume Element) consisting of masonry and mortar joints is chosen and average stress and stiffness are derived numerically. Then, they are used within a macroscopic homogeneous model, for the representation of the structural response. Applications of the mentioned approaches to masonry bridges and walls offer a further insight on the response of these structures.

Keywords: Unilateral contact · Limit analysis ·
Masonry arches-homogenization

1 Introduction

Masonry arch bridges and walls are traditional structures, consisting of stone blocks and mortar. The ability of the material and the joints to transfer compressive loading, and their inability to accept reliably tensile loading are optimally exploited. The modern approach to structural analysis of masonry arches is based on limit analysis concepts or unilateral contact mechanics, after suitable simplifications.

For common failure mechanisms revealing the formation of hinges due to loss of contact along interfaces, comparison of the two approaches with existing experimental results is satisfactory. Study of sliding failure and the influence of various strengthening

© Springer Nature Switzerland AG 2019
A. Moropoulou et al. (Eds.): TMM_CH 2018, CCIS 962, pp. 143–156, 2019.
https://doi.org/10.1007/978-3-030-12960-6_10

techniques, like FRPs bonded on the structure, may require more complicated models. First, continuous field approach based on damage mechanics can be used in order to model the contact interfaces in a smeared crack approach with less complicated models. Furthermore, numerical homogenization allows us introduce anisotropic behavior together with the unilateral contact effects into the finite element model, in the FE^2 sense. The model can be used in the sequel for collapse analysis purposes.

Mechanics of masonry structures and representative applications are presented in this paper, based on recent research and previous work of the authors and their co-workers. Further information related to unilateral contact and limit analysis of masonry bridges can be found in [1], on unilateral contact analysis for monuments in [2], on damage mechanics – field approach for the description of contact interfaces and application on the collapse prediction of masonry arches in [3] and on numerical homogenization techniques for cracked and masonry structures in [4].

The methods outlined here are suitable for the analysis of ancient and megalithic structures, Fig. 1, and of more modern masonry structures, and can be combined with modern photogrammetry and laser scanning techniques for a semi-automatic and quick structural evaluation of monuments [5].

Fig. 1. Olympia, a monument with contact interfaces, like many other masonry and megalithic structures.

In this article, three fundamental categories of numerical analysis, appropriate for the investigation of masonry structures and monuments, are presented. First, classical limit analysis is adopted for a fast evaluation of masonry arch bridges. A second approach relies on the development of macroscopic non-linear finite element models, capable of depicting collapse mechanism, limit load, displacements and stresses of the investigated structures. Finally, a third approach related to numerical homogenization and the concept that a Representative Volume Element (RVE) is initially simulated to derive the average material properties used in the macroscopic model, is presented.

2 Principles of Limit Analysis

For the simplest case of frictionless contact, the structural analysis problem can be formulated as a potential energy minimization which includes the unilateral contact inequality constraint (non-penetration) [6, 7]. This is given by the relation:

$$min_{\mathbf{u}} \left(\frac{1}{2} \mathbf{u}^T \mathbf{k} \mathbf{u} - \mathbf{P}^T \mathbf{u} \right)$$

$$\mathbf{N}_n \mathbf{u} - \mathbf{g} \leq 0$$

(1)

For the quadratic minimization problem, the Karush-Kuhn-Tucker (KKT) optimality conditions lead to the linear complementarity problem (LCP) of relations

$$\mathbf{Ku} + \mathbf{N}_n^T \mathbf{r}_n = \mathbf{P}_o + \lambda \mathbf{P}$$
$$\mathbf{N}_n \mathbf{u} - \mathbf{g} \leq 0$$
$$\mathbf{r}_n \geq 0$$
$$(\mathbf{N}_n \mathbf{u} - \mathbf{g})^T \mathbf{r}_n = 0$$

(2)

The first equation expresses the equilibrium of the discretized unilateral contact problem without friction, where $\mathbf{K}$ is the stiffness matrix and $\mathbf{u}$ the displacement vector. $\mathbf{P}_o$ denotes the self-weight of the structure and $\mathbf{P}$ represents the live load vector, multiplied by a scalar load multiplier λ. $\mathbf{N}_n$ is an appropriate geometric transformation matrix and vector $\mathbf{g}$ contains the initial gaps for the description of the unilateral contact joints. The next relations represent the constraints of the unilateral contact problem for the whole discretized structure. For the consideration of the constraints, the vector $\mathbf{r}_n$ representing Lagrange multipliers is used to depict contact pressure. The problem described above is a non-smooth parametric linear complementarity problem (LCP) parametrized by the one-dimensional load parameter λ. Values for solutions in the interval $0 \leq \lambda \leq \lambda_{\text{failure}}$ are investigated.

To obtain a solution to the mentioned general description, when applications to masonry arch bridges are discussed, Limit State Ring software is used in this article.

3 Finite Element Analysis in the Macroscopic Scale

Another tool that can be used for the structural evaluation of masonry arches and monuments, is classical non-linear finite element analysis using commercial software. The main idea, is the development of discrete models, simulating all the unilateral contact-friction interfaces which arise between the stones. For the solution of the mentioned, non-linear problem, classical methods for optimization, such as Lagrange multipliers, Augmented Lagrangian and Penalty method, are used. An incremental iterative numerical scheme is used for the solution of the mentioned, non-linear problem. Newton-Raphson is adopted in this work.

In respect to limit analysis, finite element analysis results in more computational heavy problems. On the other hand, more details for the structural response are provided, for instance displacements, principal stresses and force-displacement diagrams.

4 Numerical Homogenization

The third approach that can be used for the investigation of masonry monuments is based on homogenization. Several different homogenization procedures are proposed in the literature. Analytical methods can be more accurate in the description of the micro structure [asymptotic] but are usually applicable in simpler models. On the other hand, numerical homogenization may be used for the simulation of complex patterns of micro models, over a statistically defined representative amount of material [8].

Numerical homogenization can be extended to cover nonlinear effects, like contact, debonding, damage and plasticity, as well as dynamic behaviour and limit analysis [8]. On these methods, a unit cell is explicitly solved and the results are then used by a macroscopic constitutive law [9]. From another point of view, multi-level computational homogenization incorporates a concurrent analysis of both the macro and the microstructure, in a nested multi-scale approach [10, 11]. A non-linear Representative Volume Element (RVE) can be considered within this method.

The numerical homogenization adopted in this work is applied for the calculation of homogenized in-plane elastic properties of masonry walls. Furthermore, the results are used in a structural (macroscopic) level. The presented scheme can be integrated into a finite element program in a FE^2 sense and the extension to nonlinear problems is straightforward [12].

A theoretical description of the mentioned scheme is given in the next lines. According to the Hill-Mandel condition or energy averaging theorem, the macroscopic volume average of the variation of work equals to the local work variation, on the RVE:

$$\sigma^M : \epsilon^M = \frac{1}{V_m} \int_{V_m} \sigma^m : \epsilon^m dV_m \tag{3}$$

Three type of loading states, which satisfy the above condition, can be applied to the RVE: (a) prescribed linear displacements, (b) prescribed tractions, (c) periodic boundary conditions. In the present study linear displacements are applied to the RVE, thus:

$$\mathbf{u}|_{\partial V_m} = \epsilon^M \mathbf{x} \tag{4}$$

According to the above relation, a loading strain ε^M is applied, as a linear displacement boundary condition, to the boundaries ∂Vm of the RVE. With $\mathbf{x}$ is denoted the matrix with the undeformed coordinates of the boundary nodes of the RVE.

It follows the estimation of the effective material properties, thus the effective elasticity tensor $\mathbf{E}^*$ which is finally used at a structural level, for the investigation of the macroscopic behaviour of the structure. In order to estimate this effective tensor, the following relation between average stresses and strains of the RVE, is properly used:

$$\langle\sigma\rangle_{V_m} = \mathbf{E}^*\langle\epsilon\rangle_{V_m} \tag{5}$$

The sign < > represents the average of a quantity. The next step of the proposed procedure is related to the calculation of the averages quantities of both the microscopic strain and the stress. Generally, these are given by the following averaging relations:

$$\langle\epsilon\rangle_{V_m} = \frac{1}{V_m}\int_{V_m}\epsilon^m dV_m, \langle\sigma\rangle_{V_m} = \frac{1}{V_m}\int_{V_m}\sigma^m dV_m \tag{6}$$

However, there are some steps which reduce the computational cost and make the homogenization concept easier. Thus, according to the average strain theorem [8], in case linear boundary conditions are applied to the RVE (relation (4)) and the material is perfectly bonded, then:

$$\langle\epsilon\rangle_{V_m} = \epsilon^M \tag{7}$$

which implies that the average microscopic strain is equal to the known, loading macroscopic strain ε^M.

Furthermore, a formulation for the average stress, close to the one described in [11, 13] has been chosen here:

$$\langle\sigma\rangle_{V_m} = \frac{1}{V_m}\mathbf{fx} = \sigma^M \tag{8}$$

where $\mathbf{f}$ is the matrix of the external forces in the boundary nodes of the RVE, which are obtained in the end of the finite element analysis of the RVE.

In this study, plane stress conditions have been considered. In addition, the effective elasticity tensor is initially regarded as anisotropic, indicating that the unknown effective tensor has nine unknown terms. Consequently, three linearly independent loadings in the form of Eq. (4) are applied to the RVE. Each of them results in three average stress components, as well as in three equations (see (5)), for the determination of the effective elasticity tensor. In total, nine equations are developed for the esti-mation of the 9 unknown components of the effective tensor. This system of equations is applied to relation (5) and the solution is found.

Specifically, for the estimation of the solution in each of the three finite element analysis models of the RVE, the Lagrange multiplier method is used. Thus, the loading relation (4) is considered as an equality constraint and the linear problem is solved in a classical way.

In the following sections of the paper, details of the finite element model, together with the values of the masonry-mortal material properties for the examined problem are given. Moreover, the effective material properties calculated by the above mentioned numerical scheme, are presented.

5   Applications of the Described Schemes on Masonry Structures

5.1   Limit Analysis of Prestwood Bridge

The limit analysis software Limit State RING is initially used, for the representation of the structural performance of masonry arch bridges. As a concrete example for the demonstration of the collapse analysis, one real masonry arch bridge which had experimentally been tested, is simulated in this work using limit analysis. Prestwood Bridge [14, 15] had a span equal to 6.55 m, a rise to 1.43 m with a single ring of bricks. The static loading was applied at the one quarter of its span. Results of other bridges have been presented in [16–18]. In Tables 1 and 2 the geometric and material properties of this arch are shown. In Fig. 2 the geometry of the bridge is given.

Table 1. Geometrical parameters of Prestwood bridge [15].

Shape of intrados	Span (m)	Rise at mid span (m)	Thickness (mm)	Total width (m)	Number of units per ring
Segmental	6.55	1.43	220	3.80	60

Table 2. Material properties of Prestwood bridge ([15]).

Ring brick/stone Bulk density (kg/m^3)	Crushing strength (N/mm^2)	Modulus of elasticity (N/mm^2)	Surfacing material thickness (mm)	Fill material Density (kg/m^3)	Depth of fill at crown (mm)
1800	7.7	2200	100–300	1890	203

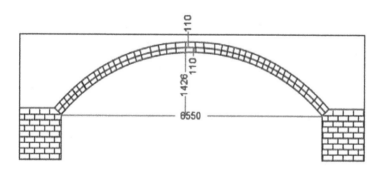

Fig. 2. Prestwood bridge modelled as a double ring arch barrel.

In software Limit State RING, a 1KN single axial vehicle load was applied across the bridge, creating different load cases. Furthermore, it was assumed that the friction coefficient is 0.5 and inter-ring coefficient is 0.6 [19]. The ultimate load, collapse mechanism and influence of backfill properties are then investigated.

To investigate the influence of backfill, a short parametric investigation with different material properties for it has been conducted. For this, the Prestwood bridge was firstly analysed with no backfill material.

According to Table 3, a good comparison between the experimental and numerical failure load is obtained. Figures 3 and 4 show both the numerical and experimental four-hinge collapse mechanism. The hinges (marked as red solid points) indicate this failure mechanism. The solid blue line in the arch barrel represents the thrust line. A similar collapse mechanism is shown in both figures.

Table 3. Comparison of experimental results with numerical results.

Experimental failure load (kN)	Limit state: RING failure load (kN)	Failure mode
228	239	4 hinge mechanism

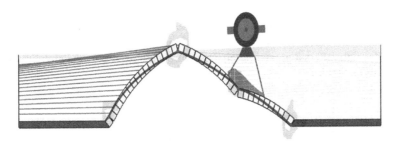

Fig. 3. Prestwood bridge failure mechanism (limit analysis).

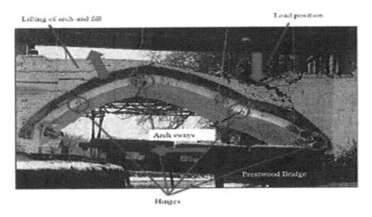

Fig. 4. Prestwood bridge failure mechanism (experimental investigation).

Figure 5 shows the change in collapse load as the position of the applied load varies across the length of the bridge. It is worth noticing that the collapse load is being reduced as the load is moving, reaching a minimum value at quarter span. But then is again increased.

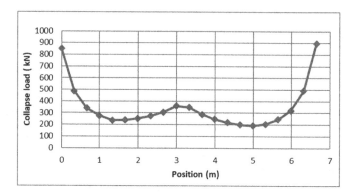

Fig. 5. Diagram illustrating the change in ultimate load as a function of position of the load application for the Prestwood bridge.

From Fig. 6, it can be noted that the ultimate load carrying capacity increases as the unit weight of the backfill material increases. This is because backfill material with a higher unit weight offers a greater passive pressure which restrains the arch bridge, thus making it more stable.

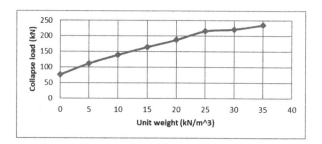

Fig. 6. Collapse load vs unit weight of the backfill.

5.2   Static Non-linear Finite Element Analysis of Masonry Bridges

A finite element analysis contact-friction model of the Prestwood bridge arch barrel was developed in Ansys commercial software, using material and geometric properties shown in Tables 1 and 2. The finite element analysis model consists of plane strain elements and total number of 2704 elements. A detail of the mesh of the arch is shown in Fig. 7.

The arch is considered to be fixed to the ground. The width of the arch barrel is taken equal to 220 mm and no fill is considered over the bridge. A load of 8000 N was then applied to the model as a concentrated point load at a quarter of the span of the arch barrel. For this model, the self-weight was included in the loading conditions. The model was analysed and the force displacement graph plotted.

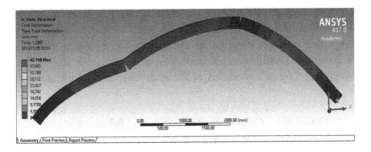

Fig. 7. Failure mechanism (commercial finite element software Ansys).

In Fig. 7 the four-hinges mechanism is shown (some hinges cannot easily be recognize due to their small size).

Figure 8 presents the change in displacement as the load is applied to the arch barrel. The arch barrel starts failing at a critical load corresponding to a displacement of 5.1 mm as shown. The diagram illustrates that there is a reduction of the load carrying capacity of the structure. Initially, in small displacement approximately equal to 1 mm, a linear distribution is obtained, indicating that no damage appears. As the load is increased, the inclination of the diagram changes, depicting reduction of the stiffness of the system. The final, almost horizontal variation represents the failure of the arch, a small increase of the load leads to a big increase of the displacements.

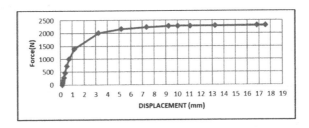

Fig. 8. Force - displacement diagram.

At this final stage of the finite element analysis, numerical problems in the system of equilibrium equations arise, resulting in loss of convergence of the simulation. Analysis is then terminated.

5.3   Dynamic Non-linear Finite Element Analysis of Masonry Bridges

To further exploit the potential of studying masonry arches using macroscopic finite element analysis, some examples are presented from a dynamic simulation conducted in a Greek masonry arch bridge, Plaka bridge, which has unfortunately been collapsed.

The main idea is to numerically test the structure, using non-linear time history finite element analysis. The unilateral contact-friction interfaces connecting the stones are still used to simulate possible damage on the bridge. Goal of this investigation is to depict influence of the backfill material properties, under dynamic loads. The following relations between the Young's modulus of the arch and the backfill, are considered in this analysis:

$$\text{Modulus of elasticity: } E_A(\text{arch}), \; E_F(\text{fill material})$$
$$\text{Model III} \qquad E_F = E_A$$
$$\text{Model III} - a \quad E_F = 1/100 \; E_A$$
$$\text{Model III} - b \quad E_F = 1/50 \; E_A$$
$$\text{Model III} - c \quad E_F = 1/10 \; E_A$$

Results are given in Fig. 9.

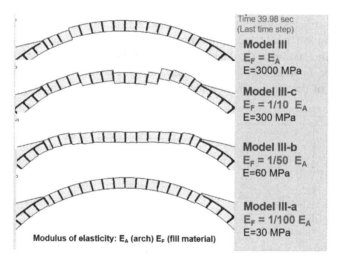

Fig. 9. Damage pattern for different backfill material properties.

In Figs. 10, 11 and 12 simulation of the whole arch bridge and damage between the interfaces are shown. The model has a number of potential interfaces, with contact and friction, and is subjected to earthquakes. Depending on the magnitude and direction of the earthquake, remaining deformation can be observed at the end of loading sequence or, for more severe loadings, collapse.

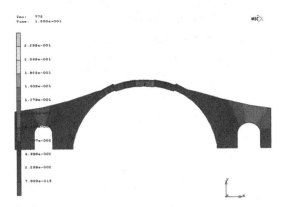

Fig. 10. A remaining deformation scheme, appearing at the end of the loading sequence.

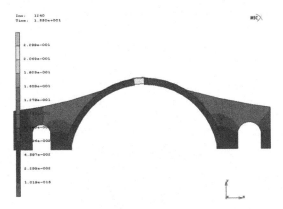

Fig. 11. The model with fill material.

Fig. 12. A collapse mechanism appearing at the end of the loading sequence.

5.4   Numerical Homogenization for Masonry Walls

A brickwork composed of regular bricks and mortar is considered. The mechanical properties of the constitutive materials are given in the following table (Table 4):

Table 4. Material properties.

Material	E (N/mm²)	v	p (kgr/m³)
Brick	4865	0,09	2403
Mortar	1180	0,06	2162

For the validation of the results obtained from the homogenization procedure described in the previous section, three finite element models with varying mesh density have been developed for the study of the RVE. In the first model, 493 nodes and 986 degrees of freedom (two degrees of freedom per node) have been considered. In the second model, a denser mesh of 777 nodes is used, while in the third model a dense mesh of 1767 nodes is regarded. For every model, first order, full integration, rectangular elements are used. The dimensions of the RVE is 260 mm × 130 mm. The width of the mortar is 10 mm. In Fig. 13, the mesh of the Representative Volume Element is given.

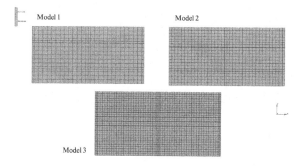

Fig. 13. Mesh and constitutive materials of the masonry RVE.

The effective material properties obtained from the three mesh density cases, are given in the following table:

Table 5. Effective material properties obtained from the homogenization procedure.

	E_{xx}	E_{yy}	v_{xy}	v_{yx}	G_{xy}
Less dense mesh (Model 1)	3992.20	3261.80	0.06703	0.08204	1500.75
More dense mesh (Model 2)	3989.90	3260.60	0.06712	0.08213	1499.77
Most dense mesh (Model 3)	3987.10	3258.10	0.06724	0.08229	1498.49

According to this table, there is no significant variation of the results, for a corresponding variation of the mesh density, on the RVE. In addition, the effective materials demonstrate an orthotropic behaviour. In the next Figure, the displacements U_x and U_y for one of the RVE models and for one loading, are shown (Fig. 14).

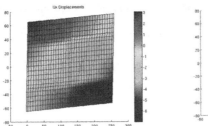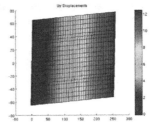

Fig. 14. Displacements U_x and U_y (mm) of the one RVE, for one strain loading, for the middle density mesh.

The values of the effective material properties of Table 5, are incorporated in a finite element model representing a masonry wall, in the structural level. Comparison between the results of this numerical homogenization procedure and a direct macro analysis of the heterogeneous structure, is then considered.

6 Conclusions

Methods which can be used to numerically assess the structural performance of masonry bridges and walls, under static and dynamic loading, are presented in this article. Limit analysis, non-linear finite element analysis and numerical homogenization are briefly described and applications to real masonry bridges and walls are presented.

Using these methods, the difficulty to deal with the heterogeneous, anisotropic nature of masonry is overcome. Collapse mechanisms and ultimate loads are in agreement with experimental research and literature results.

Numerical homogenization and finite element analysis can further be extended to account for severe loading, cracks which cross the boundaries of the microscopic masonry structure and localization phenomena.

References

1. Drosopoulos, G.A., Stavroulakis, G.E., Massalas, C.V.: Limit analysis of a single span masonry bridge with unilateral frictional contact interfaces. Eng. Struct. **28**, 1864–1873 (2006)
2. Leftheris, B.P., Stavroulaki, M.E., Sapounaki, A.C., Stavroulakis, G.E.: Computational mechanics for heritage structures. WIT Press, Southampton (2006)
3. Betti, M., Drosopoulos, G.A., Stavroulakis, G.E.: Two non-linear finite element models developed for the assessment of failure of masonry arches. Comptes Rendus Mecanique **336** (1–2), 42–53 (2008)
4. Drosopoulos, G.A., Giannis, K., Stavroulaki, M.E., Stavroulakis, G.E.: Metamodelling-assisted numerical homogenization for masonry and cracked structures. J. Eng. Mech. (ASCE) **144**, 04018072 (2018)

5. Stavroulaki, M.E., Riveiro, B., Drosopoulos, G.A., Solla, M., Koutsianitis, P., Stavroulakis, G.E.: Modelling and strength evaluation of masonry bridges using terrestrial photogrammetry and finite elements. Adv. Eng. Softw. **101**, 136–148 (2016)
6. Ferris, M.C., Tin-Loi, F.: Limit analysis of frictional block assemblies as a mathematical program with complementarity constraints. Int. J. Mech. Sci. **43**, 209–224 (2001)
7. Stavroulakis, G.E., Panagiotopoulos, P.D., Al-Fahed, A.M.: On the rigid body displacements and rotations in unilateral contact problems and applications. Comput. Struct. **40**, 599–614 (1991)
8. Zohdi, T.,I., Wriggers, P.: An Introduction to Computational Micromechanics. Springer, The Netherlands (2008). https://doi.org/10.1007/978-3-540-32360-0
9. Nguyen, V.P., Stroeven, M., Sluys, L.J.: Multiscale continuous and discontinuous modeling of heterogeneous materials: a Review on recent developments. J. Multiscale Model. **3**, 1–42 (2011)
10. Smit, R.J.M., Brekelmans, W.A.M., Meijer, H.E.H.: Prediction of the mechanical behaviour of non-linear heterogeneous systems by multi-level finite element modeling. Comput. Methods Appl. Mech. Eng. **155**, 181–192 (1998)
11. Kouznetsova, V.G.: Computational homogenization for the multi-scale analysis of multiphase materials. PhD thesis, Technical University Eindhoven, The Netherlands (2002)
12. Drosopoulos, G.A., Wriggers, P., Stavroulakis, G.E.: A Multi-scale computational method including contact for the analysis of damage in composite materials. Comput. Mater. Sci. **95**, 522–535 (2014)
13. Miehe, C., Koch, A.: Computational micro-to-macro transitions of discretized microstructures undergoing small strains. Arch. Appl. Mech. **72**, 300–317 (2002)
14. Page, J.: Masonry arch bridges. TRL State of Art Review. Her Majesty's Stationary Office, London (1993)
15. Page, J.: Load test to collapse on masonry arch bridges. Transport Research Laboratory. In: Melbourne, C. (Ed.) Arch Bridges, pp. 289–298. Thomas Telford Ltd., Bolton (1995)
16. Drosopoulos, G.A., Stavroulakis, G.E., Massalas, C.V.: FRP reinforcement of stone arch bridges: unilateral contact models and limit analysis. Compos. B **38**(2), 144–151 (2006)
17. Drosopoulos, G.A., Stavroulakis, G.E., Massalas, C.V.: Influence of the geometry and the abutments movement on the collapse of stone arch bridges. Constr. Build. Mater. **22**(3), 200–210 (2008)
18. Drosopoulos, G.A.: Non-linear analysis of stone arch bridges with the usage of a unilateral contact-friction model. PhD Thesis, Department of Material Science and Technology, University of Ioannina, Ioannina, Greece (2006)
19. Gilbert, M. (Ed.): Ring theory and modelling guide. University of Sheffield. LimitState Ltd., Sheffield (2005)

The Timber-Roofed Basilicas of Troodos, Cyprus (15th–19th Cen.). Constructional System, Anti-seismic Behaviour and Adaptability

Marios Pelekanos[✉]

Department of Architecture, Frederick University Cyprus,
7, Y. Frederikou Street, 1036 Nicosia, Cyprus
art.pem@frederick.ac.cy

Abstract. The timber-roofed basilicas of Troodos mountain area in Cyprus, which were built between the 15th and the 19th century, present a remarkable uniqueness, due to their cleverly designed constructional system. The main characteristic of this roof type is the existence of two distinctive, but co-operating, wooden parts, the Inner and the Outer Roof. The Inner Roof has a three-dimensional long triangular shape with great stiffness. The Outer Roof is literally suspended from the Inner Roof and it carries the heavy load of the solid brick flat roof tiles. A comprehensive constructional analysis showed that most of the characteristics of this roof type are similar to what is called "the eastern type roof". Timber roofs in the seismic area of the Eastern Mediterranean are constructed mainly to cope successfully with the frequent earthquake phenomenon. This constructional system presents a remarkable degree of adaptability to various types of structures. The first type is the common single-aisled basilica and the wider three-aisled basilica. The second type refers to domed churches which received a timber roof as a protective cover. The third known type is that of the converted timber-roofed basilicas, which seems to have been developed in parallel with the evolution of the original basilicas. The basic conversion was the replacement, at some point of their life, of the vaulted stone roofing with a completely new timber roof. In every case, this constructional system presented an ability to adapt each time to the particular needs and restrictions.

Keywords: Timber roof · Constructional system · Church ·
Anti-seismic behavior · Adaptability

1 Introduction

The timber-roofed basilicas of Troodos mountain area in Cyprus, which were built between the 15th and the 19th century, present a remarkable uniqueness, due to their cleverly designed constructional system. Until today, many researchers state that this particular type is unique in the European and Mediterranean area. Morphologically, the steep inclination of the roof was associated with the western European roofs and was attributed to the influence of the Franks, who ruled the island from the 13th to the 16th century [1, 2]. There are more than 130 churches of this type and, as far as typology is

© Springer Nature Switzerland AG 2019
A. Moropoulou et al. (Eds.): TMM_CH 2018, CCIS 962, pp. 157–176, 2019.
https://doi.org/10.1007/978-3-030-12960-6_11

concerned, most of them are single-aisled (75%), or three-aisled (20%) [3]. There are also some two-aisled churches (5%), most probably being the result of an extension of a single-aisled church. Almost all basilicas (over 95%) were built within the boundaries of the Troodos Rock Range [4] (Fig. 1).

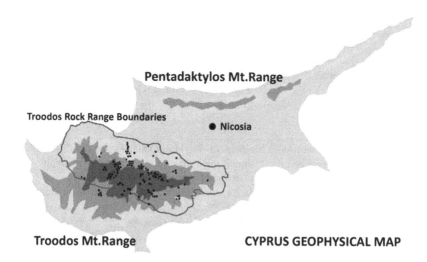

Fig. 1. The location of all timber-roofed basilicas (black dots) and the relation to the boundary of the Troodos Rock Range (red line). (Map by M. Pelekanos)

The stone walls of the basilicas are constructed of the volcanic stone of Troodos. The two long walls, which bear the load of the roof, are usually short, about 2.5–3.0 m, while the west wall rises to a pointed triangular shape, up to 6 m in single-aisled (Fig. 2) and 8 m in three-aisled churches. To the east is found the apse of the sanctuary, which sometimes is clearly visible from the exterior and in other cases it is inscribed in the general rectangular shape of the church [5].

The original openings are few, usually a door to the west (Figs. 3 and 4) and sometimes another one on the long walls, two small windows high up on the west an east walls and a small slot-opening on the apse of the sanctuary. Inside the church, the walls are coated with plaster and in many cases covered with frescoes. The foundations of the walls are usually very shallow, and when there is an inclined terrain, they are founded at different levels.

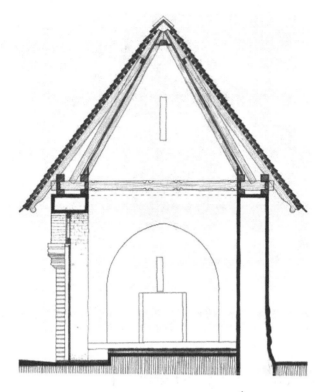

Fig. 2. Saints Sergios and Vakhos (Kalopanagiotis village-18th century), cross section facing east [5].

Fig. 3. A single-aisled timber-roofed church, dated in the 15th–16th century (Saint Nikolaos, Galata village). West façade, with the low-height main entrance. (Photo: M. Pelekanos)

Fig. 4. A three-aisled timber-roofed church, dated in the 17th century (Ayia Marina, Filousa Kelokedarwn village). West façade, with a pitched roof shelter over the main entrance. (Photo: M. Pelekanos)

2   The Constructional System of the Roof

These basilicas differ strongly from the other churches in Cyprus, not only morphologically, but also due to the uniqueness of the construction system of their wooden roof. The unique constructional feature of these basilicas is the existence of two, distinctive, but cooperating parts of the roof. The Inner Roof can be conceived as a rigid triangular prism which is based on the southern and northern walls. The Outer Roof, which carries the heavy tiles, is actually suspended from the Inner Roof, and it overhangs the two long walls, protecting in this way their upper part from the rainwater. In many basilicas the roof covers totally the sanctuary niche to the east, while on the west the roof overhangs the wall by 40–50 cm [4].

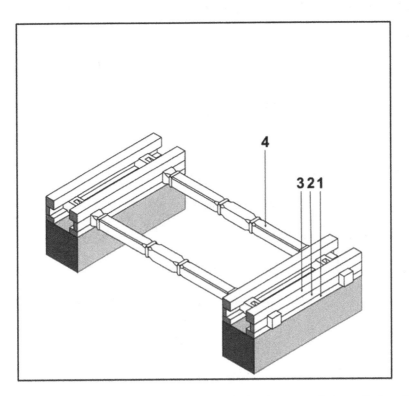

Fig. 5. The construction of the Inner Roof starts with the placement of two wall-plates (1) on which tie-beams (4) are connected. Between tie beams, pieces of wood in an upright position (2) are placed to keep the right distances. Two square shaped beams (3) are then completing and integrating the two composite "box-like" beams, on the north and south walls, to the transverse tie-beams (M. Pelekanos) [4].

The construction of the Inner Roof starts with the placement of two wall-plates (Fig. 5) (1) on which tie-beams (4) are connected. Between tie beams, pieces of wood in an upright position (2) are placed to keep the right distances. Two square shaped beams (3) are then completing the two composite "box-like" beams, on the north and south walls, and integrate it with the transverse tie-beams.

Rafters (Fig. 6) (5) are placed diagonally every 30–40 cm and hold the connective ridge-purlin ("Carina") (6) on the top of the Inner Roof. Several auxiliary wooden sections (7, 9) are placed in order to close firmly the gaps of the interior cover. A special triangular shaped wooden section (8) is penetrating the surface of the rafters, in order to connect them firmly to the wooden boards (Fig. 7) (10).

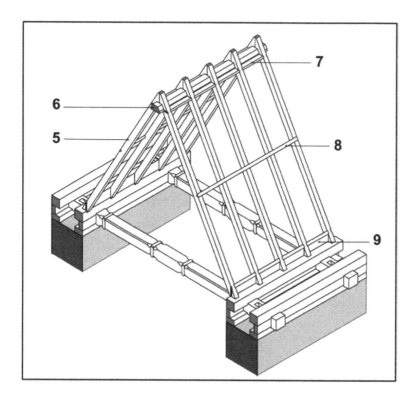

Fig. 6. Rafters (5) are placed diagonally every 30–40 cm and hold the connective ridge-purlin ("Carina") (6) on the top of the Inner Roof. Several auxiliary wooden sections (7, 9) are placed in order to close firmly the gaps of the interior cover. A special triangular shaped wooden section (8) is penetrating the surface of the rafters, in order to connect them firmly to the wooden boards (10) (M. Pelekanos) [4].

The Inner Roof is completed with the placement of the wooden boards (10) in parallel to and over the rafters. A special horizontal beam (11), nailed temporarily to the rafters, is supported by inclined strut-supports (12). Each support's lower edge is founded in a special-shaped notch on every tie-beam. In this way the heavy load of the Outer Roof is directed mainly on the strong transverse tie-beams.

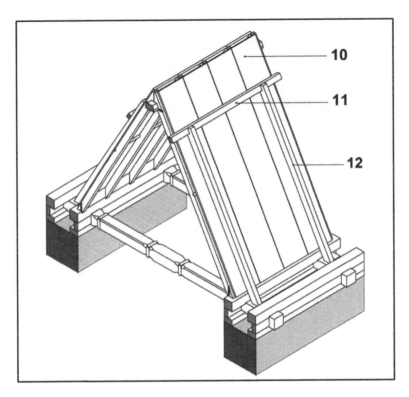

Fig. 7. The Inner Roof is completed with the placement of the wooden boards (10) along and over the rafters. A special horizontal beam (11), nailed temporarily to the rafters, is supported by inclined strut-supports (12). Each support's lower edge is founded in a special-shaped notch on every tie-beam. In this way the heavy load of the Outer Roof is directed mainly on the strong transverse tie-beams (M. Pelekanos) [4].

After the Inner Roof is completed, the Outer Roof is ready to be suspended from it, using the special horizontal beam (11). Rafters (Fig. 8) (13) are placed every 35–45 cm and overhang the walls by 50–60 cm. On top of the rafters wooden purlins (14) are nailed every 10 cm, to hold the coming tiles. Every tile covers the lower one by its two thirds, and all joints are crossed, achieving in this way the full waterproofing of the Roof.

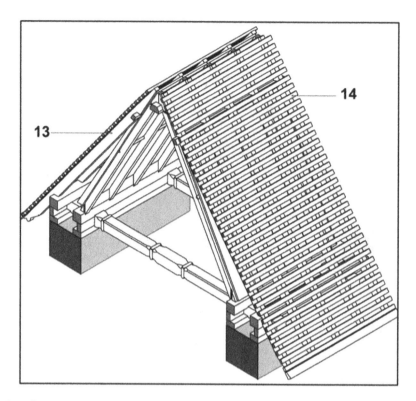

Fig. 8. After the Inner Roof is completed, the Outer Roof is ready to be suspended from it, using the special horizontal beam (11). Rafters (13) are placed every 35–45 cm and overhang the walls by 50–60 cm. On top of the rafters wooden purlins (14) are nailed every 10 cm, to hold the coming tiles. Every tile covers the lower one by its two thirds, and all joints are crossed, achieving in this way the full waterproofing of the Roof (M. Pelekanos) [4].

3 Anti-seismic Behaviour

The load of the roof is totally carried, through the two composite beams, on the southern and northern walls. In relation to the maximum compressive strength of the walls, the static load from the roof is relatively small. An important question is the way, in which the entire structure behaves, when the dynamic load of an earthquake is applied.

It is well known to engineers that masonry walls cannot resist significant loads perpendicular to their plane. In terms of vulnerability, the northern and southern walls present the lower (due to their low height and their direct connection to the roof), followed by the wall of the Sanctuary. The lower part of the eastern wall presents significant resistance to out of plane loads, due to the hemi-cylindrical construction of the niche.

The highest level of vulnerability is presented by the western wall, due to its height, the absence of ribs, the poor connection to the roof and the lack of a significant static load. To confront this weakness, the builders placed two tie-beams in contact with the western wall, one internal and one external. Despite their awareness, about the problems caused by the exposure of the external wooden beam in severe weather conditions, the decision was considered critical in order to achieve the full co-operation between the wooden roof and the masonry [6] (Fig. 9).

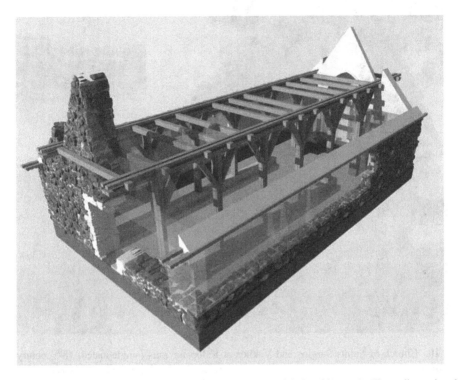

Fig. 9. Church of Panayia (St. Mary) at Kourdali (three-aisled, 16th cent.). Three-dimensional simulation showing the horizontal part of the Inner Roof, with the series of tie-beams in the middle aisle. Two pairs of these tie-beams are embracing the western and the eastern walls, achieving in this way the co-operation of the wooden roof with the stone masonry. Furthermore, these four tie-beams, along with the two composite beams over the series of wooden posts, form a belting system to secure the safe movement of the building during an earthquake [6].

Finally, the tie-beams at the western and eastern end of the church are forming, along with the two composite beams on the northern and southern walls, a full timber circumferential binding, well known from the ancient and byzantine times as "Imantosis" (from the Greek word "imantas" which means "belt"). The important role of these unique tie-beams for the whole structure can be clearly shown in cases, where the failure of the beam leads to a significant crack on the masonry's corner (church of St. Sergios and Vakhos at Kalopanayiotis village) [5] (Fig. 10).

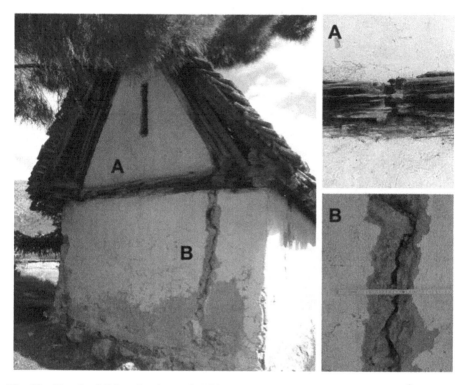

Fig. 10. Church of Saints Sergios and Vakhos at Kalopanayiotis (single-aisled, 18th century). The exposed tie-beam on the western wall was gradually decayed and finally cracked. Subsequently the stone masonry cracked very near the corner connection, due to the lateral thrusts, that were no longer counterbalanced by the tie-beam. It is worth noting that the width at the highest point of the crack on the stone wall has the same width as the crack on the wooden beam. In this case there was no internal tie-beam near the western wall [5].

A comprehensive constructional analysis showed that most of the characteristics of this type of construction are similar to the eastern type roofs: The timber circumferential binding ("Imantosis"), the "box-like" integral behaviour of the entire structure, the uniform distribution of the loads throughout the structure, the relative flexibility of the connections and its hyper-static behaviour [7]. Timber roofs in the seismic area of the Eastern Mediterranean are constructed in such a way, mainly to cope successfully

Fig. 11. Ayia Marina in Filousa Kelokedaron. Three- dimensional model showing the horizontal structure of the Inner Roof, with the series of tie-beams, and how this structure embraces the western wall, creating at the same time a timber circumferential bracing [8].

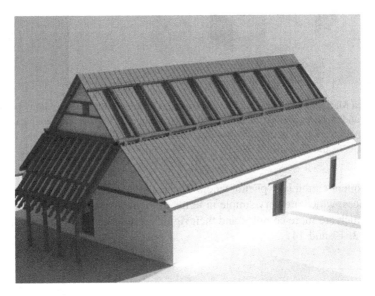

Fig. 12. Ayia Marina. The horizontal beam on the Inner Roof, supported by inclined strut-supports. On these special beams the Outer Roof will be totally suspended. Every couple of inclined strut-supports corresponds to a horizontal tie-beam of the Inner Roof [8].

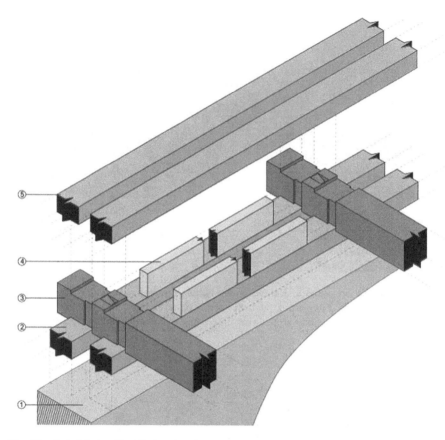

Fig. 13. Ayia Marina. Detail showing the construction of the composite "box-like" beam along the southern and northern walls. The horizontal tie-beams are locked by three couples of beams, without any nailing. The system stays firm due to the special interconnection of the beams and the heavy loads of the Outer Roof [8].

with the frequent earthquake phenomenon. These characteristics are applied on single-aisled basilicas, which are very simple in terms of their layout, but also on three-aisled basilicas, which are more complex and their rigidity is more difficult to be achieved [8] (Figs. 11, 12, 13 and 14).

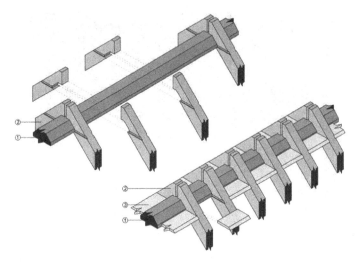

Fig. 14. Ayia Marina. Detail showing the construction of the highest part of the Inner Roof, where rafters hold the connective ridge-purlin ("Carina"). Horizontal auxiliary wooden sections are placed between the rafters in order to close firmly the gaps of the interior cover when the wooden boards are nailed on top [8].

4 Adaptability of the Constructional Type

There are at least three main cases, where this constructional type was used to serve as a roofing system. The first and basic type refers to the basilicas, that were originally built from the mid-15th century until the 19th, using this particular roofing system. Most of them are single-aisled and small, more like chapels than churches. The first three-aisled basilica appeared in the early 16th century, around 50 years after the construction of the first single-aisled basilica.

The second type was used in churches which, at some stage of their life, received a timber roof as a protective cover over their existing masonry roof (Panagia Asinou, Saint Nicholas of the "Stegi", Agios Ioannis Lampadistis, Panagia tou Araka) (Figs. 15 and 16). Despite its external morphological similarity with the basic type, i.e. that of the timber-roofed basilica, this type is different in terms of its construction, since it is just a basic post-and-beam construction and not an integrated, three-dimensional, earthquake-resistant, system. On the other hand, the Outer Roof is exactly the same as at the original basilicas and many parts of the structure are similar to the ones of the basic type [9].

The on-going investigation has recently revealed the existence of a third type, that of the 'converted timber-roofed basilica', which seems to have developed in parallel with the evolution of the original type. The basic conversion was the replacement at some point, of the vaulted stone roofing, with a completely new timber roof. The construction of the particular roof type seems to have been deemed necessary, due to

Fig. 15. Church of Panayia (St. Mary) of Arakas, at Lagoudhera village. The 12th century byzantine church was covered, between the 15th and the 17th century, by a protective timber roof [9].

Fig. 16. Church of Panayia (St. Mary) of Arakas, at Lagoudhera village. The space between the stone vault of the narthex at the west of the basilica and the timber roof [9].

the partial or total collapse of the original masonry roof, possibly caused by an earthquake or abandonment. A thorough constructional analysis reveals that distinct parts of these churches are remnants of earlier monuments, which have been incorporated into the new structure [10, 11] (Figs. 17, 18 and 19).

Fig. 17. Church of Saints Ioakeim and Anna, Kalliana village. Exterior view from the south-west. The central aisle was constructed between the 15th and 16th century. The south aisle was added later. The campanile was erected at the beginning of the 20th century, most probably at the same period when the west extension was constructed [11].

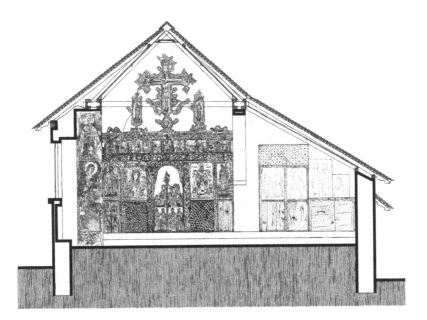

Fig. 18. Church of Saints Ioakeim and Anna. Cross- section facing east. The north wall (left on the section) belongs to the 12th century byzantine church. The main aisle was rebuilt, as a single-aisled basilica, between the 15th and 16th century. To the south (right on the section), the second aisle is an addition, most probably between the 17th and 18th century [11].

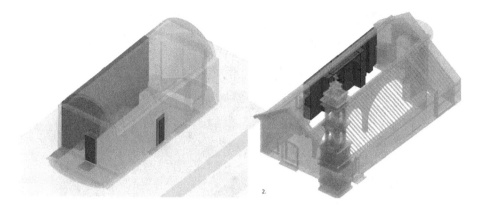

Fig. 19. Church of Saints Ioakeim and Anna. To the left, a general 3d-simulation of the 12th century byzantine church. The part which exists until today is shown darker. To the right, a 3d-simulation of the basilica as it is today, showing the part of the old church incorporated into the new timber-roofed basilica of the 15th–16th century [11].

Based on the above remarks, it is evidently clear that the constructional system of the basilicas has incorporated some specific provision principles, in order to adapt easily to the needs of every particular case.

First, the roofing system could easily extend itself to the south and/or to the north, by covering the corresponding new aisles, without any major changes to the basic triangular prism of the Inner Roof. To the west, the extension would be much easier, to any length decided.

Second, the same system could also be applied to a newly build three- aisled basilica, with some minor adjustments on the prototype, but by keeping at the same time the main core unaltered.

Third, the system could be used in order to construct a new protective roof over an existing vaulted church, usually by constructing only the Outer Roof. Finally, it could easily be applied over a ruined church without roofing, by adjusting the width of the box-like beams, by adding a second wall plate or by adjusting the length of the tie-beams, according to the existing remnants.

5   Conclusion

The timber-roofed basilicas of Troodos present a remarkable uniqueness, mainly due to their cleverly designed constructional system. The main constructional feature is the existence of two, distinctive, but cooperating parts of the roof. The constructional analysis showed that these basilicas clearly belong to the eastern type roofs which, in the seismic area of the Eastern Mediterranean, are constructed in such a way, mainly to cope successfully with the frequent earthquake phenomenon.

There are at least three main cases, where this constructional type was used to serve as a roofing system, during these four centuries: (a) The originally built one or three-aisled basilicas (Figs. 20 and 21), (b) as a protective roofing over existing masonry structures (Fig. 22) and (c) as a new timber roof over a ruined and roofless church (Fig. 23).

Fig. 20. Church of Panagia Theotokos at Galata village. Single-aisled basilica (15th century). (Photo: M. Pelekanos).

Fig. 21. Church of Panagia Chrysokourdaliotissa at Kourdali village. Three-aisled basilica (Beginning of 16th century). (Photo: M. Pelekanos).

Fig. 22. Church of Panagia Asinou near Nikitari village. Single-aisled byzantine (12th century) church with narthex (14th century) and protective timber roof (most probably of the 15th century or later). (Photo: Christina Georgiou).

Fig. 23. Church of Stavros Agiasmati near Platanistasa village. Single-aisled byzantine church, which received at a later stage a timber roof over the whole structure, forming also a covered aisle on its perimeter. (Photo: M. Pelekanos).

In order to fulfill this requirement, the design concept of this constructional system has incorporated specific provision principles, in order to adapt easily to the needs of every particular case. This particular design concept seems to be valuable even today and could be adopted when designing contemporary structures, with a long-term life objective and a need for a high level of adaptability.

References

1. Soteriou, G.: Early Christian and Byzantine Monuments of Cyprus, Athens, p. 477 (1931)
2. Fereos, H.: Timber- roofed Franco-Byzantine Architecture of Cyprus, p. 233 (1999)
3. Papageorgiou, A.: The Timber-Roofed Churches of Cyprus, Nicosia, pp. 4–22 (1975)
4. Pelekanos, M.: The Role of a Post-Byzantine Timber Roof Structure in the Seismic Behavior of a Masonry Building—The Case of a Unique Type of Timber-Roofed Basilicas in Cyprus (15th–19th Century) Published in "Historical Earthquake-Resistant Timber Framing in the Mediterranean Area (HEaRT 2015)". Springer, Switzerland (2016). ISBN 978-3-319-39491-6
5. Pelekanos, M. (Team Leader and Scientific coordinator). Constructional Analysis of the Church of Saints Sergios and Vakhos at Kalopanayiotis. Frederick University Cyprus (2013)

6. Pelekanos, M.: (Team Leader and Scientific coordinator). Constructional Analysis of the Church of Panayia at Kourdali. Frederick University Cyprus (2012)
7. Touliatos, P.: The box–framed entity and function of the structures: the importance of wood's role. In: Cestari, C.B. (Ed.), Proceedings of Culture 2000 Project: Italian Action, pp. 163–181. Elsevier, New York (2001)
8. Pelekanos, M.: (Team Leader and Scientific coordinator). Constructional Analysis of the Church of Ayia Marina at Filousa Kelokadaron. Frederick University Cyprus (2014)
9. Pelekanos, M.: (Team Leader and Scientific coordinator). Constructional Analysis of the Church of Panagia Araka at Lagoudhera. Frederick University Cyprus (2015)
10. Pelekanos, M.: The converted timber-roofed basilicas, as part of the constructional evolution of the churches in Troodos area. The First Annual Conference of Byzantine and Medieval Studies. The Byzantine Society of Cyprus, Nicosia (2016)
11. Pelekanos, M.: (Team Leader and Scientific coordinator). Constructional Analysis of the Church of Saints Ioakeim and Anna at Kalliana. Frederick University Cyprus (2017)

Alternative Dome Reconstruction Method for Masonry Structures

Argyris Fellas[(⊠)]

Frederick University, Nicosia, Cyprus
art.fa@frederick.ac.cy

Abstract. Domes first appeared in architecture in small structures such as round huts and tombs. As the need to accommodate more people became increasingly larger through the ages and the enhanced sense of symbolism of the dome became ever more important so the dome spans expanded. The construction materials switched from solid earth mounds to masonry hemispheres [1]. This growth in size has always been an engineering challenge and in many cases of masonry domed structures the dome, located at the highest point of the building, is considered to be one of the most vulnerable parts of the structure. Decreasing the weight concentrated on the highest part of the masonry structure increases the strength and durability of the structure. The aim of this paper is to present an alternative method for constructing a domed structure that was developed using computer aided design methods. This type of dome could serve as a replacement of a collapsed stone dome, usually of a church, or be fitted on a new structure.

Keywords: Church · Restoration · Reconstruction · Timber dome

1 Introduction

Reducing the mass of the architectural structures as their height increases is a common practice among architects with the intention to minimize the moment of inertia of buildings and assign to them better anti seismic qualities. In the case of churches made of stone there is a multitude of cases in which the stone domes have sustained severe damages or total collapse under the stresses developed usually in the cases of earthquakes.

Characteristic of the difficulty in building sufficiently durable masonry domes, able to withstand the earthquakes challenging the eastern part of the Mediterranean, is the multiple collapses of the Hagia Sophia dome at different years from 553 AD to 1436 AD.

The construction technology that has been developed in the last century allows architects to consider alternative methods of designing domes that could have the same appearance but significantly lighter in weight than traditional ones. Although there have been some examples in later years of light weight domes constructed from steel or wood frames these methods have not yet found their way in majority of orthodox churches and neither are they considered In most cases when a dome needs to be completely reconstructed after its collapse (Fig. 1).

© Springer Nature Switzerland AG 2019
A. Moropoulou et al. (Eds.): TMM_CH 2018, CCIS 962, pp. 177–187, 2019.
https://doi.org/10.1007/978-3-030-12960-6_12

Fig. 1. Hagia Marina dome (before restoration) in Nicosia, Cyprus.

2   Selecting the Appropriate Construction Method

2.1   A Brief History of Dome Structures

To approach the matter of a new dome construction or replacement first we have to assess its history and the reasons, if there are none, to continue on building stone domes based on the current needs (structural, aesthetic or financial) of these structures.

The first stage of dome construction technology started with the solid earth mound system (usually dug underground caves) and moved to building domes with pieces of small branches or even bones.

Around two thousand years ago the technology of masonry domes started flourishing. The increasing amount of people gathering in religious buildings created a need for a constant growth of the dome's size. This growth in size has always been an engineering challenge and in many cases of masonry domed structures the dome, located at the highest point of the building, is considered to be one of the most vulnerable parts of the structure [2]. Despite the repetitive recorded failures of large stone domes in the last millennium many times the method and technology of construction remains the same (Fig. 2).

Fig. 2. Collapsed Dome in Famagusta Area, Cyprus.

2.2 Short Literature Review

One of the earliest methodical approaches to reduce the weight of the dome is that applied in the Pantheon in Rome around 126 AD by Apollodorus of Damascus [3]. In the case of the Pantheon the materials mixed in with the cement to construct the dome were gradually reduced in density as the dome's construction was moving upwards. This method combined with the molds used that when removed left the characteristic empty rectangular spaces gave the dome the durability needed to withstand the test of time for almost 2000 years.

Another attempt made to strengthen the domes structure was later made by Filippo Brunelleschi 1436 AD in the dome of Florence Cathedral [4]. In this case the Architect chose to design an arc shaped dome so that the trusts in the domes perimeter would be reduced and strengthen the dome by adding wooden rings between the double masonry walls that composed the dome.

Another example of a different kind of dome construction is that of the addition made to the original domes of San Marco Basilica [5]. The church originally had low-profile domes that were raised by an exterior wooden shell around 1260 AD. In this case a lighter structure was chosen as to reduce the additional load to the structure need to visually change the dome's exterior.

In the 17 century another alternative to dome construction was widely employed in Russia with the construction of onion shaped domes that usually had a wooden frame as their basic structure. The wooden frame was not only easier and lighter to build but also was economical as pine tree was readily available in northern Russia.

In later years we can find examples of architects that experimented with different methods of dome construction. Such an example is Imre Makovecz, a Hungarian architect that was active in Europe for the late 1950s onward [6]. A characteristic example of a wooden dome structure by this architect is the Forest Culture Center in Visegrád, Hungary, that was constructed in 1984. In this example wood was used as the frame and interior cladding for most of the building as well as the dome that covers it as to emphasize the connection that the visitor should have with nature.

2.3   Causes of Failure in Masonry Domes

One of the main issues causing total or partial failure of masonry domes is the increased weight concentrated on the higher level of the building adding static load to the structural parts beneath and increasing the momentum generated in the event of an earthquake. The second main issue is the extended thrust around the domes perimeter. In the case of total or partial collapse of domes in churches the implications go beyond the area of the dome as the structural integrity of the rest of the system is compromised due to the open shape that is left behind after the collapse.

So the restoration of collapsed domes has to be immediate, something that is not always possible due to the process of competitions, tenders and various other governmental approvals, regulations and processes that take time until completion. Then there is the question of the method and materials that should be used to restore a collapsed dome.

2.4   Parameters for Selecting a Construction Method for a Dome

Before proceeding to alternative methods of construction we should rule out the cases in which domes could only be constructed with the traditional stone method. Firstly when the dome structure is small (less than 2 m in diameter) then there is no reason to look into alternative methods because the weight of the dome [7], estimated less than 2 tons with solid limestone as the base of calculation, is small enough not to cause a major problem to the rest of the structure. Also in cases where the dome is requested to be without an internal coating and the stone is meant to be seen for the interior of the structure an alternative method of construction is out of the question. So we are left with the cases of domes that are large in span (more than 2 m in diameter) and are covered with an internal coating, which is the usual practice in churches so that the dome can be covered with hagiography. In these cases the construction material of the dome is not visible neither for the inside, due to the hagiography, nor form the outside due to waterproofing insulation of any type which cannot be avoided. So since the internal structure is not visible any type of construction is possible and can be selected based on its durability, weight and compatibility with the rest of the structure.

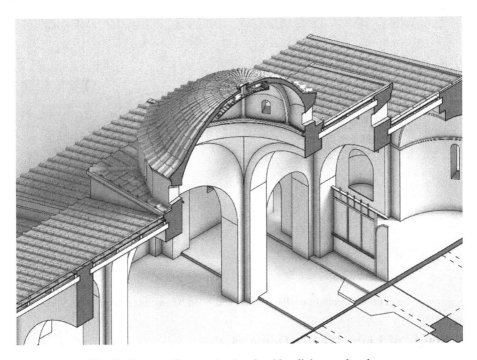

Fig. 3. Section of a sample church with a light wooden dome

Decreasing the weight concentrated on the highest part of a masonry structure and reducing the thrusts around the dome's perimeter would surely increase the strength and durability of the building. The example that will be described will suggest a light weight dome construction method that would serve both the purposes mentioned above. The technological advances of later years in laminated wood production [8] as well as the readily available stainless steel sections make the construction of such a dome possible. The ability to assemble the dome on the ground and then lift it in place also reduces the risk associated with manual labor at a height. The proposed materials and building techniques are completely compatible with masonry structures and these type of dome can serve as replacement to a totally destroyed masonry dome or can be fitted on a new masonry structure. The dome construction method to be described consists of a laminated wood frame with metallic connections that is also visible from the internal part of the building to give an aesthetic note to the dome but can also be covered completely according to the desired finished effect. A section of the whole system can be viewed on Fig. 3. The example used here is a new church to be constructed with a dome diameter of six meters. It is obvious just by observation that the section is much thinner at the part of this light dome compared to a traditional built made entirely of stone. In the case of a stone dome taking limestone with a weight of 2700 kg/m^3 for a thickness of 25 cm would weigh approximately 30 tons (30,000 kg) for a span of 6 m. In the case of the example of this light weight dome and with the basis of wood with a weigh of 600 kg/m^3 the weight comes down to less than 3 tons (3,000 kg).

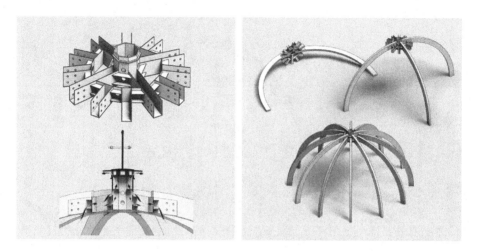

Fig. 4. Central metallic bracket and main beams installation

3   Stages of Light Weight Dome Method

3.1   Construction of the Central Connecting Bracket

Starting with the construction of the dome a metal bracket made from stainless steel is prepared to accept the twelve wooden beams that will compose the main frame (see Fig. 4). It can also be constructed from galvanized steel for cost reduction and the top final cap only can be of stainless steel. This bracket is predrilled so that the bolts can run though the beams and the bracket. The bracket has holes in the central core so that the system is breathable, avoiding moisture concentration at the top part of the dome. Also in this way the interior space can have a small scale chimney effect as to improve the buildings bioclimatic characteristics (see Fig. 4). Even with such small openings the hot air usually trapped at the highest part of the dome would be expelled based on the difference in pressure due to temperature difference between the bottom and top layer of air in the building.

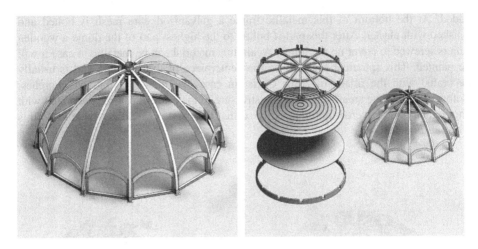

Fig. 5. Frame's perimeter completion and central spherical cap addition

3.2   Adding the Main Beams

The benefit of building a dome in this way is that it can be easily set up on the ground without the use of heavy lifting machinery. The first two wooden beams are connected to the preassembled metallic beam bracket after setting the correct distance between them and relative to the central axis. Then a third beam is bolted to the bracket so that the system can be set upright. The system of beams is completed with rest of the identical nine beams. During this stage an aerial platform should be used for the workers to be able to lift one side of the wooden beam and bolt it to the bracket.

3.3   Fixing the Geometry

The next step is to give rigidity and fixed geometry to the dome. This will serve as the layout to the next steps but also prevents the hemispherical structure to obtain an egg shape once lifted from the ground. At the bottom of the beams small brackets are bolter to each one that later will be fixed to the rest of the structure which usually ends in a rectangular stone base with a circular hole. The space between the beams as seen in the diagram (see Fig. 5) is completed with a wooden frame and galvanized steel links. In this example wood is chosen for aesthetic purposes since it will be visible for the interior space. If the intention is for this part is to be covered up or painted then metallic parts can be chosen.

3.4   Adding the Central Spherical Cap

Now the central finished interior surface of the dome is ready to be fitted. This part consists of four layers as shown in Fig. 5. The first layer also in order of construction is a metallic frame that consists of twelve bent metallic rectangular profiles following the domes radius and two rims. The central rim serves the increase of rigidity and gives a smaller bridging distance for the next layer. If the distance is larger extra rims can be added. At the bottom of this metallic frame a galvanized wire mesh is bolted and finished with plaster. After this part if bolted to the highest part of the dome a wooden ring is screwed to cover up the joints. Again this material can be metallic in case it will be painted. This spherical cap will be the centerpiece of the structure and is usually decorated with the image of Jesus Christ in case of Christian Orthodox churches. Before attaching the small spherical cap structure to the dome it can me decorated with hagiography and covered up temporarily so that the painter would not have to do this work at a height later.

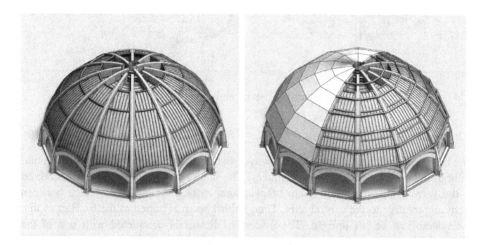

Fig. 6. Completing the interior shell and adding the insulation layer

3.5   Adding the Interior Spherical Triangles

After the spherical cap is attached the twelve spherical triangles are ready to be added as shown in Fig. 6. These parts are composed of thin metal sections that are bend to the appropriate radius to match the domes geometry. The metal sections are then covered with the same galvanized steel mesh of the spherical cap previously mentioned. Then they are covered with plaster to produce the final interior surface. The final surface can also be achieved with attaching thin gypsum boards that are cut based on the net of the curved triangles. The seams left in the gypsum boards after they are screwed into place can be taped and covered up with plaster. These spherical triangles can also be pre-assembled on a fixed location rather than in situ and follow the process of hagiography before being fixed to the main wooden beams.

3.6   Insulating the Dome

At this point the interior of the dome is finished. The step that follows is adding the thermal insulation. Secondary small wooden beams are screwed on the main beams and are covered by plywood sheets. The space between the plywood sheets and the metal frame and interior shell can be filled with mineral wool to increase the thermal and sound insulation properties of the dome. On top of the first plywood layer a second layer is added composed of small wooden beams, extruded polystyrene sheets and plywood finish (Fig. 7).

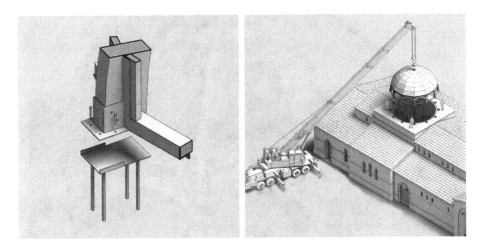

Fig. 7. Base connections and dome installation

3.7 Lifting and Fixing the Dome to the Building

With the thermal insulation of the dome completed the structure is ready to be lifted and positioned in the circular opening of the building. Twelve metal brackets are anchored into the circular perimeter of the opening in advance to serve both as landing zones for the dome and as the main connection points between the dome and the structure. At this stage a crane is needed to lift the dome into place. After the dome is positioned the final connection is made through bolting the two metal parts.

3.8 Finishing the Domes Cladding and Exterior Walls

Once the dome is fixed to the building the next step is to complete the polygonal side wall that encloses the dome. The wall can be built from any material, usually for aesthetic purposes the same as the building beneath. There will be a triangular space left between the peripheral wall and the dome and it can be filled with cement grout as to provide the appropriate slope for the water to escape freely from the roof. Covering the plywood, grout and wall surface a coating of water resistant membrane is applied. The final surface in this example is accomplished with byzantine roof tiles but it can be constructed of any other tile of even metallic roof finish (Figs. 8 and 9).

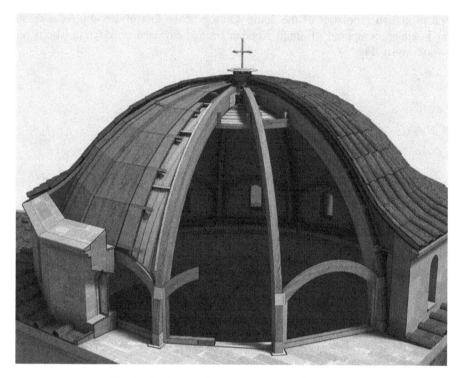

Fig. 8. Section of all the components of the dome

Fig. 9. View of the dome from the interior of the church

4   Conclusion

The example of the lightweight dome analyzed above can accept many variations both in terms of its final interior appearance as well as the construction materials of the different parts. Besides the aesthetic decisions and the various methods that can be chosen to construct such a dome the main advantage of building a dome with this design philosophy is that the weight of the dome is reduced by a great extent. This could end up determining the preservation of the interior of these structures that despite their architectural value many times carry valuable pieces of art, sculpture and decoration in their interior spaces but most importantly it could determine the safety of the people that gather in these religious buildings.

References

1. Kalopissi, S., Panayotidi, M.: Multilingual Illustrated Dictionary of Byzantine Architecture and Sculpture Terminology. Crete University Press, Herakleion (2010)
2. Campbell, D.: Stone in Traditional Architecture. Schiffer Publishing, Pennsylvania (2010)
3. Como, M.: Statics of Historic Masonry Constructions. SSSSM, vol. 5. Springer, Cham (2016). https://doi.org/10.1007/978-3-319-24569-0
4. King, R.: Brunelleschi's Dome: The Story of the Great Cathedral in Florence. Penguin Books, New York (2001)
5. Howard, D.: The Architectural History of Venice. Yale University Press, New Haven (2010)
6. Misztal, B.: Wooden Domes: History and Modern Times. Springer, Cham (2018). https://doi.org/10.1007/978-3-319-65741-7
7. Filippou, A.: The Byzantine Architecture of Cyprus. Andreas Philippou, Nicosia (2013)
8. Breyer, D., Fridley, K., Cobden, K., Pollock, D.: Design of Wood Structures. McGraw Hill, New York (2007)

Cultural Heritage Structures Strengthened by Ties Under Seismic Sequences and Uncertain Input Parameters: A Computational Approach

Angelos A. Liolios[✉]

Department of Civil Engineering,
Democritus-University of Thrace, Xanthi, Greece
aliolios@civil.duth.gr

Abstract. The seismic analysis of existing Cultural Heritage framed structures that have been damaged and upgraded by using cable elements (tension-ties) is numerically investigated. Special attention is given to uncertainty concerning input parameters for the structural elements behaviour. A double discretization, in space by the Finite Element Method and in time by a direct approach, is applied. The unilateral behaviour of the cable elements that undertake only tension stresses is strictly taken into account. Damage indices are computed for the accumulating damage due to seismic sequences. The presented numerical approach is applied to a typical reinforced concrete (RC) frame-building of the recent Greek Cultural Heritage.

Keywords: Computational structural mechanics · Seismic sequences · Upgrading by ties · Input parameters uncertainty

1 Introduction

As well-known [1], besides the usual historic monumental structures (churches, monasteries, old masonry buildings etc.), the recent built Cultural Heritage (CH) includes also existing framed structures. Examples of such CH structures are old steel and stone bridges, as well as old industrial buildings of reinforced concrete (RC), e.g. old factory premises. Concerning their global seismic behavior, it often arises the need for their seismic upgrading. Certainly this upgrading must be realized by using materials and methods in the context of the Sustainable Construction and in the frame of the current Civil Engineering praxis [1–9].

Especially for RC structures which belong to recent built Cultural Heritage, some traditional methods for their seismic upgrading (e.g. RC mantles) are available, see [4–9]. Recently, the use of cable-like members (tension-ties) has been considered as an alternative strengthening method [10–12]. As well-known, ties have been used effectively in monastery buildings and churches arches. The ties-strengthening approach has the advantages of "cleaner" and "more lenient" operation, avoiding as much as possible the unmaking, the digging, the extensive concreting and "nuisance" functionality of the existing building. The ties can undertake tension but buckle and become slack and

© Springer Nature Switzerland AG 2019
A. Moropoulou et al. (Eds.): TMM_CH 2018, CCIS 962, pp. 188–199, 2019.
https://doi.org/10.1007/978-3-030-12960-6_13

structurally ineffective when subjected to a sufficiently large compressive force. Thus the governing conditions take equality as well as an inequality form and the problem becomes a highly nonlinear one [10–14].

At this point it must be emphasized that for the seismic analysis of such old RC structures an estimation of the input parameters must be realized taking into account a lot of uncertainties. The later mainly concern the holding properties of the old materials which had been used for the building of such structures, e.g. the remaining strength of the concrete and steel, as well as the cracking effects etc. For the quantification of such uncertainties, probabilistic methods have been proposed [15–20].

Moreover, as concerns the seismic upgrading of existing RC structures, modern seismic design codes adopt exclusively the use of the isolated and rare 'design earthquake', whereas the influence of repeated earthquake phenomena is ignored. But as the results of recent research have shown [21, 22], multiple earthquakes generally require increased ductility design demands in comparison with single isolated seismic events. Especially for the seismic damage due to multiple earthquakes and to pounding [23–25], this is accumulated and so it is higher than that for single ground motions.

The present research treats with a computational probabilistic approach for the seismic analysis of Cultural Heritage existing industrial RC frame-buildings, which have to be strengthened by cable elements and to be subjected to seismic sequences. Special attention is given for the estimation of the uncertainties concerning structural input parameters. For this purpose, Monte Carlo techniques are used. Damage indices are computed for the seismic assessment of such historic and industrial structures and in order the optimum cable-bracing strengthening version to be chosen. Finally, an application is presented for a simple typical example of a two-bay two-story industrial RC frame strengthened by bracing ties under multiple earthquakes.

2 The Probabilistic Computational Approach

The probabilistic approach for the seismic analysis of Cultural Heritage existing RC frame-buildings may be obtained through Monte Carlo simulations. As well-known, see e.g. [26–33], Monte Carlo simulation is simply a repeated process of generating deterministic solutions to a given problem. Each solution corresponds to a set of deterministic input values of the underlying random variables. A statistical analysis of the so obtained simulated solutions is then performed. Thus the computational methodology consists of solving first the deterministic problem for each set of the random input variables and finally realizing a statistical analysis.

Details of the methodology concerning the deterministic problem and the probabilistic aspects are given in the next sections.

2.1 Mathematical Treatment of the Deterministic Problem

The mathematical formulation and solution of the deterministic problem concerning the seismic analysis of Cultural Heritage existing RC frame-buildings strengthened by ties has been developed in [10, 34]. Details of the developed numerical approach are summarized briefly herein.

A double discretization, in space and time, is used as usually in structural dynamics [12, 35, 36]. So, first, the structural system is discretized in space by using frame finite elements. Non-linear behavior is considered as lumped at the two ends of the RC frame elements, where plastic hinges can be developed. Pin-jointed bar elements are used for the cable-elements. The unilateral behavior of these tie-elements and the non-linear behavior of the RC structural elements can include loosening, elastoplastic or/and elastoplastic-softening-fracturing and unloading - reloading effects. All these non-linear characteristics, concerning the ends of frame elements and the cable constitutive law, can be expressed mathematically by the subdifferential relation [10–13]:

$$s_i \,(d_i) \in \hat{\partial} \, S_i(d_i). \tag{1}$$

Here s_i and d_i are generalized stress and deformation quantities. For the case of tie-elements, these quantities are the tensile force (in [kN]) and the elongation (in [m]), respectively, of the i-th cable element. $\hat{\partial}$ is the generalized gradient and S_i is the superpotential function, see Panagiotopoulos [13].

Next, dynamic equilibrium for the assembled structural system with cables is expressed by the usual incremental matrix relation:

$$\mathbf{M}\Delta\ddot{\mathbf{u}} + \mathbf{C}(\Delta\dot{\mathbf{u}}) + \mathbf{K}(\Delta\mathbf{u}) = \Delta\mathbf{p} + \mathbf{A}\Delta\mathbf{s}. \tag{2}$$

Here $\mathbf{u}$ and $\mathbf{p}$ are the displacement and the load time dependent vectors, respectively, and $\mathbf{s}$ is the cable stress vector. $\mathbf{M}$ is the mass matrix and $\mathbf{A}$ is a transformation matrix. The damping and stiffness terms, $\mathbf{C}(\dot{\mathbf{u}})$ and $\mathbf{K}(\mathbf{u})$, respectively, concern the general non-linear case. Dots over symbols denote derivatives with respect to time. For the case of ground seismic excitation $\mathbf{x_g}$, the loading history term $\mathbf{p}$ becomes $\mathbf{p} = -\mathbf{M}\,\mathbf{r}\,\ddot{\mathbf{x}}_g$, where $\mathbf{r}$ is the vector of stereostatic displacements [35].

The above relations (1)–(2), combined with the initial conditions, consist the problem formulation, where, for given p and/or $\ddot{\mathbf{x}}_g$, the vectors $\mathbf{u}$ and $\mathbf{s}$ have to be computed. From the strict mathematical point of view, using (1) and (2), we can formulate the problem as a dynamic hemivariational inequality one by following [13] and investigate it.

In Civil Engineering practical cases, a numerical treatment of the above problem is realized by applying to the above constitutive relations (1) a piecewise linearization based on experimental investigations. So, simplified stress-deformation constitutive diagrammes are used, which are piece-wise linearized. Based on the above piecewise linearization of the constitutive relations, and applying a direct time-integration scheme, in each time-step a relevant non-convex linear complementarity problem [10] of the following matrix form is solved:

$$\mathbf{v} \geq \mathbf{0}, \quad \mathbf{Dv} + \mathbf{a} \leq \mathbf{0}, \quad \mathbf{v}^T \cdot (\mathbf{Dv} + \mathbf{a}) = 0. \tag{3}$$

So, the nonlinear Response Time-History (RTH) can be computed for a given seismic ground excitation.

An alternative approach for treating numerically the problem is the incremental one based on Eq. (2). On such incremental approaches is based the structural analysis software Ruaumoko [36], which is applied hereafter.

As concerns multiple earthquakes, it is reminded [21, 22] that current seismic codes suggest the exclusive adoption of the isolated and rare "design earthquake", while the influence of repeated earthquake phenomena is ignored. This is a significant drawback for the realistic design of building structures in seismically active regions, because, as it is shown in [21], real seismic sequences have accumulating effects on various damage indices.

The decision about a possible strengthening for an existing RC structure, damaged by a seismic event, can be taken after a relevant assessment. This is obtained by using in situ structural identifications [16] and evaluating suitable damage indices [37, 38]. Further, a comparative investigation of structural responses due to various seismic excitations can be used. So, the system is considered for various cases, with or without strengthening by cable-bracings.

Among the several response parameters, the focus is on the overall structural damage index (OSDI) [37, 38]. This is due to the fact, that this parameter summarises all the existing damages on columns and beams of reinforced concrete frames in a single value, which is useful for comparison reasons.

In the OSDI model after Park/Ang [38], the global damage is obtained as a weighted average of the local damage at the section ends of each structural element or at each cable element. First the _local_ damage index DI_L is computed by the following relation:

$$DI_L = \frac{\mu_m}{\mu_u} + \frac{\beta}{F_y d_u} E_T \tag{4}$$

where: μ_m is the maximum ductility attained during the load history, μ_u the ultimate ductility capacity of the section or element, β a strength degrading parameter, F_y the yield force of the section or element, E_T the dissipated hysteretic energy, and d_u the ultimate deformation.

Next, the dissipated energy E_T is chosen as the weighting function and the _global_ damage index DI_G is computed by using the following relation:

$$DI_G = \frac{\sum_{i=1}^{n} DI_{Li} E_i}{\sum_{i=1}^{n} E_i} \tag{5}$$

where: DI_{Li} is the local damage index after Park/Ang at location i, E_i is the energy dissipated at location i and n is the number of locations at which the local damage is computed.

2.2   Numerical Treatment of the Probabilistic Problem

In order to calculate the random characteristics of the response of the considered RC buildings, the Monte Carlo simulation is used [26–33]. As mentioned, the main element of a Monte Carlo simulation procedure is the generation of random numbers from a specified distribution. Systematic and efficient methods for generating such random numbers from several common probability distributions are available. The random variable simulation is implemented using the technique of Latin Hypercube Sampling (LHS) [15, 20]. The LHS is a selective sample technique by which, for a desirable accuracy level, the number of the sample size is significantly smaller than the direct Monte Carlo simulation.

In more details, a set of values of the basic design input variables can be generated according to their corresponding probability distributions by using statistical sampling techniques. The generated basic design variables are treated as a sample of experimental observations and used for the system deterministic analysis to obtain a simulated solution as in Subsect. 2.1 is described As the generation of the basic design variables is repeated, more simulated solutions can be determined. Finally, statistical analysis of the simulated solutions is then performed. The results obtained from the Monte Carlo simulation method depend on the number of the generated basic design variables used.

Such design variables for the herein considered RC buildings are the uncertain quantities describing the plastic-hinges behavior and the spatial variation of input old materials parameters.

Concerning the plastic hinges in the end sections of the frame structural elements, a typical normalized moment- normalized rotation backbone is shown in Fig. 1, see [19]. This backbone hardens after a yield moment of a_{My} times the nominal, having a non-negative slope of a_h up to a corner normalized rotation (or rotational ductility) μ_c where the negative stifiness segment starts. The drop, at a slope of a_c, is arrested by the residual plateau appearing at normalized height r that abruptly ends at the ultimate rotational ductility μ_u. The normalized rotation is the rotational ductility $\mu = \theta / \theta^{yield}$.

The above six backbone parameters shown in Fig. 1, namely a_h, a_c, μ_c, r, μ_u and a_{My}, $= M/M_y$ are assumed to vary independently from each other according to Normal distribution.

As regards the random variation of parameters for the old materials, which had been used for the building of old RC structures, their input estimations concern mainly the remaining strength of the concrete and the steel and the elasticity modulus. According to [20] and JCSS (Joint Committee Structural Safety), see [28], concrete strength and elasticity modulus follow the Normal distribution, whereas the steel strength follows the Lognormal distribution.

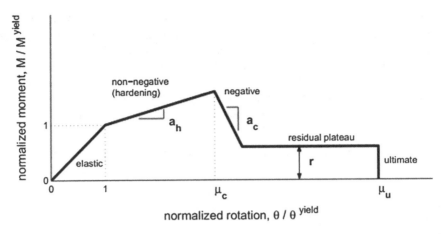

Fig. 1. Representative normalized moment - normalized rotation backbone diagramme for plastic hinges [19].

3   Numerical Example

3.1   Description of the Considered Cultural Heritage RC Structural System

The old industrial reinforced concrete frame F0 of Fig. 2 is considered to be upgraded by ties and subjected to a multiple ground seismic excitation.

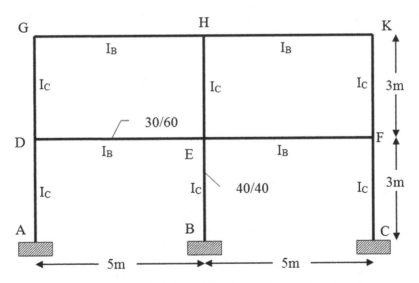

Fig. 2. System F0: the old industrial RC frame without cable-strengthening

The estimated concrete class is C12/15 and the steel class is S220. As mentioned, according to [20] and JCSS (Joint Committee Structural Safety), see [28], concrete strength and elasticity modulus follow Normal distribution and the steel strength follows the Lognormal distribution. So the statistical characteristics of the input random variables concerning the building materials are estimated to be as shown in Table 1. The mean/median values of the random variables correspond to the best estimates employed in the deterministic model according to Greek code KANEPE [8].

Table 1. Statistical data for the building materials treated as random variables

	Distribution	Mean	COV
Compressive strength of concrete	Normal	8.0 MPa	15%
Yield strength of steel	Lognormal	191.3 MPa	10%
Initial elasticity modulus, concrete	Normal	26.0 GPA	8%
Initial elasticity modulus, steel	Normal	200 GPA	4%

Further, for the frame assessment, probabilistic section models are derived by using the data of the Table 3. The procedure is shown qualitatively in Fig. 3.

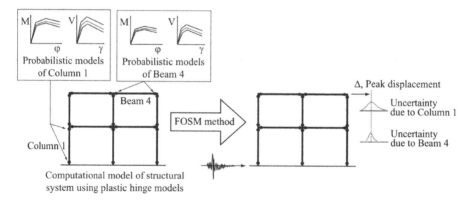

Fig. 3. System evaluation procedure using probabilistic section models

Due to various extreme actions (environmental etc.), corrosion and cracking has been taken place, which has caused a strength and stiffness degradation. The stiffness reduction due to cracking [39] results to effective stiffness of 0.60 I_g for the two external columns, 0.80 I_g for the internal columns and 0.40 I_g for the beams, where I_g is the gross inertia moment of their cross-section. Using Ruaumoko software [36], the columns and the beams are modeled using prismatic frame elements. Nonlinearity at the two ends of RC members is idealized using one-component plastic hinge models, following the Takeda hysteresis rule. Interaction curves (M-N) for the critical cross-sections of the examined RC frame have been computed.

The frame-system F0 of Fig. 2 was initially without cable-bracings. After the seismic assessment, it was decided the frame F0 to be strengthened by ties. The X-cable-bracings system, shown in Fig. 4, has been proposed as the optimal one in order the frame F0 to be seismically upgraded. The system of the frame with the X-bracing diagonal cable-elements shown in Fig. 4 is denoted as system F4.

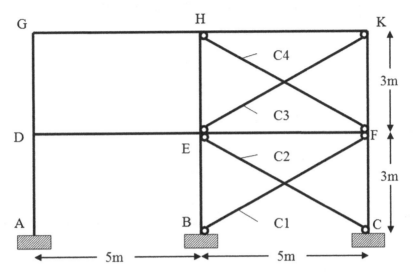

Fig. 4. System F4: the industrial RC frame cable-strengthened with X-bracings

The cable constitutive law, concerning the unilateral (slackness), hysteretic, fracturing, unloading-reloading etc. behavior, is depicted in Fig. 5. The cable elements have a cross-sectional area $F_c = 18$ cm^2 and they are of steel class S220 with elasticity modulus $E_c = 200$ GPa.

3.2   Earthquakes Sequence Input

A list of multiple earthquakes, downloaded from the strong motion database of the Pacific Earthquake Engineering Research (PEER) Center [21, 40], appears in Table 2.

3.3   Representative Probabilistic Results

After application of the herein proposed computational probabilistic approach, some representative results are shown in Table 3. These concern the Coalinga case of the seismic sequence of Table 2.

In column (2) of the Table 3, the Event E_1 corresponds to Coalinga seismic event of 0.605 g PGA, and Event E_2 to 0.733 g PGA, (g = 9.81 m/s^2). The sequence of events E_1 and E_2 is denoted as Event ($E_1 + E_2$).

In the table column (3) the mean value and in column (4) the coefficient of variation COV of the Global Damage Indices DI_G are given. Similarly, in the column (5) and in the column (6) the mean value and the coefficient of variation COV of the absolutely maximum horizontal top roof displacement u_{top}, respectively, are given.

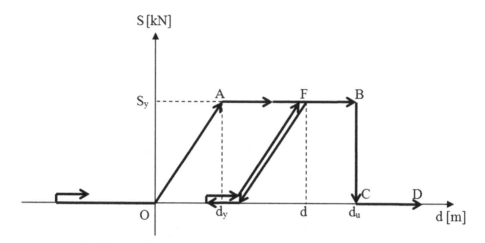

Fig. 5. The diagramme for the constitutive law of cable-elements.

Table 2. Multiple earthquakes data

No	Seismic sequence	Date (Time)	Magnitude (M_L)	Recorded PGA (g)	Normalized PGA (g)
1	Coalinga	1983/07/22 (02:39)	6.0	0.605	0.165
		1983/07/25 (22:31)	5.3	0.733	0.200
2	Imperial valley	1979/10/15 (23:16)	6.6	0.221	0.200
		1979/10/15 (23:19)	5.2	0.211	0.191
3	Whittier narrows	1987/10/01 (14:42)	5.9	0.204	0.192
		1987/10/04 (10:59)	5.3	0.212	0.200

Table 3. Representative probabilistic response quantities for the systems F0 and F4

System	Events	DI_G		u_{top} [cm]	
		Mean value	COV	Mean value	COV
(1)	(2)	(3)	(4)	(5)	(6)
F0	Event E_1	0.176	14.8%	2.48	12.7%
	Event E_2	0.385	13.7%	3.87	11.8%
	Event ($E_1 + E_2$)	0.442	15.7%	4.64	13.4%
F4	Event E_1	0.082	14.4%	1.36	12.9%
	Event E_2	0.107	12.8%	1.74	11.8%
	Event ($E_1 + E_2$)	0.128	15.2%	1.88	14.7%

As the values in the Table 3 show, multiple earthquakes generally increase response quantities, especially the damage indices. On the other hand, the strengthening of the frame F0 by X-bracings (system Frame F4 of Fig. 4) improves the response behaviour.

4 Concluding Remarks

As the results of the numerical example have shown, the herein presented computational approach can be effectively used for the probabilistic numerical investigation of the seismic inelastic behaviour of Cultural Heritage old RC framed structures strengthened by cable elements and subjected to multiple earthquakes. The probabilistic estimation of the uncertain input parameters is effectively realized by using Monte Carlo simulation. Finally, the optimal cable-bracing scheme can be selected among investigated alternative ones by using computed damage indices.

References

1. Asteris, P.G., Plevris, V. (eds.): Handbook of Research on Seismic Assessment and Rehabilitation of Historic Structures. IGI Global, Hershey (2015)
2. Moropoulou, A., Labropoulos, K.C., Delegou, E.T., Karoglou, M., Bakolas, A.: Non-destructive techniques as a tool for the protection of built cultural heritage. Constr. Build. Mater. **48**, 1222–1239 (2013)
3. Moropoulou, A., et al.: NDT investigation of Holy Sepulchre complex structures. In: Radonjanin, V., Crews, K. (eds.) Proceedings of Structural Faults and Repair 2012, Proceedings in CD-ROM (2012)
4. Bertero, V.V.: Seismic upgrading of existing buildings. In: Proc. Simposio Internacional de Ingeniería Civil, a los 10 Años de los Sismos de 1985, Sociedad Mexicana de Ingeniería, Sísmica, AC Mexico, D.F., Mexico (1995)
5. Dritsos, S.E.: Repair and Strengthening of Reinforced Concrete Structures. University of Patras, Greece (2001). (in Greek)
6. Fardis, M.N.: Seismic Design, Assessment and Retrofitting of Concrete Buildings: Based on EN-Eurocode 8. Springer, Heidelberg (2009). https://doi.org/10.1007/978-1-4020-9842-0
7. FEMA P440A: Effects of strength and stiffness degradation on the seismic response of structural systems. U.S. Department of Homeland Security, Federal Emergency Management Agency (2009)
8. Greek Retrofitting Code-(KANEPE): Greek Organization for Seismic Planning and Protection (OASP), Greek Ministry for Environmental Planning and Public Works. Athens, Greece (2013). (in Greek). www.oasp.gr
9. Penelis, G.Gr., Penelis, Gr.G.: Concrete Buildings in Seismic Regions. CRC Press, Boca Raton (2014)
10. Liolios, A., Chalioris, C.: Industrial reinforced concrete buildings strengthened by cable elements: a numerical investigation of the response under seismic sequences. In: Moropoulou, A. (ed.) Proceedings of Scientific Conference on "Scientific Support for Decision-Making on Sustainable and Compatible Materials and Interventions for the Preservation and Protection of Cultural Heritage", Thalis Project, NTUA, Athens, pp. 244–257 (2015)

11. Liolios, A.: A computational investigation for the seismic response of RC structures strengthened by cable elements. In: Papadrakakis, M., Papadopoulos, V., Plevris, V. (eds.) Proceedings of COMPDYN 2015: Computational Methods in Structural Dynamics and Earthquake Engineering, 5th ECCOMAS Thematic Conference, Crete Island, Greece, 25–27 May 2015, vol. II, pp. 3997–4010 (2015)

12. Liolios, A., Chalioris, C.: Reinforced concrete frames strengthened by cable elements under multiple earthquakes: a computational approach simulating experimental results. In: Proceedings of 8th GRACM International Congress on Computational Mechanics, Volos, 12–15 July 2015

13. Panagiotopoulos, P.D.: Hemivariational Inequalities. Applications in Mechanics and Engineering. Springer, Heidelberg (1993). https://doi.org/10.1007/978-3-642-51677-1

14. Leftheris, B., Stavroulaki, M.E., Sapounaki, A.C., Stavroulakis, G.E.: Computational Mechanics for Heritage Structures. WIT Press, Southampton (2006)

15. Papadrakakis, M., Stefanou, G. (eds.): Multiscale Modeling and Uncertainty Quantification of Materials and Structures. Springer, Cham (2014). https://doi.org/10.1007/978-3-319-06331-7

16. Strauss, A., Frangopol, D.M., Bergmeister, K.: Assessment of existing structures based on identification. J. Struct. Eng. ASCE **136**(1), 86–97 (2010)

17. Vamvatsikos, D., Cornell, C.A.: Incremental dynamic analysis. Earthquake Eng. Struct. Dynam. **31**, 491–514 (2002)

18. Vamvatsikos, D., Cornell, C.A.: Direct estimation of the seismic demand and capacity of oscillators with multi-linear static pushovers through IDA. Earthquake Eng. Struct. Dynam. **35**(9), 1097–1117 (2006)

19. Vamvatsikos, D., Fragiadakis, M.: Incremental dynamic analysis for estimating seismic performance sensitivity and uncertainty. Earthquake Eng. Struct. Dynam. **39**(2), 141–163 (2010)

20. Thomos, G.C., Trezos, C.G.: Examination of the probabilistic response of reinforced concrete structures under static non-linear analysis. Eng. Struct. **28**, 120–133 (2006)

21. Hatzigeorgiou, G., Liolios, A.: Nonlinear behaviour of RC frames under repeated strong ground motions. Soil Dyn. Earthq. Eng. **30**, 1010–1025 (2010)

22. Liolios, As., Liolios, A., Hatzigeorgiou, G.: A numerical approach for estimating the effects of multiple earthquakes to seismic response of structures strengthened by cable-elements. J. Theor. Appl. Mech. **43**(3), 21–32 (2013). https://doi.org/10.2478/jtam-2013-0021

23. Maniatakis, C.A., Spyrakos, C.C., Kiriakopoulos, P.D., Tsellos, K.P.: Seismic response of a historic church considering pounding phenomena. Bull. Earthq. Eng. **16**(7), 2913–2941 (2018)

24. Spyrakos, C.C., Maniatakis, Ch.A.: Retrofitting of a historic masonry building. In: 10th National and 4th International Scientific Conference on Planning, Design, Construction and Renewal in the Construction Industry (iNDiS 2006), Novi Sad, 22–24 November 2006, pp. 535–544 (2006)

25. Spyrakos, C.C., Maniatakis, C.A.: Seismic protection of monuments and historic structures – the SEISMO research project. In: Proceedings of the VII European Congress on Computational Methods in Applied Sciences and Engineering ECCOMAS 2016, 5–10 June 2016, Crete Island, Greece (2016)

26. Lee, T.H., Mosalam, K.M.: Probabilistic seismic evaluation of reinforced concrete structural components and systems. Report 2006/04, Pacific Earthquake Engineering Research Center. University of California, Berkeley, USA Google Scholar (2006)

27. Melchers, R.E., Beck, A.T.: Structural Reliability Analysis and Prediction, 3rd edn. Wiley, New York (2018)

28. JCSS: Probabilistic Model Code-Part 1: Basis of Design (12th draft). Joint Committee on Structural Safety, March 2001. http://www.jcss.ethz.ch/
29. Georgioudakis, M., Stefanou, G., Papadrakakis, M.: Stochastic failure analysis of structures with softening materials. Eng. Struct. **61**, 13–21 (2014)
30. Ang, A.H., Tang, W.H.: Probability Concepts in Engineering Planning and Design, vol. 2: Decision, Risk, and Reliability. Wiley, New York (1984)
31. Casciati, F., Augusti, G., Baratta, A.: Probabilistic Methods in Structural Engineering. CRC Press, Boca Raton (2014)
32. Kottegoda, N., Rosso, R.: Statistics, Probability and Reliability for Civil and Environmental Engineers. McGraw-Hill, London (2000)
33. Dimov, I.T.: Monte Carlo Methods for Applied Scientists. World Scientific, Singapore (2008)
34. Liolios, A., Moropoulou, A., Liolios, As., Georgiev, K., Georgiev, I.: A computational approach for the seismic sequences induced response of cultural heritage structures upgraded by ties. In: Margenov, S., Angelova, G., Agre, G. (eds.) Innovative Approaches and Solutions in Advanced Intelligent Systems. SCI, vol. 648, pp. 47–58. Springer, Cham (2016). https://doi.org/10.1007/978-3-319-32207-0_4
35. Chopra, A.K.: Dynamics of Structures: Theory and Applications to Earthquake Engineering. Pearson Prentice Hall, New York (2007)
36. Carr, A.J.: RUAUMOKO - Inelastic Dynamic Analysis Program. Department of Civil Engineering, University of Canterbury, Christchurch, New Zealand (2008)
37. Mitropoulou, C.C., Lagaros, N.D., Papadrakakis, M.: Numerical calibration of damage indices. Adv. Eng. Softw. **70**, 36–50 (2014)
38. Park, Y.J., Ang, A.H.S.: Mechanistic seismic damage model for reinforced concrete. J. Struct. Div. ASCE **111**(4), 722–739 (1985)
39. Paulay, T., Priestley, M.J.N.: Seismic Design of Reinforced Concrete and Masonry Buildings. Wiley, New York (1992)
40. PEER: Pacific Earthquake Engineering Research Center. PEER Strong Motion Database (2011). http://peer.berkeley.edu/smcat

Masonry Compressive Strength Prediction Using Artificial Neural Networks

Panagiotis G. Asteris[1](✉), Ioannis Argyropoulos[1], Liborio Cavaleri[2],
Hugo Rodrigues[3], Humberto Varum[4], Job Thomas[5],
and Paulo B. Lourenço[6]

[1] Computational Mechanics Laboratory, School of Pedagogical
and Technological Education, Heraklion, Athens, Greece
panagiotisasteris@gmail.com
[2] Department of Civil, Environmental,
Aerospace and Materials Engineering (DICAM),
University of Palermo, Palermo, Italy
[3] RISCO, Department of Civil Engineering,
Polytechnic Institute of Leiria, Leiria, Portugal
[4] Civil Engineering Department,
Faculty of Engineering of the University of Porto, Porto, Portugal
[5] Department of Civil Engineering,
Cochin University of Science and Technology, Cochin, Kerala, India
[6] Department of Civil Engineering, ISISE,
University of Minho, Guimarães, Portugal

Abstract. The masonry is not only included among the oldest building materials, but it is also the most widely used material due to its simple construction
and low cost compared to the other modern building materials. Nevertheless,
there is not yet a robust quantitative method, available in the literature, which
can reliably predict its strength, based on the geometrical and mechanical
characteristics of its components. This limitation is due to the highly nonlinear
relation between the compressive strength of masonry and the geometrical and
mechanical properties of the components of the masonry. In this paper, the
application of artificial neural networks for predicting the compressive strength
of masonry has been investigated. Specifically, back-propagation neural network
models have been used for predicting the compressive strength of masonry
prism based on experimental data available in the literature. The comparison of
the derived results with the experimental findings demonstrates the ability of
artificial neural networks to approximate the compressive strength of masonry
walls in a reliable and robust manner.

Keywords: Artificial Neural Networks (ANNs) ·
Back-Propagation Neural Networks (BPNNs) · Building materials ·
Compressive strength · Masonry · Masonry unit · Mortar ·
Soft-computing techniques

© Springer Nature Switzerland AG 2019
A. Moropoulou et al. (Eds.): TMM_CH 2018, CCIS 962, pp. 200–224, 2019.
https://doi.org/10.1007/978-3-030-12960-6_14

1 Introduction

Masonry, as it is constructed with the use of natural materials, is one of the oldest building systems known to Humanity and is believed to have been in use for over 6,000 years. Also, masonry is the most widely used construction type, not only in the poverty-stricken countries, due to its low cost compared to the other modern materials, but also in developed countries, due to the aesthetic value that it provides when used in modern constructions. Furthermore, masonry structures exhibit a very good seismic behavior for usual low-rise houses, as well as, excellent thermal properties, as it can keep the structure cool in the summer and warm in the winter [1]. The fact that the masonry material is the oldest building material explains that the majority of monuments are masonry structures, meaning stone (or/and brick) elements joined together through the use of mortars.

In the light of the above, it can be clearly seen why masonry structures are so popular among the civil engineering community. Masonry structures, as an important part of our cultural and historical identity, have attracted the interest of many researchers since the early part of 20th century. A dominant position among the first researchers involved in masonry structures is Engesser, who, in 1907, proposed in his work entitled *Überweitgespanntewölbbrücken,* the first formulae for the estimation of masonry compressive strength considering the mortar and unit strengths [2].

Despite the plethora of research work in the last three decades, the mechanics of masonry structures remains an open issue and, at the same time, a challenge for the practicing civil engineer. This is mainly related to the complicated, inhomogeneous and anisotropic nature of this particular structure type [3–9]. Furthermore, there not yet exists a robust quantitative method, available in the literature, which can reliably predict its strength, based on its components and its geometric and mechanical characteristics.

Masonry is composed of different materials, namely: the masonry units and the mortar phase. Masonry units may be either solid or hollow and may be made of a wide variety of materials. Clay bricks, blocks of stone, concrete blocks, pressed earth bricks, calcium silicate bricks, soft mud bricks etc. are some examples of masonry units used in masonry construction. The two material phases in masonry are joined by a weak interface and hence masonry is generally weak in tension. Masonry structures are therefore expected to resist only compressive forces [10]. The conventional design practice emphasizes that masonry structures are subjected to compressive stresses alone [11, 12] and hence an accurate determination of compressive strength is extremely important. Empirical values for the masonry strength are suggested in SP: 20 [13] for the design of masonry based on the unit strength and properties of mortar. Alternatively, masonry specimens can be tested to obtain a more accurate value of their compressive strength. The results of proposed equations available in the literature have a large dispersion, as will be shown in the next section. The mechanical properties of masonry structures exhibit a strong nonlinear nature derived from the parameters involved in their structure; it is this nonlinear behaviour that makes the development of an analytical formula for the prediction of the mechanical properties using deterministic methods a rather difficult task.

Based on the above non-linear nature of the parameters involved in the mechanical behavior of masonry, as well as on the limitations of deterministic methods to give a reliable and robust prediction regarding the compressive strength, the last three decades, soft-computing techniques, such as artificial neural networks (ANNs), have come to contribute to the solution of the problem.

ANNs have emerged over the last decades as an attractive meta-modelling technique applicable to a vast number of scientific fields, including material science. An important characteristic of ANNs is that they can be used to build soft-sensors, i.e., models with the ability to estimate critical quantities without having to measure them [14]. In particular, such surrogate models can be constructed after a training process with only a few available data, which can be used to predict pre-selected model parameters, reducing the need for time- and money-consuming experiments.

To the authors' knowledge, there is no many attempts to apply Neural Network (NN) for the prediction of masonry behavior in general. NNs have been used in many ways for various other problems in the literature. Thus far, the literature includes studies in which ANNs were used for predicting the mechanical properties of concrete materials [15–23]. In the study of Asteris et al. [23] ANNs were used to estimate the compressive strength of self-compacting concrete through a training process, involving eleven input parameters and one output parameter, which is compressive strength of concrete. Moreover, similar methods such as fuzzy logic and genetic algorithms have also been used for modelling the compressive strength of concrete [24–32]. A detailed state-of-the-art report can be found in earlier literatures [33–38].

In this context, in the work presented herein, the modelling of the mechanical characteristics of masonry structures materials has been investigated in-depth using soft-computing techniques such as surrogate models. In particular, this study investigates the application of Artificial Neural Networks (ANNs) models for the prediction of the compressive strength of masonry prisms. Specifically, for the development and the training of NN models, a database consisting of 232 specimens, taken from the literature, was utilized. The compressive strength of masonry unit, the compressive strength of mortar, the height-to-thickness ratio of masonry prism, the volume fraction of masonry unit and the volume ratio of bed joint were used as input parameters, while the value of compressive strength was used as output parameter. The optimum NN model developed in this study has proven to be very successful, exhibiting very reliable predictions.

2 Literature Review

For the determination of the masonry wall compressive strength (f_{wc}), several semi-empirical expressions are available in the literature [2, 10, 39–53]. Common feature of these expressions, global effects contributing to the system resistance, such as buckling-effects or local-compression resistance are not considered.

The majority of these proposals [2, 10, 39–42, 46–53], in the form of the first formula proposed by Engesser in 1907 [2], take under consideration only the compressive strengths of brick and mortar, while only a few proposals pay attention to the height-to-thickness ratio of masonry prism (Fig. 1), and only two proposals [12, 43] take into account the volume fraction of masonry unit and the volume ratio of bed joint mortar which are used as input parameters. In Table 1, the most representative formulae (from the literature) are being presented, concerning the estimation of masonry prism compressive strength.

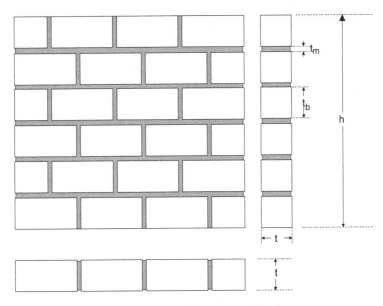

Fig. 1. Geometry of masonry wall prism.

Recently, Thaickavil and Thomas [12] have been proposed a formula that takes into account the majority of the parameters that affect the masonry compressive strength. The authors [12] carried out regression analysis on a plethora of test data (232 datasets) corresponding to the masonry unit strength of 3.1 to 127.0 MPa, mortar strength of 0.3 to 52.6 MPa and h/t ratio of 1.15 to 5.75, have proposed the following formula:

$$f_{wc} = \frac{0.54 \times f_{bc}^{1.06} \times f_{mc}^{0.004} \times VF_{b}^{3.3} \times VR_{mH}^{0.6}}{\left(h/t\right)^{0.28}} \quad (1)$$

where VF_b is the volume fraction of brick, while VR_{mH} is the volume ratio of bed joint to mortar and are defined by the following equations:

$$VF_b = \frac{V_u}{V_p} \quad (2)$$

where V_u is the volume of masonry units and V_p is the volume of prism.

$$VR_{mH} = \frac{V_{mH}}{V_{mH} + V_{mV}} \qquad (3)$$

where V_{mH} is the volume fraction of mortar in horizontal joints and V_{mV} is the volume fraction of mortar in vertical joints. More specifically, the volume fraction is obtained by dividing the respective volume with the corresponding total volume in the prism. The above proposed analytical formula (Eq. 3) seems to be the most reliable for the determination of masonry compressive strength among a plethora of proposed equations available in the literature [2, 10, 39–53].

Table 1. Formulae for the estimation of masonry compressive strength.

No.	Reference		Formula
1	Engesser	[2]	$f_{wc} = \frac{1}{3}f_{bc} + \frac{2}{3}f_{mc}$
2	Bröcker	[39]	$f_{wc} = 0.68f_{bc}^{1/2}f_{mc}^{1/3}$
3	Mann	[40]	$f_{wc} = 0.83f_{bc}^{0.66}f_{mc}^{0.18}$
4	Henry and Malek	[41]	$f_{wc} = 0.317f_{bc}^{0.531}f_{mc}^{0.208}$
5	Dayaratnam	[42]	$f_{wc} = 0.275f_{bc}^{0.5}f_{mc}^{0.5}$
6	Rozza	[43]	$f_{wc} = (v_u f_{bc} + 0.8v_m f_{mc})/10$
7	Bennett et al.	[44]	$f_{wc} = 0.3f_{bc}$
8	AS 3700	[45]	$f_{wc} = K_h K_m f_{bc}^{0.5}$
9	Dymiotis and Gutlederer	[46]	$f_{wc} = 0.3266f_{bc} \times (1 - 0.0027f_{bc} + 0.0147f_{mc})$
10	Eurocode 6	[47]	$f_{wc} = Kf_{bc}^{0.7}f_{mc}^{0.3}$
11	Kaushik et al.	[10]	$f_{wc} = 0.317f_{bc}^{0.866}f_{mc}^{0.134}$
12	Gumaste et al.	[48]	$f_{wc} = 0.63f_{bc}^{0.49}f_{mc}^{0.32}$
13	Christy et al.	[49]	$f_{wc} = 0.35f_{bc}^{0.65}f_{mc}^{0.25}$
14	Garzón-Roca et al.	[50]	$f_{wc} = 0.53f_{bc} + 0.93f_{mc} - 10.32$
15	Sarhat and Sherwood	[51]	$f_{wc} = 0.886f_{bc}^{0.75}f_{mc}^{0.18}$
16	Lumantarna et al.	[52]	$f_{wc} = 0.75f_{bc}^{0.75}f_{mc}^{0.31}$
17	Kumavat	[53]	$f_{wc} = 0.69f_{bc}^{0.6}f_{mc}^{0.35}$

f_{wc} is the masonry compressive strength; f_{bc} is the brick compressive strength; f_{mc} is the mortar compressive strength; v_u is the relative volume of unit; v_m is the relative volume of mortar;

K is a constant in Eurocode 6 formula, modified according to the National Annex for different countries. The value of this constants in the UK is 0.52 [54] while in Greece 0.20 to 1.00 depending on brick/block unit properties and their arrangement;

K_h is a factor in Australian AS 3700 [45]code that accounts for the ratio of unit height to mortar joint thickness (1.3 for blocks of 190 mm high blocks and mortar joints with 10 mm thickness);

K_m is also a factor in Australian AS 3700 [45] code that accounts for bedding type (1.4 for full bedding and 1.6 for face-shell bedding).

3 Artificial Neural Networks

This section summarizes the mathematical and computational aspects of artificial neural networks. In general, ANNs are information-processing models configured for a specific application through a training process. A trained ANN maps a given input onto a specific output. The main advantage of a trained ANN over conventional numerical analysis procedures (e.g., regression analysis) is that the results can be produced with much less computational effort [23, 55–62].

3.1 General

The concept of an artificial neural network is based on the concept of the biological neural network of the human brain. The basic building block of the ANN is the artificial neuron, which is a mathematical model trying to mimic the behaviour of the biological neuron. Information is passed into the artificial neuron as input and processed with a mathematical function leading to an output that determines the behaviour of the neuron (similar to fire-or-not situation for the biological neuron). Before the information enters the neuron, it is weighted in order to approximate the random nature of the biological neuron. A group of such neurons consists of an ANN in a manner similar to biological neural networks. In order to set up an ANN, one needs to define: (i) the architecture of the ANN; (ii) the training algorithm, which will be used for the ANN learning phase; and (iii) the mathematical functions describing the mathematical model. The architecture or topology of the ANN describes the way the artificial neurons are organized in the group and how information flows within the network. For example, if the neurons are organized in more than one layers, then the network is called a multilayer ANN. Regarding the training phase of the ANN, it can be considered as a function minimization problem, in which the optimum value of weights need to be determined by minimizing an error function. Depending on the optimization algorithms used for this purpose, different types of ANNs exist. Finally, the two mathematical functions that define the behaviour of each neuron are the summation function and the activation function. In the present study, back-propagation neural network (BPNN) is used, which is described in the next section.

3.2 Architecture of BPNN

A BPNN is a feed-forward, multilayer network [55], i.e., information flows only from the input towards the output with no feedback loops, and the neurons of the same layer are not connected to each other, but they are connected with all the neurons of the previous and subsequent layer. A BPNN has a standard structure that can be written as

$$N - H_1 - H_2 - \cdots - H_{NHL} - M \qquad (4)$$

where N is the number of input neurons (input parameters); H_i is the number of neurons in the i-th hidden layer for $i = 1, \ldots, NHL$; NHL is the number of hidden layers; and M is the number of output neurons (output parameters). Figure 2 depicts an example of a BPNN composed of an input layer with 5 neurons, two hidden layers with 8 and 3 neurons respectively, and an output layer with 2 neurons, i.e., a 5-8-3-2 BPNN.

A notation for a single node (with the corresponding R-element input vector) of a hidden layer is presented in Fig. 3.

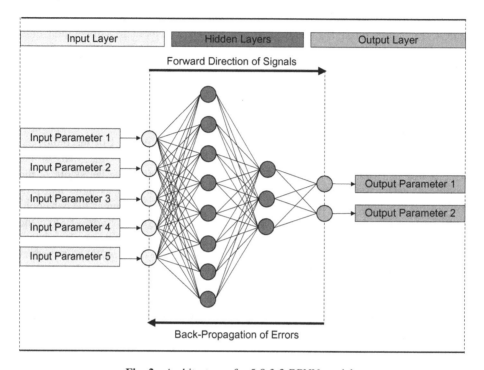

Fig. 2. Architecture of a 5-8-3-2 BPNN model.

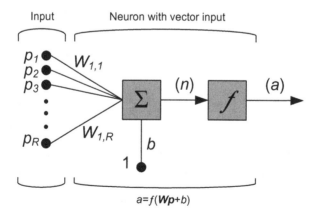

Fig. 3. A neuron with a single R-element input vector.

For each neuron i, the individual element inputs $p_1, \ldots, p_R$ are multiplied by the corresponding weights $w_{i,1}, \ldots, w_{i,R}$ and the weighted values are fed to the junction of the summation function, in which the dot product $(W \cdot p)$ of the weight vector $W = [w_{i,1}, \ldots, w_{i,R}]$ and the input vector $p = [p_1, \ldots, p_R]^T$ is generated. The threshold b (bias) is added to the dot-product forming the net input n, which is the argument of the transfer function f:

$$n = W \cdot p = w_{i,1}p_1 + w_{i,2}p_2 + \ldots + w_{i,R}p_R + b \tag{5}$$

The choice of the transfer (or activation) function f may strongly influence the complexity and performance of the ANN. Although different type of functions are available, sigmoidal transfer functions are the most commonly used. In the present study, the Logistic Sigmoid and the Hyperbolic Tangent transfer functions were found to be appropriate for the problem investigated. During the training phase, the training data are fed into the network which tries to create a mapping between the input and the output values. This mapping is achieved by adjusting the weights in order to minimise the following error function:

$$E = \sum (x_i - y_i)^2 \tag{6}$$

where x_i and y_i are the measured value and the prediction of the network, respectively, within an optimization framework. The training algorithm used for the optimization plays a crucial role in building a quality mapping, thus an exhaustive investigation was performed in order to find the most suitable for this problem. The most common method used in the literature is the back-propagation technique, in which, as stated by its name, the information propagates to the network in a backward manner in order to adjust the weights and minimize the error function. To adjust the weights properly, a general method called gradient descent is applied, in which the gradients of the error function with respect to the network weights is calculated. Further discussion on the training algorithms is given in the numerical example section.

4   Results and Discussion

This section presents the process for tuning optimum ANNs used for the prediction of the compressive strength of masonry walls, based on experimental data available in the literature.

4.1   Experimental - Database

A prerequisite for the successful function of artificial neural networks is the use of an extended and reliable database, capable of training the system. In the case of masonry walls this presents several difficulties due to several factors.

First of all, it is obvious that the production of a very large number of specimens can be a rather difficult, exorbitant and time-consuming procedure. Apart from that, state of the art facilities and laboratories are needed in order for the properties of the samples to be as homogenous as possible, while the time between the production of the specimen and its actual measurement, the samples must be properly stored and cured, a process that demands plenty of space with specific requirements. It has also been observed that homogeneity of samples that were created inside a specific laboratory with the same staff and under the same conditions cannot be always maintained, resulting in variations in their properties. Namely, for the case of compressive strength of masonry walls, deviations greater than ±20% has been observed in the literature [63]. For all these reasons, researchers face a major difficulty obtaining an adequate amount of experimental data capable of training the ANNs, yet this issue can be addressed by using additional data from other published databases.

In addition, the mortars and masonry units are composite materials, consisting of a binder material and aggregates, while in most cases additives are used, either natural or manufactured or both. Thus, mortars are produced through the mix of water with different natural/manufactured raw materials. During the data collection process, it is important to distinguish the necessary mix parameters, while it is of crucial importance to be accurate regarding the type of raw materials used, in order to train the system appropriately. For example, the use of Ordinary Portland cement (OPC) in the mortar mix will have a different effect on the final compressive strength values when compared to the use of white cement or high alumina cement; thus, if different types of cement were used in the mortars of masonry specimens included in the database, it must be appropriately described to the ANN in order to account for its influence.

Similarly, the bricks that formulate the masonry units may have a wide range of properties that is based on their production procedures along with the materials used. Hollow or solid bricks can be found as masonry units in various researches, that might have been manufactured using common burnt clay, sand lime, concrete, fly ash clay etc. having vastly varying properties. Hence, their compressive strength or deformation characteristics when subjected to external load or compressive stresses will be varying accordingly.

To recapitulate, it is common for adequate experimental databases to be compiled through the accumulation of smaller databases acquired by different researchers and available relevant literature. During the process of compiling the database, the reliability of each individual database has to be examined, while the raw materials that were used must be adequately described (dimensions and properties of each material along with the geometry of the masonry units). Furthermore, it is important that the same standards have been followed during the experimental procedure, in order for the results to be comparable and the comparison to be meaningful. An adequate number of specimens must have tested in order for the values to be statistically acceptable; a small amount of tested specimens, regardless of credibility, cannot give a result that can be considered reliable. When training an ANN, in addition to the reliability of the database, it is crucial that the values of the input parameters (mortar mix synthesis parameters and specimen age) cover all possible value ranges of the parameters.

It is no exaggeration to state that the reliability of the optimum developed neural network is crucially dependent on the reliability of the experimental data, thus confirming the famous expression in the field of informatics Garbage In, Garbage Out (GIGO). Predictive analytics begins with good data; more data doesn't necessarily mean better data. A successful predictive analytics project requires, first and fore-most, relevant and accurate data.

In the light of the above a vast database has been composed that consists of 232 experimental data sets (Table 2) that have been obtained from 21 different experimental published works [10, 12, 48, 52, 64–80]. During the experimental data selection from the literature it was taken under consideration the compressive strength of the masonry brick (f_b), the compressive strength of mortar (f_m), the height-to-thickness ratio of the masonry prism (h/t), the volume fraction of the masonry brick (VF_b) and the volume ratio of bed joint mortar (VR_{mH}).

Thaickavil and Thomas [12] published 64 data sets, in which 32 of them were made out of cement stabilized pressed earth units and 32 out of them using burnt clay units, with an average compressive strength of 4.56 and 6.68 MPa respectively. In each category, 8 different configurations were formed implementing 4 different mortar proportions (M1 (1:6), M2 (1:5), M3 (1:4) and M4 (1:3)) in each configuration that were prepared using Ordinary Portland cement (OPC) and river sand conforming to Zone II of IS: 383 [81]. The compressive strength of the mortars varied between 13.60 and 35.50 MPa, while they were determined through testing cubes of 50 cm^2 face area as per IS: 2250 [82]. Based on the 64 produced configurations, 192 brick masonry prisms were created in total (3 masonry units per configuration). The dimensions of half of the masonry units were 190 × 113 × 100 mm^3 (from cement stabilized pressed earth bricks) while the dimensions of the rest were 210 × 96 × 50 mm^3 (from burnt clay bricks). These brick masonry units were later capped with a 1–2 mm thin layer of dental plaster in order to level the contact surface between the specimen face and platens of the testing machine. The results from the performed analyses enabled the researchers to draw useful conclusions concerning the effect of (a) the strength of the brick, (b) the strength of the mortar, (c) the height-to-thickness ratio of the prism, (d) the volume fraction of brick and (e) the volume ratio of bed joint to mortar, on the compressive strength of the masonry prisms (f_{wc}).

In the research work of Francis et al. [80], the effect of joint (mortar) thickness on the compressive strength of brickwork was investigated. 33 data sets of four-high stack-bonded prisms were created, in which the thickness of the mortar varied, while using 2 types of bricks (solid and perforated) with the mean dimensions taken from a significantly large sample of individual measurements. The properties of the mortar derived from a 1:1:6 mix (Portland cement: lime: sand) with its compressive strength varied between 0.30 and 52.60 MPa. Following the same pattern, the compressive strength of the bricks presented a wide dispersion since the recorded values were spread between 3.10 and 127.00 MPa after being measured from 12 single bricks. In each type of brick, six prisms were created with mortar joints of approximately 10.00 mm and 15.00 mm thickness and four prisms of each brick type with 25.00 mm mortar, while four more prisms in each brick type were created with as thin joints as possible. In each case a layer of mortar was placed at the top and bottom of each prism and then the prisms were cured for 14 days in air inside the laboratory. It was observed that thinner joints make the brick units stronger. Hence, it was concluded that the bond strength is of paramount importance when bending or eccentricity of load produces tensile stresses.

Vermeltfoort [78] stated that strength and stiffness of both brick and mortar is a crucial factor influencing the compressive strength of masonry units. Through the implementation of 29 data sets, 170 masonry specimens were finally created and tested. A large variation in the test configuration was achieved by altering the mechanical

Table 2. Data from experiments published in literature.

No.	Reference		Type of unit	Number of datasets	Compressive strength [MPa]
1	Thaickavil and Thomas	[12]	PEB & BCLB	64	0.70–3.20
2	Ravula and Subramaniam	[64]	SFCLB	2	5.80–8.00
3	Singh and Munjal	[65]	BCB	12	2.10–11.60
4	Zhou et al.	[66]	HCB	12	10.20–27.00
5	Balasubramanian et al.	[67]	CLB	1	2.80–2.80
6	Vindhyashree et al.	[68]	SCB	3	4.00–4.80
7	Lumantarna et al.	[52]	SLCB	14	6.51–30.79
8	Nagarajan et al.	[69]	BCLB	3	1.90–2.40
9	Thamboo	[70]	HCB	4	6.90–10.10
10	Vimala and Kumarasamy	[71]	STMB	6	0.70–1.60
11	Reddy et al.	[72]	PEB	4	3.20–3.90
12	Kaushik et al.	[10]	CLB	12	2.90–8.50
13	Gumaste et al.	[48]	TMB & WCB	6	1.30–10.00
14	Mohamad et al.	[73]	HCB	6	7.50–11.70
15	Brencich and Gambarotta	[74]	SLCB	2	3.90–13.50
16	Bakhteri and Sambasivam	[75]	SLCB	6	9.10–16.90
17	Ip	[76]	FS & SLCB	4	11.00–41.00
18	Hossein et al.	[77]	BCLB	1	18.20–18.20
19	Vermeltfoort	[78]	SFCLB, PFB, WCB, CLSLB	29	3.90–39.80
20	McNary and Abrams	[79]	SMP & MCU	8	19.70–48.20
21	Francis et al.	[80]	SLB & PFB	33	7.80–21.90
	Total			**232**	**0.70–48.20**

PEB: Pressed Earth bricks
SFCLB: Soft Clay bricks
SCB: Solid Concrete bricks
CLB: Clay bricks
TMB: Table Mounted brick
FS: Flagstone
CLSLB: Calcium Silicate bricks
MCU: Modular Cored unit
BCLB: Burnt Clay brick
HCB: Hollow Concrete bricks
STMB: Stabilized Mud blocks
SLCB: Solid Clay bricks
WCB: Wire Cut bricks
PFB: Perforated bricks
SMP: Standard Modular paver
SLB: Solid bricks

properties of bricks, using 3 soft mud brick categories (PO, VE, ER), 2 wire cut brick (JB, JG) and 1 calcium silicate (CS), with their compressive strength varying between 27.00 and 127.00 MPa. In addition, the use of mortars of different compressive strength (4.00 to 48.00 MPa) greatly facilitated towards the preparation of the afore-mentioned configurations. Finally, the preparation of 3 different geometries (wide, narrow and high) resulted into specimens with dimensions of $430 \times 100 \times 340 \text{ mm}^3$, $200 \times 100 \times 340 \text{ mm}^3$ and $100 \times 100 \times 500 \text{ mm}^3$ (L $\times$ D $\times$ H) respectively. The researcher concluded that the equation of the characteristic compressive strength of masonry that is found on CEN-EC6 [83] gives an approximation of the actual/measured value with a deviation of 8% for groups of more than 3 specimens.

Based on the above database, each input training vector p is of dimension 1×5 and consists of five parameters, namely, the compressive strength of masonry unit, the compressive strength of mortar, the height-to-thickness ratio of masonry prism, the volume fraction of masonry unit and the volume ratio of bed joint. The corresponding output training vectors are of dimension 1×1 and consist of the value of the compressive strength of the masonry prism. Their mean values together with the minimum and maximum values, as well as the standard deviation (STD) values are listed in Table 3.

Table 3. The input and output parameters used in the development of BPNNs.

Variable	Units	Type	Data used in NN models			
			Min	Average	Max	STD
Volume fraction of masonry unit (brick)	–	Input	0.76	0.89	0.98	0.04
Compressive strength of masonry unit (brick)	MPa	Input	3.10	29.27	127.00	28.67
Compressive strength of mortar	MPa	Input	0.30	13.85	52.60	10.79
Height-to-thickness ratio of masonry prism	–	Input	1.15	3.27	5.75	0.99
Volume ratio of bed joint mortar	–	Input	0.66	0.96	1.00	0.09
Masonry compressive strength	MPa	Output	0.70	9.97	48.20	9.37

4.2 Training Algorithms

For the training of the BPNN models the use of a large set of training function such as quasi-Newton, Resilient, One-step secant, Gradient descent with momentum and adaptive learning rate and Levenberg-Marquardt backpropagation algorithms has been investigated. From all these algorithms the best prediction for the non-linear behaviour of the compressive strength of masonry prism of SCC is achieved, when the Levenberg-Marquardt implemented by Levmar [84]. This algorithm appears to be the fastest method for training moderate-sized feedforward neural networks (up to several hundred weights) as well as non-linear problems. It also has an efficient implementation in MATLAB® software, because the solution of the matrix equation is a built-in function, so its attributes become even more pronounced in a MATLAB environment.

4.3   BPNN Model Development

In this work, a large number of different BPNN models have been developed and implemented. Each one of these ANN models was trained over 155 data-points out of the total of 232 data-points, (66.81% of the total number) and the validation and testing of the trained ANN were performed with the remaining 77 data-points. More specifically, 39 data-points (16.81%) were used for the validation of the trained ANN and 38 (16.38%) data-points were used for the testing.

The development and training of the ANNs occurs with a number of hidden layers ranging from 1 to 2 and with a number of neurons ranging from 1 to 30 for each hidden layer. Each one of the ANNs is developed and trained for a number of different activation functions, such as the Log-sigmoid transfer function (logsig), the Linear transfer function (purelin) and the Hyperbolic tangent sigmoid transfer function (tansig) [85–91].

The reliability and accuracy of the developed neural networks were evaluated using Pearson's correlation coefficient R and the root mean square error (RMSE). RMSE presents information on the short term efficiency which is a benchmark of the difference of predicated values about the experimental values. The lower the RMSE, the more accurate is the evaluation. The Pearson's correlation coefficient R measures the variance that is interpreted by the model, which is the reduction of variance when using the model. R values range from 0 to 1 while the model has healthy predictive ability when it is near to 1 and is not analyzing whatever when it is near to 0. These performance metrics are good measures of the overall predictive accuracy.

Furthermore, the following new engineering index, the a20-index, is proposed for the reliability assessment of the developed ANN models:

$$a20 - index = \frac{m20}{M} \tag{7}$$

where M is the number of dataset sample and $m20$ is the number of samples with value of rate Experimental value/Predicted value between 0.80 and 1.20. It should be noted that for a perfect predictive model, the values of the *a20-index* are expected to be equal to 1.00. The proposed *a20-index* has the advantage that its value has a physical engineering meaning which declares the amount of the samples that satisfy the predicted values with a deviation of ±20%, compared to experimental values.

Based on the above, a total of 3,600,000 different BPNN models have been developed and investigated in order to find the optimum NN model for the prediction of the compressive strength of masonry walls. Namely, five different cases of NN architectures that were based on the number of input parameters were investigated as it is presented in Table 4. For each case a total of 720,000 different BPNN models were developed and investigated.

The developed ANN models were sorted in a decreasing order based on the *a20-index* value. Based on this ranking, the optimum BPNN model for the prediction of the compressive strength for each one from the five cases of NN architectures based on Input Parameters used are presented in Table 5.

Table 4. Cases of NN architectures based on the number of input parameters that were used.

Case	Number of input parameters	Input parameters				
		VF_b	f_b	f_m	h/t	VR_{mH}
I	2		√	√		
II	3		√	√	√	
III	4		√	√	√	√
IV	4	√	√	√	√	
V	5	√	√	√	√	√

Table 5. Statistical indexes of the optimum BPNN for each one from the five cases of NN architectures based on input parameters used (see also Table 4).

Case of input parameters	Optimum BPNN model	Data	Indexes		
			R	RMSE	*a20-index*
I	2-28-4-1	Training	0.9744	2.0218	0.5677
		Validation	0.9513	2.8845	0.4872
		Test	0.9607	3.0847	0.3947
		All	0.9671	2.3850	0.5259
II	3-8-28-1	Training	0.9975	0.6332	**0.9290**
		Validation	0.9919	1.2626	**0.7179**
		Test	0.9531	3.2991	**0.7368**
		All	0.9867	1.5227	**0.8621**
III	4-4-26-1	Training	0.9961	0.7969	0.8129
		Validation	0.9763	1.9561	0.6667
		Test	0.9700	2.6522	0.5263
		All	0.9873	1.4899	0.7414
IV	4-4-18-1	Training	0.9927	1.0871	0.7290
		Validation	0.9807	1.7796	0.5385
		Test	0.9792	2.3673	0.6053
		All	0.9872	1.4966	0.6767
V	5-7-30-1	Training	0.9973	0.6629	0.9097
		Validation	0.9776	2.0029	0.6667
		Test	0.9730	2.8492	0.5263
		All	0.9869	1.5158	0.8060

Based on these results, the optimum BPNN model is that of case II, which corresponds of BPNN architecture with three input parameters (Masonry unit compressive strength, mortar compressive strength and height-to-thickness ratio of masonry wall). Namely, the optimum BPNN model is that of architecture 3-8-28-1, which represents BPNN model with three input parameters, two hidden layers with 8 and 28 neurons and one output parameter (Fig. 4). As it is presented in Fig. 4, the transfer functions are Hyperbolic tangent sigmoid transfer function (tansig) for the first hidden layer, the Log-sigmoid transfer function (logsig) for the second hidden layer and the Linear transfer function (purelin) for the output layer.

All figures from Figs. 5, 6, 7 and 8, depict the comparison of the experimental values with the predicted values of the optimum BPNN model for the case of train, validation, test and all data. It is clearly depicted that the proposed optimum 3-8-28-1 BPNN predicted reliably the compressive strength of masonry prism. Also, Figs. 9 and 10 present a comparison between the exact experimental values with the predicted values of the optimum BPNN model. It is worth mentioning that for samples with experimental values of compressive strength greater than 4.00 MPa the predicted values have a deviation less than ±20% that means the value of the *a20-index* is 1.00 (points between the two dotted lines in Figs. 5, 6, 7 and 8).

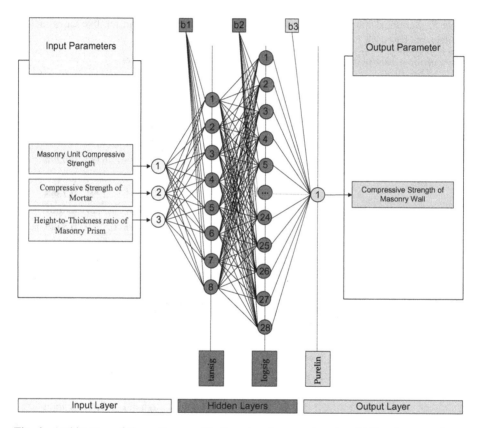

Fig. 4. Architecture of the optimum with three input parameters, two hidden layers and one output parameter 3-8-28-1 BPNN model.

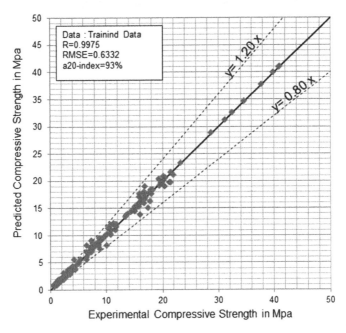

Fig. 5. Comparison of Experimental and predicted values of compressive strength for the training process.

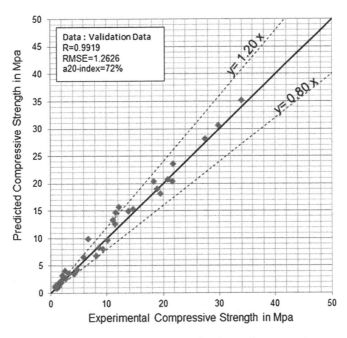

Fig. 6. Comparison of Experimental and predicted values of compressive strength for the validation process

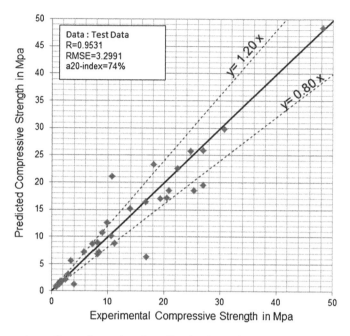

Fig. 7. Comparison of experimental and predicted values of compressive strength for the test process.

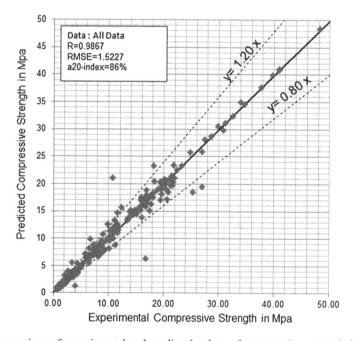

Fig. 8. Comparison of experimental and predicted values of compressive strength for all data.

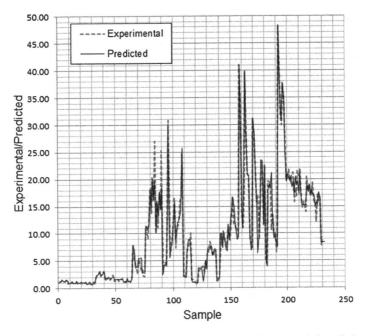

Fig. 9. Experimental vs predicted values of compressive strength for all data.

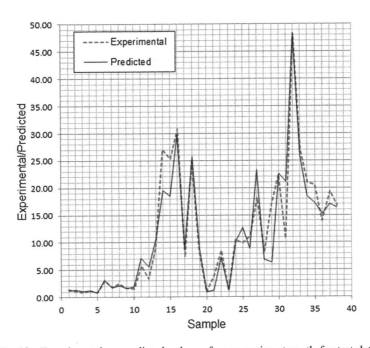

Fig. 10. Experimental vs predicted values of compressive strength for test data.

4.4   Comparisons with Code Provisions

In Table 6, a comparison among the proposed BPNN model with the available in the literature models for the prediction of masonry wall compressive strength is presented, with the models being sorted based on the *a20-index*. It is clearly shown that the optimum BPNN model with architecture 3-8-28-1 (three input parameters, two hidden layers with 8 and 28 neurons and one output parameter the masonry prism compressive strength) predict the compressive strength of masonry wall in a reliable and robust manner. Namely, the optimum BPNN model corresponds to a value of a *a20-1index* equal to 0.8621 while for the first of the available in the literature that is the proposed formula by Thaickavil and Thomas [12] is 0.4526.

Table 6. Ranking of the developed optimum BPNN models with the code provisions and other research formulae about the prediction of masonry prism compressive strength available in the literature based on the engineering *a20-index* (All Data).

No.	Model	Indexes		
		R	RMSE	*a20-index*
1	3-8-28-1 BPNN	0.9867	1.5227	0.8621
2	5-7-30-1 BPNN	0.9869	1.5158	0.8060
3	4-4-26-1 BPNN	0.9873	1.4899	0.7414
4	4-4-18-1 BPNN	0.9872	1.4966	0.6767
5	2-28-4-1 BPNN	0.9671	2.3850	0.5259
6	Thaickavil and Thomas 2018	0.8660	4.8500	0.4526
7	Bennet et al. 1997	0.8589	4.9662	0.3621
8	Dymiotis and Gutlederer 2002	0.8892	4.4770	0.3491
9	Mann 1982	0.8943	4.3206	0.2672
10	Eurocode 6 2005	0.8805	4.4416	0.2629
11	Kaushik et al. 2007	0.8841	5.1618	0.2629
12	Kumavat et al. 2016	0.8652	4.7715	0.2284
13	ACI 530.99	0.8589	5.4791	0.2241
14	Bröcker 1963	0.8580	6.6316	0.1767
15	Gumaste et al. 2007	0.8609	7.4510	0.1724
16	Engesser 1907	0.7439	11.9769	0.1724
17	Sarhat et al. 2014	0.8913	8.1023	0.1552
18	Hendry and Malek 1986	0.8922	10.7358	0.1379
19	Lumantarna et al. 2014	0.8781	11.7124	0.1336
20	Christy et al. 2013	0.8887	7.9826	0.1207
21	Garzón-Roca et al. 2013	0.7704	14.6882	0.1034
22	Dayaratnam 1987	0.7880	9.0089	0.0819

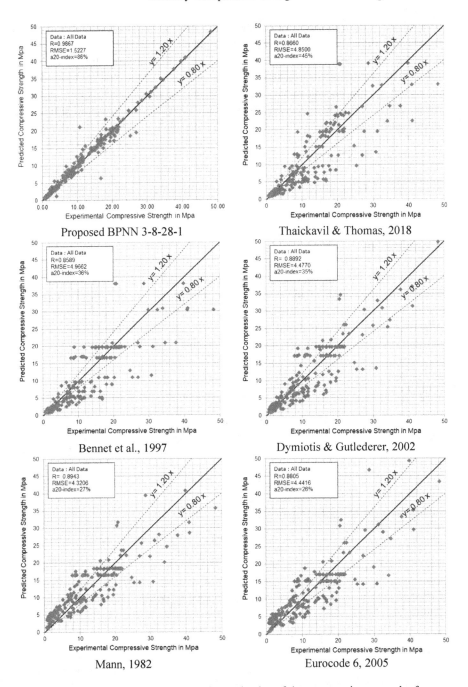

Fig. 11. Comparison of the "exact" experimental value of the compressive strength of masonry with the ones predicted by the existing empirical expressions and by the proposed BPNN model (all data).

Furthermore, the advantages of the produced BPNN model compared to the code provisions and other research formulae are depicted in Fig. 11. The importance of graphing the data and the effect of outliers on the statistical properties of a data set should also be pointed out. For example, Anscombe [92] presented four simple data sets and even though they show identical statistical properties (mean values, standard deviation, correlation factor, etc.), they were quite different when inspected graphically. For these reasons, Fig. 11 compares the "exact" experimental value of the compressive strength of masonry with the ones predicted by the existing empirical expressions and by the proposed BPNN model.

From Fig. 11, it is also clear that the proposed 3-8-28-1 BPNN provides much more reliable values for the compressive strength of masonry prism compared to those proposed by the available equations, thus confirming the validity of the proposed NN.

5 Conclusions

The present study investigates the application of Artificial Neural Networks (ANNs) models for the prediction of the compressive strength of masonry walls. The comparison between the derived results and the experimental findings, clearly demonstrates the effectiveness of ANNs to build a series of soft sensors resulting in the ability to predict, in a reliable and comprehensive manner, their compressive strength.

References

1. ACI/TMS 122R-14: Guide to Thermal Properties of Concrete and Masonry Systems, Reported by ACI/TMS Committee 122, December 2014
2. Engesser, F.: Über weitgespannte wölbbrücken. Z. Architekt. Ing.-wesen **53**, 403–440 (1907)
3. Syrmakezis, C.A., Asteris, P.G.: Masonry failure criterion under biaxial stress state. J. Mater. Civ. Eng. **13**(1), 58–64 (2001)
4. Lourenço, P.: Computations on historic masonry structures. Prog. Struct. Mat. Eng. **4**(3), 301–319 (2002)
5. Milani, G., Lourenço, P.B., Tralli, A.: Homogenised limit analysis of masonry walls, part I: failure surfaces. Comput. Struct. **84**(3–4), 66–180 (2006)
6. Asteris, P.G., Antoniou, S.T., Sophianopoulos, D.S., Chrysostomou, C.Z.: Mathematical macromodeling of infilled frames: state of the art. J. Struct. Eng. **137**(12), 1508–1517 (2011)
7. Chrysostomou, C.Z., Asteris, P.G.: On the in-plane properties and capacities of infilled frames. Eng. Struct. **41**, 385–402 (2012)
8. Asteris, P.G., Cotsovos, D.M., Chrysostomou, C.Z., Mohebkhah, A., Al-Chaar, G.K.: Mathematical micromodeling of infilled frames: state of the art. Eng. Struct. **56**, 1905–1921 (2013)
9. Asteris, P.G., et al.: Seismic vulnerability assessment of historical masonry structural systems. Eng. Struct. **62–63**, 118–134 (2014)
10. Kaushik, H.B., Rai, D.C., Jain, S.K.: Stress-strain characteristics of clay brick masonry under uniaxial compression. J. Mater. Civ. Eng. **19**(9), 728–739 (2007)

11. Thomas, J.: Concrete block reinforced masonry wall panels subjected to out-of-plane monotonic lateral loading. In: Proceedings of National Conference on Recent Advances in Structural Engineering, Hyderabad, India, pp. 123–129, February 2006

12. Thaickavil, N.N., Thomas, J.: Behaviour and strength assessment of masonry prisms. Case Stud. Constr. Mater. **8**, 23–38 (2018)

13. SP 20 (S&T): Handbook on Masonry Design and Construction. Bureau of Indian Standards, New Delhi (1991)

14. Alexandridis, A.: Evolving RBF neural networks for adaptive soft-sensor design. Int. J. Neural Syst. **23**, 1350029 (2013)

15. Dias, W.P.S., Pooliyadda, S.P.: Neural networks for predicting properties of concretes with admixtures. Constr. Build. Mater. **15**, 371–379 (2001)

16. Lee, S.C.: Prediction of concrete strength using artificial neural networks. Eng. Struct. **25**, 849–857 (2003)

17. Topçu, I.B., Saridemir, M.: Prediction of compressive strength of concrete containing fly ash using artificial neural networks and fuzzy logic. Comput. Mater. Sci. **41**, 305–311 (2008)

18. Trtnik, G., Kavčič, F., Turk, G.: Prediction of concrete strength using ultrasonic pulse velocity and artificial neural networks. Ultrasonics **49**, 53–60 (2009)

19. Waszczyszyn, Z., Ziemiański, L.: Neural networks in mechanics of structures and materials —new results and prospects of applications. Comput. Struct. **79**, 2261–2276 (2001)

20. Douma, O.B., Boukhatem, B., Ghrici, M., Tagnit-Hamou, A.: Prediction of properties of self-compacting concrete containing fly ash using artificial neural network. Neural Comput. Appl. **28**, 1–12 (2016). https://doi.org/10.1007/s00521-016-2368-7

21. Mashhadban, H., Kutanaei, S.S., Sayarinejad, M.A.: Prediction and modeling of mechanical properties in fiber reinforced self-compacting concrete using particle swarm optimization algorithm and artificial neural network. Constr. Build. Mater. **119**, 277–287 (2016)

22. Açikgenç, M., Ulaş, M., Alyamaç, K.E.: Using an artificial neural network to predict mix compositions of steel fiber-reinforced concrete. Arab. J. Sci. Eng. **40**, 407–419 (2015)

23. Asteris, P.G., Kolovos, K.G., Douvika, M.G., Roinos, K.: Prediction of self-compacting concrete strength using artificial neural networks. Eur. J. Environ. Civ. Eng. **20**, s102–s122 (2016)

24. Baykasoğlu, A., Dereli, T.U., Taniş, S.: Prediction of cement strength using soft computing techniques. Cem. Concr. Res. **34**, 2083–2090 (2004)

25. Akkurt, S., Tayfur, G., Can, S.: Fuzzy logic model for the prediction of cement compressive strength. Cem. Concr. Res. **34**, 1429–1433 (2004)

26. Özcan, F., Atiş, C.D., Karahan, O., Uncuoğlu, E., Tanyildizi, H.: Comparison of artificial neural network and fuzzy logic models for prediction of long-term compressive strength of silica fume concrete. Adv. Eng. Softw. **40**, 856–863 (2009)

27. Saridemir, M.: Predicting the compressive strength of mortars containing metakaolin by artificial neural networks and fuzzy logic. Adv. Eng. Softw. **40**(9), 920–927 (2009)

28. Eskandari-Naddaf, H., Kazemi, R.: ANN prediction of cement mortar compressive strength, influence of cement strength class. Constr. Build. Mater. **138**, 1–11 (2017)

29. Oh, T.-K., Kim, J., Lee, C., Park, S.: Nondestructive concrete strength estimation based on electro-mechanical impedance with artificial neural network. J. Adv. Concr. Technol. **15**, 94–102 (2017)

30. Khademi, F., Akbari, M., Jamal, S.M., Nikoo, M.: Multiple linear regression, artificial neural network, and fuzzy logic prediction of 28 days compressive strength of concrete. Front. Struct. Civ. Eng. **11**, 90–99 (2017)

31. Türkmen, İ., Bingöl, A.F., Tortum, A., Demirboğa, R., Gül, R.: Properties of pumice aggregate concretes at elevated temperatures and comparison with ANN models. Fire Mater. **41**, 142–153 (2017)

32. Nikoo, M., Zarfam, P., Sayahpour, H.: Determination of compressive strength of concrete using self organization feature map (SOFM). Eng. Comput. **31**, 113–121 (2015)
33. Adeli, H.: Neural networks in civil engineering: 1989–2000. Comput.-Aided Civ. Infrastruct. Eng. **16**, 126–142 (2001)
34. Safiuddin, M., Raman, S.N., Salam, M.A., Jumaat, M.Z.: Modeling of compressive strength for self-consolidating high-strength concrete incorporating palm oil fuel ash. Materials **9**, 396 (2016)
35. Mansouri, I., Kisi, O.: Prediction of debonding strength for masonry elements retrofitted with FRP composites using neuro fuzzy and neural network approaches. Compos. Part B Eng. **70**, 247–255 (2015)
36. Mansouri, I., Gholampour, A., Kisi, O., Ozbakkaloglu, T.: Evaluation of peak and residual conditions of actively confined concrete using neuro-fuzzy and neural computing techniques. Neural Comput. Appl. **29**, 1–16 (2016). https://doi.org/10.1007/s00521-016-2492-4
37. Reddy, T.C.S.: Predicting the strength properties of slurry infiltrated fibrous concrete using artificial neural network. Front. Struct. Civ. Eng. **12**, 1–14 (2017). https://doi.org/10.1007/s11709-017-0445-3
38. Salehi, H., Burgueño, R.: Emerging artificial intelligence methods in structural engineering. Eng. Struct. **171**, 170–189 (2018)
39. Bröcker, O.: Die auswertung von tragfähigkeitsversuchen an gemauerten wänden. Betonstein-Ztg. **10**, 19–21 (1963)
40. Mann, W.: Statistical evaluation of tests on masonry by potential functions. In: Proceedings of the Sixth International Brick Masonry Conference, Rome, Italy, May 1982, pp. 86–98 (1982)
41. Hendry, A.W., Malek, M.H.: Characteristic compressive strength of brickwork walls from collected test results. Mason. Int. **7**, 15–24 (1986)
42. Dayaratnam, P.: Brick and Reinforced Brick Structures. Oxford & IBH, New Delhi (1987)
43. Apolo, G.L., Matinez-Luengas, A.L.: Curso Técnicas de Intervención en El Patrimonio Arquitectonico. Consultores Tecnicos de Contstruccion (1995)
44. Bennett, R., Boyd, K., Flanagan, R.: Compressive properties of structural clay tile prisms. J. Struct. Eng. **123**(7), 920–926 (1997)
45. AS Committee 3700-2001. Masonry structures. Australian Standard Association, Sydney, 197 p. (2001)
46. Dymiotis, C., Gutlederer, B.M.: Allowing for uncertainties in the modelling of masonry compressive strength. Constr. Build. Mater. **16**(2002), 443–452 (2002)
47. EN 1996-1-1: Eurocode 6: design of masonry structures-Part 1-1: general rules for reinforced and unreinforced masonry structures. European Committee for Standardization, Brussels (2005)
48. Gumaste, K.S., Rao, K.S.N., Reddy, B.V.V., Jagadish, K.S.: Strength and elasticity of brick masonry prisms and wallettes under compression. Mater. Struct. **40**(2), 241–253 (2007)
49. Christy, C.F., Tensing, D., Shanthi, R.: Experimental study on axial compressive strength and elastic modulus of the clay and fly ash brick masonry. J. Civ. Eng. Constr. Technol. **4**(4), 134–141 (2013)
50. Garzón-Roca, J., Marco, C.O., Adam, J.M.: Compressive strength of masonry made of clay bricks and cement mortar: estimation based on neural networks and fuzzy logic. Eng. Struct. **48**(2013), 21–27 (2013)
51. Sarhat, S.R., Sherwood, E.G.: The prediction of compressive strength of ungrouted hollow concrete block masonry. Constr. Build. Mater. **58**, 111–121 (2014)
52. Lumantarna, R., Biggs, D.T., Ingham, J.M.: Uniaxial compressive strength and stiffness of field-extracted and laboratory-constructed masonry prisms. J. Mater. Civ. Eng. **26**(4), 567–575 (2014)

53. Kumavat, H.R.: An experimental investigation of mechanical properties in clay brick masonry by partial replacement of fine aggregate with clay brick waste. J. Inst. Eng. India Ser. A **97**(3), 199–204 (2016)

54. British Standards Institution (BSI): BS EN 1996 (Eurocode 6): Design of Masonry Structures, British Standards Institution, p. 128 (2005)

55. Hornik, K., Stinchcombe, M., White, H.: Multilayer feedforward networks are universal approximators. Neural Netw. **2**, 359–366 (1989)

56. Plevris, V., Asteris, P.G.: Modeling of masonry compressive failure using neural networks. In: Proceedings of the OPT-i 2014—1st International Conference on Engineering and Applied Sciences Optimization, Kos, Greece, 4–6 June, pp. 2843–2861 (2014)

57. Plevris, V., Asteris, P.G.: Modeling of masonry failure surface under biaxial compressive stress using neural networks. Constr. Build. Mater. **55**, 447–461 (2014)

58. Plevris, V., Asteris, P.: Anisotropic failure criterion for brittle materials using artificial neural networks. In: Proceedings of the COMPDYN 2015—5th ECCOMAS Thematic Conference on Computational Methods in Structural Dynamics and Earthquake Engineering, Crete Island, Greece, 25–27 May 2015, pp. 2259–2272 (2015)

59. Giovanis, D.G., Papadopoulos, V.: Spectral representation-based neural network assisted stochastic structural mechanics. Eng. Struct. **84**, 382–394 (2015)

60. Asteris, P.G., Plevris, V.: Neural network approximation of the masonry failure under biaxial compressive stress. In: Proceedings of the 3rd South-East European Conference on Computational Mechanics (SEECCM III), an ECCOMAS and IACM Special Interest Conference, Kos Island, Greece, 12–14 June 2013, pp. 584–598 (2013)

61. Asteris, P.G., Plevris, V.: Anisotropic masonry failure criterion using artificial neural networks. Neural Comput. Appl. **28**, 1–23 (2016). https://doi.org/10.1007/s00521-016-2181-3

62. Asteris, P.G., Kolovos, K.G.: Self-compacting concrete strength prediction using surrogate models. Neural Comput. Appl. 1–16 (2017). https://doi.org/10.1007/s00521-017-3007-7

63. Page, A.W.: The biaxial compressive strength of brick masonry. Proc. Instn. Civ. Engrs. **71**(2), 893–906 (1981)

64. Ravula, M.B., Subramaniam, K.V.L.: Experimental investigation of compressive failure in masonry brick assemblages made with soft brick. Mater. Struct. **50**(19), 1–11 (2017)

65. Singh, S.B., Munjal, P.: Bond strength and compressive stress-strain characteristics of brick masonry. J. Build. Eng. **9**, 10–16 (2017)

66. Zhou, Q., Wang, F., Zhu, F.: Estimation of compressive strength of hollow concrete masonry prisms using artificial neural networks and adaptive neuro-fuzzy inference systems. Constr. Build. Mater. **125**, 199–204 (2016)

67. Balasubramanian, S.R., et al.: Experimental determination of statistical parameters associated with uniaxial compression behaviour of brick masonry. Curr. Sci. **109**(11), 2094–2102 (2015)

68. Vindhyashree, Rahamath, A., Kumar, W.P., Kumar, M.T.: Numerical simulation of masonry prism test using ANSYS and ABAQUS. Int. J. Eng. Res. Technol. **4**(7), 1019–1027 (2015)

69. Nagarajan, S., Viswanathan, S., Ravi, V.: Experimental approach to investigate the behaviour of brick masonry for different mortar ratios. In: Proceedings of the International Conference on Advances in Engineering and Technology, Singapore, March 2014, pp. 586–592 (2014)

70. Thamboo, J.A.: Development of thin layer mortared concrete masonry. Ph.D. dissertation, Queensland University of Technology, Brisbane (2014)

71. Vimala, S., Kumarasamy, K.: Studies on the strength of stabilized mud block masonry using different mortar proportions. Int. J. Emerg. Technol. Adv. Eng. **4**(4), 720–724 (2014)

72. Reddy, B.V., Vyas, C.V.U.: Influence of shear bond strength on compressive strength and stress-strain characteristics of masonry. Mater. Struct. **41**(10), 1697–1712 (2008)

73. Mohamad, G., Lourenço, P.B., Roman, H.R.: Mechanics of hollow concrete block masonry prisms under compression: review and prospects. Cement Concrete Comp. **29**(3), 181–192 (2007)
74. Brencich, A., Gambarotta, L.: Mechanical response of solid clay brickwork under eccentric loading. Part I: unreinforced masonry. Mater. Struct. **38**, 257–266 (2005)
75. Bakhteri, J., Sambasivam, S.: Mechanical behaviour of structural brick masonry: an experimental evaluation. In: Proceedings of the 5th Asia - Pacific Structural Engineering and Construction Conference, Johor Bahru, Malaysia, August 2003, pp. 305–317 (2003)
76. Ip, F.: Compressive strength and modulus of elasticity of masonry prisms. Master of Engineering thesis, Carleton University, Ottawa (1999)
77. Hossain, M.M., Ali, S.S., Rahman, M.A.: Properties of masonry constituents. J. Civil Eng. Inst. Eng. Bangladesh **25**(2), 135–155 (1997)
78. Vermeltfoort, A.T.: Compression properties of masonry and its components. In: Proceedings of the 10th International Brick and Block Masonry Conference, Calgary, Canada, vol. 3, pp. 1433–1442 (1994)
79. McNary, W., Abrams, D.: Mechanics of masonry in compression. J. Struct. Eng. **111**(4), 857–870 (1985)
80. Francis, A.J., Horman, A.A., Jerrems, L.E.: The effect of joint thickness and other factors on the compressive strength of brickwork. In: SIBMAC Proceedings of 2nd International Brick and Block Masonry Conference, Stoke on Trent, pp. 31–37 (1970)
81. IS: 383: Indian Standard Specification for Coarse and Fine Aggregates from Natural Sources for Concrete. Bureau of Indian Standards, New Delhi, India (1970)
82. IS: 2250: Indian Standard Code of Practice for Preparation and Use of Masonry Mortars. Bureau of Indian Standards, New Delhi, India (1981)
83. N.N. Common unified rules for masonry structures, Eurocode No. 6, CEN
84. Lourakis, M.I.A.: A brief description of the Levenberg-Marquardt algorithm Implemened by levmar, Hellas (FORTH). Institute of Computer Science Foundation for Research and Technology (2005). http://www.ics.forth.gr/ ∼ lourakis/levmar/levmar
85. Asteris, P.G., Nozhati, S., Nikoo, M., Cavaleri, L., Nikoo, M.: Krill herd algorithm-based neural network in structural seismic reliability evaluation. Mech. Adv. Mater. Struct. (Article in press). https://doi.org/10.1080/15376494.2018.1430874
86. Asteris, P.G., Roussis, P.C., Douvika, M.G.: Feed-forward neural network prediction of the mechanical properties of sandcrete materials. Sensors (Switzerland) **17**(6), 1344 (2017)
87. Cavaleri, L., et al.: Modeling of surface roughness in electro-discharge machining using artificial neural networks. Adv. Mater. Res. (South Korea) **6**(2), 169–184 (2017)
88. Asteris, P.G., et al.: Prediction of the fundamental period of infilled RC frame structures using artificial neural networks. Comput. Intell. Neurosci. **2016**, 12 (2016). 5104907
89. Nikoo, M., Hadzima-Nyarko, M., Karlo Nyarko, E., Nikoo, M.: Determining the natural frequency of cantilever beams using ANN and heuristic search. Appl. Artif. Intell. **32**(3), 309–334 (2018)
90. Nikoo, M., Ramezani, F., Hadzima-Nyarko, M., Nyarko, E.K., Nikoo, M.: Flood-routing modeling with neural network optimized by social-based algorithm. Natural Hazards **82**(1), 1–24 (2016)
91. Nikoo, M., Sadowski, L., Khademi, F., Nikoo, M.: Determination of damage in reinforced concrete frames with shear walls using self-organizing feature map. Appl. Comput. Intell. Soft Comput. **2017**, 10 (2017). 3508189
92. Anscombe, F.J.: Graphs in statistical analysis. Am. Stat. **27**(1), 17–21 (1973)

Nominal Life of Interventions for Monuments and Historic Structures

Constantine C. Spyrakos$^{(\boxtimes)}$ and Charilaos A. Maniatakis ⓘ

Laboratory for Earthquake Engineering, National Technical University
of Athens, Polytechnic Campus, 9 Heroon Polytechniou Street,
15780 Zografos, Athens, Greece
cspyrakos@gmail.com, chamaniatakis@gmail.com

Abstract. This work presents a methodology that harmonizes the modern philosophy of seismic codes with the engineering and archaeological practice of interventions on cultural heritage structures. By introducing the notion of "nominal life of intervention, T_Λ", and use of attenuation relationships, an easy to apply procedure is proposed that allows application of less extended interventions on monuments and satisfies both the aesthetic and archaeological criteria, provided that the nominal life of these interventions satisfies an acceptable limit. Following the proposed methodology, rehabilitation measures are designed for a specific performance level that is associated to a certain nominal life, after which the structure should be re-evaluated. Utilizing attenuation equations, it arrives at a simple to apply procedure and diagrams that can be used to evaluate and design interventions in heritage structures. The methodology could serve as a stepping stone to address the challenging task of balancing public safety with acceptable interventions on historic structures in many countries with high seismicity and rich cultural heritage.

Keywords: Nominal life of interventions · Cultural heritage structures · Monuments · Seismic performance levels

1 Introduction

The maintenance of cultural heritage structures has gained the interest of researchers and authorities during the last decades in seismic prone regions following the occurrence of significant earthquake events, including Greece, Italy and Turkey, e.g., [1, 2] given the fact that many of them have been constructed with no seismic regulations while, in many cases, they have already suffered damage from past seismic activity.

Recent advances in engineering seismology have significantly increased our knowledge regarding seismic hazard, especially at small distances from active faults, in the so-called near-fault region [3–8]. These new concepts for seismic hazard suggest an increased difficulty for restoration studies of both conventional and cultural heritage structures. Similar advances have occurred in the field of computational methods [9–11], in-situ and laboratory testing [12–17], as well as in the use of new materials for maintenance and restoration of historic structures and monuments [18–22].

© Springer Nature Switzerland AG 2019
A. Moropoulou et al. (Eds.): TMM_CH 2018, CCIS 962, pp. 225–236, 2019.
https://doi.org/10.1007/978-3-030-12960-6_15

The applicability of monitoring technology allows for non- invasive assessment of monuments state for both static and seismic actions [23–26].

Despite this progress of the aforementioned fields the restoration of a monument under these constraints remains a challenging engineering issue that requires balance between safety and feasibility of implementation. Special difficulties that are not common for conventional structures are often met during the preservation and restoration of heritage structures. These difficulties mainly arise from the fact that, according to the widely adopted national standards, international regulations for the conservation and restoration of monuments, such as the Athens Charter, the Venice Charter and the Declaration of Amsterdam [27–29], and legislative documents that exist in this area, i.e., [30], only a limited number of interventions that are non-invasive and reversible can be applied. More limitations arise from aesthetic and archaeological criteria. However, these less invasive interventions rarely provide performance stan-dards compatible with the ones required for non preserved structures according to modern concepts for earthquake hazard.

In the present study, a methodology is proposed that introduces the notion of "nominal life of intervention, T_Δ", that is, the duration within which a certain quality of performance, "a performance level", is complied, given that the acceptable and feasible interventions are applied. The national code of Italy for the restoration of monuments [31] proposes a similar methodology that is applicable to countries that their national code is based on site specific seismic spectra.

Practically the method specifies a shorter duration, T_Δ, of the interventions for a preserved structure, requesting a re-assessment of the heritage structure after the T_Δ duration expires. The concept of the T_Δ can be applicable to the existing infrastructure [32], including buildings and bridges that have reached their design life and may not fully comply with the safety requirements of current codes [31].

2   Performance Levels and Categorization

For a structure with a conventional life cycle of $T_L = 50$ years, the performance levels according to Eurocode 8-Part 3 [31] are defined as: "Damage Limitation" (A), "Sig-nificant Damage" (B) and "Near Collapse" (C). Three levels of seismic risk are adopted: (a) seismic excitation with exceedance probability $P_R = 20\%$ in 50 years and an average return period $T_{RL} = 225$ years; (b) seismic excitation with $P_R = 10\%$ and $T_{RL} = 475$ years; and (c) seismic excitation with $P_R = 2\%$ and $T_{RL} = 2475$ years. The level of protection resulting from a combination of exceedance levels and seismic risk is presented in Table 1, in conjunction with Step 4 in Sect. 3.

Also, Eurocode 8-Part 1 [33] defines four importance classes, I, II, III and IV to which correspond the importance factors $\gamma_I = 0.8$, 1.0, 1.2 and 1.4, respectively. The categories I, II, III and IV are defined as follows:

I   Buildings of minor importance for public safety, e.g., agricultural buildings, etc.
II   Ordinary buildings, not belonging in the other categories.
III   Buildings whose seismic resistance is of importance in view of the consequences associated with a collapse, e.g., schools, assembly halls, cultural institutions, etc.

IV   Buildings whose integrity during earthquakes is of vital importance for civil protection, e.g., hospitals, fire stations, power plants, etc.

It should be stated that Eurocode 8 does not refer specifically to monuments or historic structures; however, current concepts regarding earthquake hazard, as reflected to its provisions, suggest a valuable means to assess the structural integrity of existing historic structures and monuments.

Also, depending on the exposure to visitors, it is also proposed that the monuments should be classified into three categories (Ci with i = 1, 2, 3) assigning a proper Ci, that is:

C1:   almost continuous presence of public or frequent presence of large groups: (i) inhabited buildings in historical city centers, (ii) monuments used as museums, and (iii) monuments continuously used for worshipping;

C2:   occasional habitation or intermittent presence of small groups: (i) monuments visited only under specific conditions, and (ii) remote and rarely visited monuments;

C3:   entrance allowed only to service-personnel with visitors standing only outside the monument.

Comparing the three importance classes defined in EC8-1 [33] or CSI [34], it follows that, as regards the exposure to visitors, C2 can be associated to the importance class II and C1 with the importance class III or IV.

By accepting that cultural heritage structures belong to categories III and IV, the following assessment and redesign objectives can be adopted: A1, A2, B1.

Table 1. Assessment or redesign objectives of the structure [30].

P_R for $T_L = 50$	Performance level		
	Damage Limitation (A)	Significant Damage (B)	Near Collapse (C)
$P_R = 10\%$ ($T_{RL} = 475$ years)	A1	B1	C1
50% ($T_{RL} = 72$ years)	A2	B2	C2

Cultural heritage buildings belong to importance class III or IV, not necessarily as being associated with continuous presence of public, but mainly because of their cultural and aesthetic value; thus, correlated to seismic actions that are characterized by a high return period [30, 33]. In order to accommodate for high seismic demand the preservation of heritage structures could most likely require invasive interventions in order to meet the "safety standards" adopted for new constructions. According to the principles of interventions on historic buildings and monuments, less intrusive interventions are imposed in most cases, which, nevertheless, would continue to ensure the safety of the monument to a certain degree, e.g., [35, 36]. In general, interventions on monuments should satisfy the following three principles [27–29]: (i) Reversibility, (ii) Durability (in-time), and (iii) Feasibility of the proposed solution.

The basic philosophy of codes and comprehensive research efforts promotes mainly two intervention alternatives: (i) Global Rehabilitation, which attempts to improve the seismic safety of the whole structure by means of non intrusive, yet, extensive interventions that are inferior to safety levels of protection for new structures and (ii) Local Rehabilitation, which improves the response of the structure through local interventions that do not affect the overall behavior of the structure. The two alternatives appear to be the most appropriate and, in fact, the most widely used to preserve historic structures and monuments. Nevertheless, in many cases interventions carried out on monuments are made without complying with a certain earthquake intensity level. Unfortunately, in many cases the interventions are made following previous experience and simplified techniques, without conducting detailed analysis to assess the effectiveness of these interventions. This trend is diminishing with time in a global sense, since previous experience is combined with advances made on computational mechanics and rehabilitation technologies for a more effective design of interventions on monuments.

3   Description of the Methodology

This section presents an approach that could bridge seismic safety demands with the implementation of interventions on monuments for either Global or Local Rehabilitation measures. The basic idea is to consider the rehabilitation measures as fulfilling a predefined limit state for a certain time duration, after which the structure with its interventions should be re-evaluated and, if necessary, appropriate measures will be taken at this later time.

The methodology also requires: (i) the introduction of the term "nominal life of an intervention T_Δ", defined as the time for which the intervention ensures a selected performance level, e.g., B2 for $P_R = 10\%$ and A1 for $P_R = 50\%$, and (ii) the use of either an attenuation relationship or site specific spectra [33, 37]; the latter case, is currently adopted in the Italian code [30]. A more detailed presentation of the methodology and an extensive attempt to address these issues is presented in previous work of the first author [38–40]. The steps of the proposed methodology are the following:

STEP 1: Define the a_{gRL} for which the structure reaches the desired performance level. This acceleration is determined by analysis using appropriate numerical modeling of the structure that takes into account both the overall and local failure mechanisms.

STEP 2: Calculate the return period T_{RL} related to the corresponding reference peak ground acceleration in stiff soil, a_{gR}, with the use of a proper attenuation relationship, as the following attenuation relations valid for the three seismic hazard zones, Zi, (i = 1, 2, 3) of Greece

$$\text{Zone Z1} : \ \log a_{gR} \approx 0.277 \log T_{RL} + 1.579 \tag{1}$$

$$\text{Zone Z2} : \ \log a_{gR} \approx 0.264 \log T_{RL} + 1.739 \tag{2}$$

$$\text{Zone Z3}: \ \log a_{gR} \approx 0.240 \log T_{RL} + 2.015 \tag{3}$$

where $a_{gR} = a_{gRL}/0.8$ is considered. For each one of the three seismic hazard zones Zi, με i = 1, 2 και 3, respectively, and for soil category A, the reference peak ground acceleration a_{gR} with a probability of exceedance 10% in 50 years is defined, as listed in Table 2.

Table 2. Reference peak ground acceleration for different seismic hazard zones (Greek National Annex of [33]).

Seismic hazard zone	Reference peak ground acceleration in stiff soil, a_{gR} (P_R = 10% in 50 years)
Z1	0.16 g
Z2	0.24 g
Z3	0.36 g

STEP 3: Calculate the nominal life T_Δ from the return period T_{RL} and the probability of occurrence P_R, adapting a Poissonian distribution for the seismic events according to the following equation:

$$T_\Delta = -\ln(1 - P_R) \times T_{RL} \approx P_R \times T_{RL} \tag{4}$$

STEP 4: Calculate $T_{\Delta R}$, given the value of the importance factor γ_I multiplying the reference peak ground acceleration a_{gR} and applying the following equation

$$\gamma_i \approx (T_{\Delta R}/ T_\Delta)^{-1/k} \tag{5}$$

where k is in the order of 3 [33]. The $T_{\Delta R}$ duration has the same meaning with T_Δ but refers to different importance categories than II ($\gamma_I \neq 1.00$).

STEP 5: The simultaneous fulfillment of the following criteria for T_Δ is examined

(i) Exceed 50 years for the performance level A2.
(ii) Approach 50 years or at least not be shorter than 20 years for the performance level B1.

After the end of the periods specified by T_Δ the structure should be re-evaluated for the corresponding performance level, that is, A2 or B1. If the selected T_Δ for the retrofitted structure does not meet the criteria (i) and (ii), the structure should be considered as belonging to category C3 and should therefore not be accessible by the public.

In Figs. 1, 2 and 3 diagrams that are obtained by applying the relations (1) to (5) are developed for the three seismic hazard zones Z1, Z2 and Z3, respectively. By using the diagrams, the calculation of the T_Δ duration is more easily achieved by providing the a_{gR} acceleration, the probability of exceedance, P_R, and the category of importance.

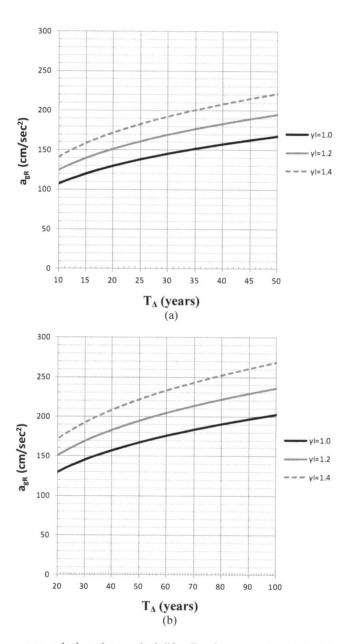

Fig. 1. Diagrams to calculate the nominal life, T_Δ, for zone of seismic risk Z1 and for probabilities of exceedance: (a) $P_R = 10\%$; (b) $P_R = 50\%$.

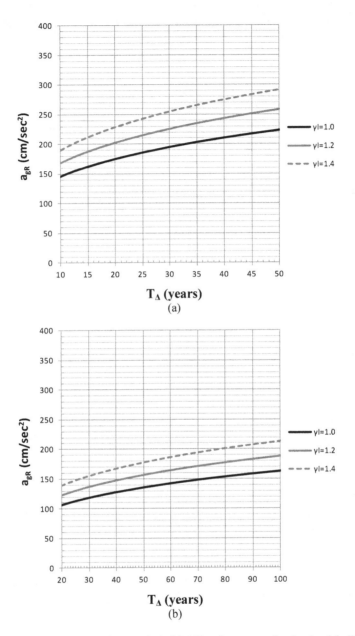

Fig. 2. Diagrams to calculate the nominal life, T_{Δ}, for zone of seismic risk Z2 and for probabilities of exceedance: (a) $P_R = 10\%$; (b) $P_R = 50\%$.

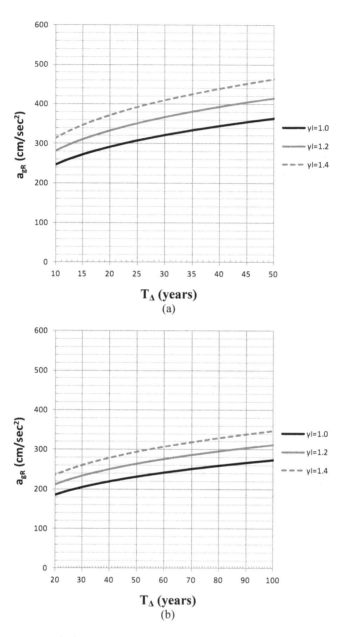

Fig. 3. Diagrams to calculate the nominal life, T_Δ, for zone of seismic risk Z3 and for probabilities of exceedance: (a) $P_R = 10\%$; (b) $P_R = 50\%$.

4 Conclusions

Earthquake protection of cultural heritage structures can be realized through a preventive knowledge of the seismic risk in order to plan mitigation strategies and schedule the necessary strengthening measures to reduce vulnerability. In order to meet the requirements of contemporary seismic codes strengthening of cultural heritage structures often requires invasive interventions that may not be applied because of relevant limitations. According to the proposed methodology time limits are set after the expiration of which the seismic risk of the structure should be re-evaluated. By means of a proper estimation of the nominal life of interventions, T_A, the application of more invasive interventions is transferred at a future time when more advanced scientific data and techniques will be available. The methodology can be easily adopted to current trends in seismology that emphasize on fault-specific attenuation relationships.

Acknowledgements.   The investigation was performed as a task of the research project "Seismic Protection of Monuments and Historic Structures – SEISMO" co-financed by the Greek Ministry of Education and the European Union under the action "THALES" within the context of the Operational Programme – Education and Lifelong Learning, NSRF 2007–2013.

References

1. Oliveira, C.S., Çaktı, E., Stengel, D., Branco, M.: Minaret behavior under earthquake loading: the case of historical Istanbul. Earthquake Eng. Struct. Dynam. **41**, 19–39 (2012)
2. GEER-EERI-ATC: Earthquake Reconnaissance January 26th/February 2nd 2014 Cephalonia, Greece events. GEER Association Report No. GEER-034 in collaboration with EERI and ATC, Version 1, 6 June 2014
3. Somerville, P.G., Smith, N.F., Graves, R.W., Abrahamson, N.A.: Modification of empirical strong ground motion attenuation relations to include the amplitude and duration effects of rupture directivity. Seismol. Res. Lett. **68**, 199–205 (1997)
4. Spyrakos, C.C., Maniatakis, C.A., Taflambas, J.: Evaluation of near source seismic records based on damage potential parameters: case study: Greece. Soil Dyn. Earthq. Eng. **28**, 738–753 (2008)
5. Spyrakos, C.C., Maniatakis, C.A., Kiriakopoulos, P., Francioso, A., Taflampas, I.M.: Performance of a post-Byzantine triple-domed basilica under near and far fault seismic loads: analysis and intervention. In: Asteris, P.G., Plevris, V. (eds.) Handbook of Research on Seismic Assessment and Rehabilitation of Historic Structures, vol. II, pp. 831–867. IGI Global Editions, Hershey (2015)
6. Alavi, B., Krawinkler, H.: Consideration of near-fault ground motion effects in seismic design. In: Proceedings of the 12th World Conference on Earthquake Engineering, article 2665. New Zealand Society for Earthquake Engineering, Upper Hutt, N.Z. (2000)
7. Maniatakis, C.A., Spyrakos, C.C.: A new methodology to determine elastic displacement spectra in the near-fault region. Soil Dyn. Earthq. Eng. **35**, 41–58 (2012)
8. Spyrakos, C.C., Maniatakis, C.A., Taflambas, J.: Critical evaluation of near field seismic records in Greece. In: Syngellakis, S. (ed.) Earthquake Ground Motion - Input Definition for Aseismic Design, pp. 1–10. Wessex Institute of Technology Press, Southampton (2015)

9. Maniatakis, C.A., Spyrakos, C.C., Kiriakopoulos, P.D., Tsellos, K.P.: Seismic response of a historic church considering pounding phenomena. Bull. Earthq. Eng. **16**(7), 2913–2941 (2018)
10. Spyrakos, C.C., Pugi, F., Maniatakis, C.A., Francioso, A.: Evaluation of the dynamic response for a historic Byzantine crossed-dome church through block joint and kinematic analysis. In: Proceedings of the 5th International Conference on Techniques Methods in Structural Dynamics and Earthquake Engineering COMPDYN 2015, pp. 2354–2364. Institute of Structural Analysis and Antiseismic Research, National Technical University of Athens, Greece (2015)
11. Spyrakos, C.C., Touliatos, P., Patsilivas, D., Pelekis, G., Xampesis, A., Maniatakis, C.A.: Seismic analysis and retrofit of a historic masonry building. In: Syngellakis, S. (ed.) Retrofitting of Heritage Structures - Design and Evaluation of Strengthening Techniques, pp. 65–74. Wessex Institute of Technology Press, Southampton (2013)
12. Spyrakos, C.C.: Finite Element Modeling in Engineering Practice. Algor Publishing Division, Pittsburg (1995)
13. Spyrakos, C.C., Raftoyannis, J.: Linear and Nonlinear Finite Element Analysis in Engineering Practice. Algor Publishing Division, Pittsburg (1997)
14. Lemos, J.V.: Discrete element modeling of the seismic behaviour of stone masonry arches. In: Pande, G., Middleton, J., Kralj, B. (eds.) Computer Methods in Structural Masonry, 4th edn, pp. 220–227. CRC Press, London (1998)
15. Lourenco, P.B., Milani, G., Tralli, A., Zucchini, A.: Analysis of masonry structures: review of and recent trends in homogenization techniques. Can. J. Civ. Eng. **34**, 1443–1457 (2007)
16. Lagomarsino, S., Penna, A., Galasco, A., Cattari, S.: TREMURI program: an equivalent frame model for the nonlinear seismic analysis of masonry buildings. Eng. Struct. **56**, 1787–1799 (2013)
17. Roca, P., Cervera, M., Gariup, G.: Structural analysis of masonry historical constructions. Classical and advanced approaches. Arch. Computat. Methods Eng. **17**(3), 299–325 (2010)
18. Oliveira, D.V., Basílio, I., Lourenço, P.B.: FRP strengthening of masonry arches towards an enhanced behavior. In: Proceedings of the International Conference on Bridge Maintenance, Safety and Management (IABMAS 2006), 3, Porto, Portugal (2006)
19. Credali, L., Ussia, G.: Composite materials technologies in constructions structural retrofitting: new developments and applications in historical buildings and applications in seismic zone. In: Proceedings of the 3rd International Conference on Computational Methods in Structural Dynamics and Earthquake Engineering COMPDYN 2011, pp. 1866–1882. Institute of Structural Analysis and Antiseismic Research, National Technical University of Athens, Greece (2011)
20. Spyrakos, C.C., Kiriakopoulos, P.D., Smyrou, E.: Seismic strengthening of the historic church of Sts Helen and Constantine in Piraeus. In: Proceedings of the 3rd International Conference on Techniques Methods in Structural Dynamics and Earthquake Engineering COMPDYN 2011, pp. 2401–2413. Institute of Structural Analysis and Antiseismic Research, National Technical University of Athens, Greece (2011)
21. Spyrakos, C.C., Francioso, A., Kiriakopoulos, P.D., Papoutsellis, S.: Seismic evaluation of the historic church of St. Nicholas in Piraeus before and after interventions. In: Proceedings of 4th International Conference on Techniques Methods in Structural Dynamics and Earthquake Engineering COMPDYN 2013, pp. 3015–3029. Institute of Structural Analysis and Antiseismic Research, National Technical University of Athens, Greece (2013)
22. Spyrakos, C.C.: Ενίσχυση κατασκευών για σεισμικά φορτία (Strengthening of structures for seismic loads). Technical Chamber of Greece, Athens (2004). (in Greek)

23. Alaggio, R., Benedettini, F., De Sortis, A., Lucarelli, V.: Structural identification of monuments in Rome using ambient vibration measurements. In: Proceedings of EVACES 2011 – 4th International Conference on Experimental Vibration Analysis for Civil Engineering Structures, Varenna, Italy (2011)

24. Russo, S.: On the monitoring of historic Anime Sante church damaged by earthquake in L'Aquila. Struct. Control Health Monit. **20**(9), 1226–1239 (2013)

25. De Stefano, A., Matta, E., Clemente, P.: Structural health monitoring of historical heritage in Italy: some relevant experiences. J. Civ. Struct. Health Monit. **6**(1), 83–106 (2016)

26. Spyrakos, C.C., Fessas, C.P., Andrikopoulos, G.A.: Health monitoring of a monument at Acropolis using an expert acquisition system and wired optical strand devices. In: Proceedings of the 6th International Conference on Techniques Methods in Structural Dynamics and Earthquake Engineering COMPDYN 2017, pp. 2651–2658. Institute of Structural Analysis and Antiseismic Research, National Technical University of Athens, Greece (2017)

27. The Athens Charter for the Restoration of Historic Monuments. First International Congress of Architects and Technicians of Historic Monuments, Athens (1931)

28. International Council on Monuments and Sites – ICOMOS: International Charter for the Conservation and Restoration of Monuments and Sites (The Venice Charter 1964). Second International Congress of Architects and Technicians of Historic Monuments, Venice (1964)

29. Congress on the European Architectural Heritage: The Declaration of Amsterdam, Amsterdam (1975)

30. Italian Building Code: Linee Guida per la valutazione del rischio sismico del patrimonio culturale allineate alle nuove norme tecniche per le costruzioni (Guidelines: Assessment and mitigation of seismic risk of cultural heritage with reference to the 2008 Italian Building Code), G.U. No. 47 (2011). (in Italian)

31. Comité Européen de Normalisation (CEN): EN 1998-3:2004. Eurocode 8: Design of structures for earthquake resistance - Part 3: Assessment and retrofitting of buildings. CEN, Brussels, Belgium (2004)

32. Steenbergen, R.D.J.M., Vrouwenvelder, A.C.W.M.: Safety philosophy for existing structures and partial factors for traffic loads on bridges. Heron **55**(2), 123–139 (2010)

33. Comité Européen de Normalisation (CEN): EN 1998-1:2004. Eurocode 8: Design of structures for earthquake resistance - Part 1: General rules, seismic action and rules for buildings. CEN, Brussels, Belgium (2004)

34. Earthquake Planning Protection Organization (EPPO), Harmonization Team of Code of Interventions to Eurocodes: Code of Structural Interventions (CSI), Final Harmonized Text, English temporary version V1 (2012)

35. Borri, A. Candela, M.: Strutturisti e Restauratori: Sicurezza Vs conservazione? Problemi, dubbi e proposte, anche alla luce di esperienze successive al terremoto dell'Aquila (Structural engineers and Restorers: Security and Conservatism? Problems, doubts and proposals, also in the light of subsequent experience after the L' Aquila earthquake). In: Proceedings of the XVI National Congress "L' Ingegneria Sismica in Italia", pp. 5–26. ANIDIS, L' Aquila (2015). (in Italian)

36. Lagomarsino, S.: The seismic prevention as a tool for the conservation of cultural heritage. Ingenio 26 (2014)

37. Papazachos, B.C., Papaioannou, Ch.A., Margaris, B.N., Theodulidis, N.P.: Regionalization of seismic hazard in Greece based on seismic sources. Nat. Hazards **8**, 1–13 (1993)

38. Spyrakos, C.C.: Seismic risk of historic structures and monuments: a need for a unified policy. In: Proceedings of the 5th International Conference on Techniques Methods in Structural Dynamics and Earthquake Engineering COMPDYN 2015, pp. 2423–2439. Institute of Structural Analysis and Antiseismic Research, National Technical University of Athens, Greece (2015)

39. Spyrakos, C.C.: Modern approaches for the earthquake protection of cultural heritage structures - review and proposal. In: Proceedings of the 4th National Conference of Restoration. ETEPAM, Thessaloniki (2015)
40. Spyrakos, C.C.: Bridging performance based seismic design with restricted interventions on cultural heritage structures. Eng. Struct. **160**, 34–43 (2018)

Cultural Heritage and Pilgrimage Tourism

Restoration and Management of World Heritage Christian Cultural Sites: Problems of Religious Tourism – Challenges - Perspectives

Alkiviadis Prepis[✉]

Department of Architectural Engineering, Democritus University of Thrace,
3, Polytechneiou str., 1st parodos, 69100 Komotini, Greece
alkisprepis@gmail.com

Abstract. Taking under consideration the exceptional value and unique religious and cultural significance of the Old City of Jerusalem, as well as that since 1981 the "Old City of Jerusalem and its Walls" are inscribed on the World Heritage List, is worth a discussion on the linkages between the World Heritage Convention and the heritage of Christian religious interest (with special reference to those connected with the image of the "New Jerusalem"): whether doctrinal texts and principles are capable to be applied in practice and in the local context of individual examples, identifying our interest on specific issues, such as: the definition, acceptance and interpretation of associated spiritual values of the past in the modern world, the question about the meaning and the limits of restoration and reconstruction, particularly in relation to the notions of authenticity and integrity, the research of the problems in the field of sustainable development (religious tourism, pilgrimage etc.), and the role of the religious community in the management of the World Heritage property.

1 Introduction

1.1 The World Heritage Convention

The Convention for the Protection of World Cultural and Natural Heritage[1] was adopted by the General Conference of UNESCO in Paris on 16 November 1972. It was established to recognize "sites of Outstanding Universal Value" which are part of the heritage of humankind as a whole, which deserve protection and transmission to future generations, and which are important for the whole of humanity.

The Operational Guidelines for the Implementation of the World Heritage Convention[2] define *monuments* as architectural works, works of monumental sculpture and

---

[1] https://whc.unesco.org/en/convention/.

[2] UNESCO, Operational Guidelines for the Implementation of the World Heritage Convention, rev. 2017, https://whc.unesco.org/en/guidelines/, IIA 45, Article 1.

Honorable Professor of Kazan State University of Architecture and Engineering, ICOMOS expert.

© Springer Nature Switzerland AG 2019
A. Moropoulou et al. (Eds.): TMM_CH 2018, CCIS 962, pp. 239–261, 2019.
https://doi.org/10.1007/978-3-030-12960-6_16

painting, elements or structures of an archaeological nature, inscriptions, cave dwellings and combinations of features, which are of outstanding universal value from the point of view of history, art or science.

The Operational Guidelines define *Outstanding Universal Value (OUV)* as being cultural and/or natural significance which is so exceptional as to transcend national boundaries and to be of common importance for present and future generations of all humanity[3].

To be deemed of Outstanding Universal Value, a property must also meet the conditions of *integrity* and/or *authenticity* and must have an adequate *protection and management system* to ensure its safeguarding[4].

The List of World Heritage includes 1073 properties forming part of the cultural and natural heritage, which the Commission considers that it has excellent universal value. These include:

- 832 monuments of cultural heritage,
- 206 monuments of natural heritage,
- 35 mixed heritage monuments,
  located in 165 States Parties.

States Parties (193) are countries which have adhered to the World Heritage Convention. They agree to identify and nominate properties and protect the World Heritage values of the properties inscribed and are encouraged to report periodically on their condition.

1.2   The Scope of the Analysis

Taking under consideration the exceptional value and unique religious and cultural significance of the Old City of Jerusalem, as well as that since 1981 the "Old City of Jerusalem and its Walls" are inscribed on the World Heritage List, is worth a discussion on the linkages between the World Heritage Convention and the heritage of Christian religious interest (with special reference to those connected with the image of the "New Jerusalem"): whether doctrinal texts and principles are capable to be applied in practice and in the local context of individual examples, identifying our interest on specific issues, such as: the definition, acceptance and interpretation of associated spiritual values of the past in the modern world, the question about the meaning and the limits of restoration and reconstruction, particularly in relation to the notions of authenticity and integrity, the research of the problems in the field of sustainable development (religious tourism, pilgrimage etc.), and the role of the religious community in the management of the World Heritage property.

In order to analyse the above mentioned subjects, we consider that selected representative Christian monuments having been included on the World Heritage List can be presented in separate groups, as follows:

---

[3] Operational Guidelines, 49.

[4] Cameron Ch (2009) The evolution of the concept of Outstanding Universal Value. In: King J, Stanley-Price N (eds) Conserving the authentic – Essays in honour of Jukka Jokilehto, ICCROM Conservation Studies, No.10, pp 127–136.

2   Religious Monuments within Historical Centers – Representations of the (Heavenly) Jerusalem and St Sophia Church in Constantinople

Several cities and individual churches in the Middle Ages were associated with the idea of representing or incorporating Jerusalem in one manner or another. For medieval Christian societies, both in the East and in the West, the centrality of the New Jerusalem idea was crucial, since the city of Jerusalem was regarded as the "centre of the world, the axis of the universe, the place of Salvation". Numerous and outstanding are medieval examples that reveal mechanisms for setting up the sacral topography of Jerusalem in other parts of the world. The *hierotopical* approach[5] has made possible more profound insights into this important issue of medieval era. It led us to awareness of the fact that translation of sacred spaces, creation of New Jerusalems and images of the Holy Land were highly significant aspects of the mediaeval culture. In this respect a wide circle of medieval Christian centers have reproduced the image of a terrestrial Jerusalem "in imitation of and as a path to the Heavenly Jerusalem": Constantinople, Rome in Lateran (S. Maria Maggiore, Sto. Stefano Rotondo, and S. Croce), "Moscow Jerusalem" (the Church of the Veil of the Virgin on the Moat (St. Basil's) and its Golgotha (Lobnoye Mesto)), etc.

The Holy Sepulcher began to represent the holy places of Jerusalem or the Holy Land even before the 11th century, in the Carolingian era, after they were no longer in Christian hands: Palatine Chapel at Aachen (Germany - 805), Sankt Michael in Fulda (Germany- 820), and San Stefano in Bologna (Italy – from 9th c.). During the following centuries, multiple medieval architectural copies of the Holy Sepulcher were made. Some were dedicated to or directly called Jerusalem. This widely attested phenomenon occurred in a large range of variants, depending on the 'type' of Jerusalem represented and the way in which the representation was made concrete, thus, creating places of salvation with a strong catechetical goal of educating people about Christianity[6].

At the turn of the 15th and 16th centuries the phenomenon of *Sacri Monti* began with the aim of creating places of prayer in Europe as an alternative to the Holy Land (Jerusalem and Palestine). At that time, access to the Holy Land was becoming more and more difficult for pilgrims owing to the rapid expansion of Muslim culture. Initially, three different locations were proposed for the "New Jerusalem": Vareallo in Valsesia (which became a model for latter constructions in Piedmont and Lombardy, due to the originality of its design and the strong presence of artistic and spiritual references), Montaione in Tuscany, and the Sanctuary of Bom Jesus do Monte in

---

[5] For the term *hierotopy* see: Lidov A (2006) Hierotopy. The Creation of Sacred Spaces as a Form of Creativity and Subject of Cultural History. In: Lidov A (ed) Hierotopy. Creation of Sacred Spaces in Byzantium and Medieval Russia, Moscow, pp 32–58.

[6] Erdeljan J (2011) New Jerusalems as New Constantinoples? Reflexions of the reasons and principles of Translatio Constantinopoleos in Slavia Orthodoxa. In: Δελτίον Χριστιανικής Αρχαιολογικής Εταιρείας 32, pp 11–18; Hoffmann A, Wolf G (ed) Jerusalem as Narrative Space/Elzählraum Jerusalem, Leiden (2012); Sible de Blaauw (2014) Translations of the Sacred City between Jerusalem and Rome. In: Goudeau J, Verhoeven M, Weijers W (eds) The Imagined and Real Jerusalem in Art and Architecture, Leiden, Chapter 6, pp 136–165.

Braga, Northern Portugal (which latter effectively inspired the Sanctuary of Bom Jesus do Matozinhos in Congonhas, Brazil). The phenomenon has spread around Europe (Switzerland, Austria, Germany, Spain, France, Hungary, Poland, Slovakia, the Netherland and Belgium).

2.1   The Holy City of Mtskheta, Georgia - the Second Jerusalem (World Heritage List - 1994)

Mtskheta has maintained its role as the spiritual and cultural centre of the country, assumed ever since the introduction of Christianity in the region. The Holy Cross Monastery of Jvari, Svetitskhoveli Cathedral and Samtavro Monastery are key monuments of medieval Georgia (Fig. 1). The present churches include the remains of earlier buildings on the same sites, as well as the remains of ancient wall paintings. The complex of the Svetitskhoveli Cathedral in the centre of the town includes the cathedral church, the palace and the gates of the Katolikos Melchizedek that date from the 11th century, built on the site of earlier churches dating back to the 5th century. The small domed church of the Samtavro Monastery was originally built in the 4th century and has since been subject to various restorations. The main church of the monastery was built in the early 11th century.

Fig. 1. Historical core of Mtskheta with Svetitskhoveli Cathedral

The Holy Capital Mtskheta and its cathedral church Svetitskhoveli are the axis of the Georgian history and the center of its "earthly paradise". St. Nino erected Svetitskhoveli church (the "Cross of Christ") on the river bank of Mtkvari at the place where the Holy Tunic, taken off from Jesus, immediately before his crucifixion, as well as the pelerine of the Prophet Elijah are buried. Here, the head of the church – Katolicos - Patriarch - and bishops are consecrated. After that, similar to Jerusalem, the worship of the holy places connected to the Savior's earthly life began: the "Great Zion", while on its opposite side – on Georgia's Golgotha – the Mtskheta's Jvari (Cross) church was built (6th c.). The mountain slopes between the Aragvi River and the church were called Gethsemane. There are also Tabor, Eleoni, a cave where St. Nino prayed and is called Bethlehem, Antioch – the cathedral of St. Stephan First Martyr (beg. 5th c.) at the rocky cape at the confluence of Aragvi and Mtkvari rivers, and the baptism place of all Georgians that was located at the left river bank of Mtskheta - "Our Jordan". These sites are functionally linked through a litany (religious procession), which represents the venerable pilgrimage to the holy city of Christendom – Jerusalem[7].

A proposed by the Georgian Church project for the rehabilitation of the "New Jerusalem" concept for the religious monuments of the city includes a new pedestrian bridge over the river Aragvi, related infrastructure with new walkways, connecting roads and look-outs at the historically important area at the banks of the Aragvi and Mtkvari rivers and by the Jvari Church. Moreover - new square and new buildings for visitors and baptistery for pilgrims are proposed near Antiokia Nunnery and an extensive landscaping of the surrounding areas including a new baptism canal and a basin for the baptism process. However, it is very problematic that a large and comprehensive project like this is being considered separately from the preparation of the Master Plan for the World Heritage property by the local authorities and the national authorities (Ministry of Culture). Clearly there is disconnect between the actions of the relevant stakeholders, as additional problems identified: increasing tourism pressure with a rehabilitation of new public spaces and developing commercial tourism facilities by the local authorities before finalizing the Master Plan process; lack of awareness by the local community of World Heritage commitment, of the benefits and constraints, of positive and celebratory measures; lack of an overall shared vision for the development of the city of Mtskheta as a guiding principle for the process of elaboration of the Urban Planning and for all decision-making[8].

2.2 Saint-Sophia Cathedral and Related Monastic Buildings - Kyiv-Pechersk Lavra, Ukraine (World Heritage List - 1990)

Serious problems on a great scale displays the urban space planning in the protected buffer zone of St. Sophia of Kiev, covering an area of 119 ha. The Cathedral, designed to rival Haghia Sophia in Constantinople, is one of the major edifices representing the culture of Eastern Christianity in the 11th century (Fig. 2). The stylistic features of the cathedral's decoration were spread throughout Kievan Russia. A complex of monastic buildings surrounds the church. The imposing walled monastery complex is the core of

---

[7] Chkhartishvili M (2009) Mtskheta as New Jerusalem: Hierotopy in the Life of St Nino. In: Lidov A (ed) New Jerusalems. Hierotopy and iconography of sacred spaces, Moscow, pp 131–150.

[8] Sidorenko A, Prepis A, Lizitsin K (2018) Historical monuments of Mtskheta – Georgia, Report on the joint UNESCO World Heritage Centre/ ICOMOS/ ICCROM Reactive Monitoring Mission (19-24/2/2018), Paris, p 61 + 158 photos.

Fig. 2. Saint-Sophia Cathedral ensemble within Kyiv historical core

the protected historic urban area, where a duality of responsibilities is recognised. The property control is exercised by the "St. Sophia Museum Foundation", maintaining to a high quality level the authenticity of the churches and historic buildings, combined with cultural – educational – research functions.

However, the extended buffer zone area, being under the control of the Municipality Technical Services, is characterized by the absence of a supporting policy for the rescue, maintenance and rehabilitation of historic buildings. In contrast, a policy of overexploitation (in volume and height) is implemented, by providing permission for constructing gigantic buildings, next to or by demolishing the traditional ones. The new buildings have concealed the city historic core, posing under risk even the historic landscape with the monasteries of Kyiv-Pechersk Lavra along the Dnieper River. The current planning of urban space is out of traditional city's scale and local character, following the requirements of a misguided "modern development": big investments in private offices, hotels or occasional touristic development, while the historic city traffic is literally chaotic. A lucrative real estate politics is practiced, in the context of big financial investments.

In this aspect, the most encouragement factor is the massive reaction of the residents who often gets inventive forms, organized in NGOs.

3   Living Religious Ensembles

The Context of the Religious Heritage of the Past in the Modern World

In our days, our views have much changed with regard to the approach of the so-called "revival" of historical monuments and notably living religious ensembles. A new interpretation of such sites is necessary, based on the concept of continuity, and its evolution to the present. The current theoretical framework and practice of conservation, as best epitomized in a values-based approach and the World Heritage concept, is based on discontinuity created between the monuments (considered to belong to the past) and the people of the present, thus seemingly unable to embrace living religious heritage sites. A "Living Heritage Approach"[9] is an innovative approach that views communities and sites as an inseparable entity: This approach brings a new insight into key concepts such as authenticity and sustainable development. The discussion generated aims to shift the focus of conservation from "preservation" towards a continual process of "creation" in an ongoing present, attempting to change the way heritage is perceived, protected and, more importantly, further created.

Historical religious monuments should not be treated as "remnants of an important historical past", but more and more in the light of the needs of contemporary society, as part of a dynamic process within the modern city, considering them as essential factors of redesigning and development both the immediate environment of the historical monument itself and the wider area.

Today the religious architectural heritage of the past is considered an important factor for the identity of local people, which can substantially contribute to the homogenization of multicultural societies and the achievement of social peace. Especially in multicultural societies the different religious historical monuments must be treated equally, in the same spirit, and thus one of the requirements of the modern era restoration principles is satisfied, while respecting the diachronic dimension of the cultural heritage.

In addition, today the process of preservation and revival of the religious monuments is treated in a different way: the process follows various paths – in many cases for the monument itself – following for the solution of the problems ways and methodologies much more interesting than their unambiguous solutions of the past; methodologies that can successfully combine monumental conservation and restoration with the redesign of a modern environment around and within the monument itself. In this way the whole process can also contribute to the satisfaction of the social need to improve the quality of life not only of the religious community but, also, of all residents.

However, key problem remains how to rehabilitate the overall original form of the monument in a way that does not compromise the authenticity of its material and aesthetic appearance.

Consequently, historic religious monuments should be treated within the frames of an integrated conservation process, which enables the continuation of the urban cultural heritage, being kept in an appropriate human setting, and particularly their assignment towards a continual process of "creation" of a new synthesis, combining their elements of authenticity and sustainable development of the site.

Pilgrimage Problems

A significant number of living religious ensembles face growing problems from the large number of visitors which is constantly showing upward trends. The number of

---

[9] Poulios, I (2014) Past in the Present: A Living Heritage Approach – Meteora, Greece, London, Ubiquity Press. doi: https://doi.org/10.5334/bak.

pilgrims who flow together especially in order to attend fest liturgies is so big that it is necessary to extend across the surrounding outdoor area and even to organize open-air mass. Infrastructure accommodation of sanctuaries is not capable to accept such big numbers of pilgrims and on these cases special care should be taken for exceptional measures. Also, there is need traffic management plans to be prepared for the transition of thousands of pilgrims through the city center to the very place of the sanctuary. Additionally - to organize the execution of the procession in the wider area around the sanctuary, with the presence of thousands of pilgrims. Usually, these factors have not been taken into consideration on the management assessment of the long-term impacts on the surrounding area of these religious ensembles.

Indicatively, it can be taken under consideration that:

- About 5 million pilgrims visit Lourdes every year and within France only Paris has more hotels than Lourdes. And about 10 million pilgrims visit Our Lady of Guadalupe each year, where each mass can accommodate up to 40,000 people. Thus each decade, just Lourdes and Guadalupe amount to over one hundred million Catholic pilgrimages, based on Marian apparitions to two people on two remote hilltops.
- The Sanctuary of our Lady of Fátima also attracts a large number of Roman Catholics, and every year pilgrims fill the country road that leads to the shrine with crowds that approach one million on May 13 and October 13 - the significant dates of Fátima apparitions. Overall, about four million pilgrims visit the basilica every year.

3.1 Meteora Rock Monasteries, Greece (World Heritage List - 1988, Cultural + Natural Heritage)

In a region of almost inaccessible sandstone peaks, monks settled on these "columns of the sky" from the 11th century onwards (Fig. 3). Twenty-four of these monasteries were built, despite incredible difficulties, at the time of the great revival of the eremitic ideal in the 15th century. Their 16th-century frescoes mark a key stage in the development of post-Byzantine painting. At Meteora, within the context of the living heritage approach, the primary role would be given to the monastic communities, who would be encouraged to continue the creation of the site, with the involvement of the conservation professionals and the broader community.

Within the domestic market, and within the "See, Sun and Culture" model, Meteora is by far the most popular monastic site in Greece. Its popularity is additionally eased by the fact that Meteora is also open to women, is much easier to access by a well-organized transportation system, and there are no special entry procedures or restrictions in the number of visitors. In this context, the benefits are most significant for the Greek state, while rather limited for the local community, with the exception of a few restaurants, souvenir shops, and hotels. The benefits for the Meteora monastic communities, as the ones who control the access to the monasteries, are significant.

The positioning of a monastery in relation to its landscape, as well as the arrangement of its internal space is defined by the monastery's two-fold function: as a place of worship and as place that sustain the life of the monastic community.

Fig. 3. Roussanou nunnery (1545) after the construction of the new buildings and the new road

However, developing mass tourism at Meteora during the last decades led to compromising results:

– Widening of the road network, with considerable impact on the sensitive landscape.
– The site, although remained a monastic one, at the same time has clearly developed into a major tourist attraction. The external space of the monasteries was used by the visitors, whilst the monastic communities were restricted to their monasteries, the internal space of which was divided between the monastic communities and the visitors. Therefore, the monastic communities have been limited in mostly secondary and peripheral areas and buildings of the monasteries (St. Stephen monastery).
– New buildings have been erected because of the continuing need for more space for everyday monastic needs. Thus, the Roussanou monastery is now in three parts: the original monastery on the top of the rock, the five storey building next to the rock, with a tower-lift connecting it with the original monastery, and a two-storey building, close to the five storey one[10].

3.2 Stari Ras and Sopočani Monastery, Serbia (World Heritage List - 1979)

A phenomenon with a growing tendency in the Balkans in recent years is the rehabilitation of living religious ensembles, of exceptional historical significant, under the pressing needs of modern society and political decisions, which may lead to possible alteration or to jeopardize the authenticity of the monument. In particular, following the

---

[10] Poulios, I., op.cit.

Fig. 4. Sopočani monastery (1265) in Stari Ras – Novi Pazar, Serbia

recent wars in the Balkans, the Orthodox Church seeks to restore or rebuild old monasteries, supporting cultural, religious and ethnic identities. Unfortunately, sometimes the new additions are incompatible with the size, morphology or construction traditions of the old monasteries. Thus, restoration problems move to another level: that of restoring religious life to the new building around the original monument, within the context of the historical environment. In many cases the imitation of the old forms and the volumetric composition is followed, in order to create a new, adequate building environment on the ruins of the original: such are the cases of the planned reconstruction of the wings at the Sopočani Monastery (1265, World Heritage List -1979) (Fig. 4) and the additions and reconstruction of the cells to the Monastery of Djurdjevi Stupovi (1166, World Heritage List -1979) both in Stari Ras – Novi Pazar, Serbia.

3.3   Saint Catherine Area, Egypt (World Heritage List - 2002)

The Greek-Orthodox Monastery of St Catherine stands at the foot of Mount Horeb where, the Old Testament records, Moses received the Tablets of the Law. The mountain is known and revered by Muslims as Jebel Musa. The entire area is sacred to three world religions: Christianity, Islam, and Judaism. The monastery, founded in the 6th century, is the oldest Christian monastery still in use for its initial function. Its walls and buildings are of great significance to studies of Byzantine architecture and the monastery houses outstanding collections of early Christian manuscripts and icons. The rugged mountainous landscape, containing numerous archaeological and religious sites and monuments, forms a perfect backdrop to the monastery (Fig. 5).

A unique example - to be underlined with emphasis - of coexistence of Christianity and Islam and of religious tolerance, which is part of the history of the monastery, is the concession by the monks to convert the old monastery refectory into a Moslem temple in order to satisfy the religious needs of Bedouins – constant and historical servants of the monastery.

It should be emphasized that the monastery itself hosts an extremely large number of visitors ranging from 300 to 3,000 people per day depending on the season. Works undertaken secure good living conditions for the monks' community and contribute to the better organization of the visitors' circulation, among others with a control gate to monitor the number of tourists and the establishment of a museum with representative and some of the most valuable religious artifacts within the monastery walls.

Fig. 5. Greek-Orthodox Monastery of St Catherine in Sinai, Egypt

Numerous and positive actions are undertaken for the management of the property, regarding both the landscape and archaeological values. Those actions, focused particularly on the tourist infrastructure, include the building and furnishing of a Visitor Centre at a distance of 1 km from the monastery with an organized parking area; the establishment of a First Aid Unit and training of its staff; the building of restrooms in Gebel Musa Mountain as well as the installation of several water points on the trail for the visitors. Much work has been done for the overall maintenance of the site, including the arrangement of trails to the surrounding mountains, as well as in several *wadis* (valleys). Problems, however, continue with the control point, presently installed at the

Holy Summit, to be transferred to the lower and wider Prophet Elijah plateau, together with the infrastructure and accommodation facilities for the big number of visitors.

4 Religious Monuments Inscribed on the List of World Heritage in Danger

The destruction of Cultural Heritage worldwide is a topic that receives growing attention: Cultural Heritage is threatened in armed conflicts, through climate change and environmental influences, and through neglect. Wars, earthquakes and other natural disasters, pollution of urban environment, contamination, illegal trade and non-controlled touristic development provide great problems on World Heritage monuments. These negative factors can affect the Outstanding Universal Value under which a property was inscribed on World Heritage List.

Fifty four properties the World Heritage Committee has decided to be included on the List of World Heritage in Danger in accordance with Article 11 (4) of the *Convention*. Six of them are religious monuments.

4.1 Medieval Monuments in Kosovo (World Heritage List - 2006)

The policy of eliminating the presence of the "other" and its historical memory was expressed with extreme severity in the recent nationalist wars that followed the collapse of Yugoslavia (1991–1999) and the creation of the new national states (more precisely: the re-establishment of the national states after World War I) in the Balkans. The religious monuments of the heterodox (Muslim) and/or the coreligionist (Christian Orthodox or Catholic) - although until a while ago of the same nationality - have been the subject of targeted destruction attacks to the extent that they were symbols of a different national identity[11].

Thus, during the recent wars in the Western Balkans actions of intentional destruction of the religious cultural heritage took place: mosques, churches and hundreds of historic buildings were burned or blown up, religious and cultural symbols were eliminated, graveyards excavated. The postwar peace policy imposed by the United Nations considered the restoration/reconstruction of religious monuments as a condition *sine qua non* for the consolidation of peaceful coexistence in the region. Many monuments have not only be repaired, but almost totally rebuilt often in a brisk manner and with modern methods. However, unsolved problem remains the restoration of the interior historical decoration of the monuments, which is lost forever.

---

[11] Prepis A (2015) The management of historic monuments in the process of creating national identities in the Balkans (19th-20th c.). In: International Symposium - Culture and Space in the Balkans, 17ᵗʰ -20ᵗʰ c. University of Macedonia – Department of Balkan, Slavic and Eastern Studies, Thessaloniki, pp 535–556; Αλκιβιάδης Πρέπης, « Η διαχείριση των ιστορικών μνημείων στη διαδικασία συγκρότησης των εθνικών ταυτοτήτων στα Βαλκάνια (19ος-20ος αι.) » , Διεθνές Συμπόσιο Πολιτισμός και Χώρος στα Βαλκάνια, 17ος -20ος αιώνας, Πανεπιστήμιο Μακεδονίας – Τμήμα Βαλκανικών, Σλαβικών και Ανατολικών Σπουδών, Θεσσαλονίκη 2015, σελ. 535–556.

Fig. 6. Cathedral of Our Lady Ljeviška in Prizren, after 2004 riots

If the question of material reality is fundamental, it is also the symbolical meaning of the monument equally important, as well. Whenever we make interventions in a "heavy" way, the authenticity of the character of the historic building is under threat to be lost. The 16th-century *Mostar Bridge* was destroyed as a political act during the civil war in the ex-Yugoslavia in 1993. After the destruction of the bridge, the original parts that remained in situ were kept, but the arch of the bridge was entirely rebuilt new on the original site with the support of UNESCO, Aga Khan Trust for Culture and World Monuments Fund. Although the site has certainly lost part of its authenticity, the World Heritage Committee inscribed the site on the World Heritage List emphasizing its significance as: *"a symbol of reconciliation, international cooperation and of the coexistence of diverse cultural, ethnic and religious communities"*. UNESCO has pointed out that the intangible component is as important as the physical: the previous is certainly the main issue concerning the Outstanding Universal Value of this site, already having been accepted as *"renaissance of the monument"* by the people[12].

The case of the *Bridge of Mostar* gave rise to other communities to express their will for full reconstruction of religious monuments - community's unity symbols. The colorful decorated *Aladža mosque in Foča* (1551) and *St. Trinity Orthodox Cathedral in Mostar* (1873), the largest Christian church in Bosnia-Herzegovina, were leveled

---

[12] Old Bridge Area of the Old City of Mostar, https://whc.unesco.org/en/list/946/documents/ Advisory Body Evaluation (ICOMOS), 2005, Mostar (Bosnia and Herzegovina) No 946 rev, file:///C:/ Users/user/Downloads/946rev-ICOMOS-1107-en.pdf, p. 183.

over the war events of 1992. The Muslim community, as well as the Serb minority in the city, considers the complete reconstruction of the monuments will contribute to the resettlement of refugees to their homeland, affirming their right to free expression of their identity and their religious beliefs. And, indeed, today both monuments are under full reconstruction process by the local religious communities.

What makes a completely reconstructed building an asset of architectural heritage? Most certainly it is an immaterial component: the cultural significance of the building, fragments of collective memory or the unwillingness of the people to accept the fact of destruction. This is the social dimension of the reconstruction, compared with restoration of the tangible dimensions of these religious monuments.

In 2004, in an outbreak of new nationalist vandalism, Christian monasteries, churches, cemeteries, religious and historical monuments suffered severe disasters. The Joint Committee of UNESCO and Council of Europe group of experts to whom I participated recorded a total of 48 Christian, 14 Islamic and 13 monuments of traditional architecture, more or less destroyed. The monasteries of the Serbian Patriarchate of Peć, of Gračanica and Dečani, as well as the Cathedral of Our Lady Ljeviška in Prizren (Fig. 6), are the four most important monuments of medieval Serbia representing the mature phase of the Palaeologian Renaissance in architecture and wall painting in Kosovo. The monuments were included in the UNESCO World Heritage List in 2006. Today the first three are preserved with minor disasters as religious enclaves, kept constantly under the protection of Kosovo Police. The fourth has almost completely been burned inside with no conservation and restoration works having been carried out yet, and still remaining closed behind a tall, wire barbed wall, under protection of the NATO-led Kosovo Stabilization Force, KFOR. There are great difficulties to monitor the property and for pilgrims to access the religious ensembles due to political instability, post-conflict situation (visits under the Kosovo Stabilization Force/United Nations Interim Administration Mission in Kosovo (KFOR/UNMIK) escort) and lack of guards and security[13].

4.2   Bagrati Cathedral, Kutaisi, Georgia

The question about the meaning and limits of reconstruction is certainly one of the issues that can and should be raised, particularly in relation to the notions of authenticity and integrity. *Bagrati Cathedral* (constructed end of the 10th century and completed in 1003) was badly damaged during wars in 1691 and in 1770. The repair of the ruin, a national symbol of Georgia's revival, already started in the '50 s. It was inscribed on the World Heritage List in 1994.

In 2004, ICOMOS noted that any reconstruction must be carried out in keeping with the Outstanding Universal Value of the property and its authenticity and therefore

---

[13] Prepis A, Lefantzis M, Johnson D (2004) Technical Assessment Report on the Religious Buildings/Ensembles and Cultural Sites Damaged in March 2004 in Kosovo, 1st – 2nd – 3rd expert visit. In: Integrated Rehabilitation project Plan/Survey of the Architectural Heritage, Council of Europe; Prepis A, Johnson D, Bianchi A, Gödicke H, Šurdić B, Hoxha G, Montanes F, Mills A, Wik S (2005) Protection and Conservation of Cultural Heritage in Kosovo – Consolidated Summary, UNESCO - Council of Europe - UNMIK, Paris.

Fig. 7. The controversial restoration of 11th c. Bagrati cathedral led to remove from UNESCO World Heritage List

it would be more appropriate to retain the property as a ruin. In January 2008, the President of Georgia and the Georgian Orthodox Church initiated the reconstruction project of the Bagrati Cathedral with the intention of restoring the initial religious use and functions of the Cathedral. In 2008 the UNESCO World Heritage Committee urged the State Party not to carry out any reconstruction work which may adversely affect the Outstanding Universal Value and its authenticity and not to commence any constructions before consideration of the project by the World Heritage Committee. In 2009 the Georgian Ministry of Culture approved the Bagrati Cathedral Rehabilitation Plan envisaging a reinforcement of the existing structure with the aim of complete reconstruction – a project a group of Georgian experts and ICOMOS Georgia protested against in vain. In 2010 the World Heritage Committee placed Bagrati Cathedral on the List of World Heritage in Danger in view of *"irreversible interventions carried out on the site as part of a major reconstruction project"* and in 2013 came to the conclusion that *"due to the inappropriate rehabilitation the authenticity of Bagrati Cathedral has been irreversibly compromised and that it no longer contributes to the justification for the criterion for which the property was inscribed"* (Fig. 7)[14].

---

[14] Thirty-Sixth Session of the UNESCO World Heritage Committee, Bagrati Cathedral and Gelati Monastery (Georgia) (C710), Decision: 37 COM 7A.32, point 4, https://whc.unesco.org/en/decisions/5009/.

4.3   The Old City of Jerusalem and Its Walls (World Heritage List - 1981)

The Old City of Jerusalem and its Walls (Fig. 8) was inscribed, as a holy city for Judaism, Christianity and Islam, on the World Heritage List in 1981; the site was proposed by Jordan taking under consideration the special status of Jerusalem (corpus separatum according to the 1947 partition plan of the United Nations).

Fig. 8. The Old City of Jerusalem and its Walls

Serious problems affecting the property and identified in reports, remain unsolved up today, concerning: natural risk factors; threats of destruction due to urban development plans, governance and management processes; alteration of the urban and social fabric; impact of persistent archaeological excavations; modification of juridical status of the property diminishing the degree of its protection; deterioration of monuments due to lack of conservation policy; significant loss of historical authenticity; traffic, access and circulation problems, as well as the disastrous impact of tourism on the protection of the monuments; tunneling, works, projects and other practices in and around the Old City of Jerusalem, which are illegal under international law and are violations which are not in conformity with the provisions of the relevant UNESCO conventions, resolutions and decisions.

Consequently, the World Heritage Committee has been further inscribed since 1982 the holy city on the List of World Heritage in Danger, recalling that the Old City of Jerusalem must be safeguarded in its entirety as a coherent whole and that the threats to any one of the elements of which it is composed endanger the property as such, as well as its authenticity and its specific character.

5   Restoration – Rehabilitation - Reconstruction of Religious Sites: Re-Writing of the History? - Innovative Restoration Practices

In 1982, the Declaration of Dresden was approved, being specially devoted to the reconstruction of monuments destroyed by war. It was based on the existing theoretical principles and stated: *"The complete reconstruction of severely damaged monuments must be regarded as an exceptional circumstance which is justified only for special reason resulting from the destruction of a monument of great significance by war. Such reconstruction must be based on reliable documentation of its condition before destruction"*[15].

The reconstruction of historical monuments reappears with particular intensity during the current period of globalization as a complex cultural phenomenon by hiring local or national ideological characteristics. Political, symbolic or economic incentives can be decisive for the monuments of cultural heritage, while claiming the authentic reconstruction based on the availability of accurate and rich documentation.

International charters (Venice Charter, 1964[16]; Nara Document on Authenticity, 1994[17]) are overcome by the reality and the fact that the conditions expressed in these as doctrines are impossible to limit the reversals of the rules established by the principles themselves. According to Venice Charter it was clear that restoration or renovation would be possible or desirable only under certain preconditions, or perhaps must be strictly rejected - that was the official doctrine defended by the official "school" of restoration till '70s. However, it was realized that in practice there were other different approaches "except rule". Nevertheless another important factor ought to be considered – the dialogue with society: we should listen to the social demand.

In 2000 "The Riga Charter on Authenticity and Historical Reconstruction in Relationship to Cultural Heritage" was approved, stated that the need for reconstruction should be established *"through full and open consultations among national and local authorities and the community concerned"*[18].

Therefore, it was concluded that the process of reconstruction is not *a priori* fatal and unacceptable, but crucial is the way in which it takes place.

"The valid contributions of all periods to the building of a monument must be respected, since unity of style is not the aim of a restoration"[19].

---

[15] Declaration of Dresden on the "Reconstruction of Monuments Destroyed by War (1982) https://www.icomos.org/en/charters-and-texts/179-articles-en-francais/ressources/charters-and-standards/184-the-declaration-of-dresden, Article 8.

[16] International Charter for the Conservation and Restoration of Monuments and Sites (The Venice Charter, 1964), https://www.icomos.org/charters/venice_e.pdf;.

[17] The Nara Document on Authenticity (1994): https://www.icomos.org/charters/nara-e.pdf.

[18] The Riga Charter on Authenticity and Historical Reconstruction in Relationship to Cultural Heritage (2000), point 6: http://www.vilagorokseg.hu/_upload/editor/UNESCO_hatteranyagok/Riga_Charter_2000.pdf.

[19] Venice Charter, 1964, article 11.

"Interpretation should explore the significance of a site in its multi-faceted historical, political, spiritual, and artistic contexts. It should consider all aspects of the site's cultural, social, and environmental significance and values. The public interpretation of a cultural heritage site should clearly distinguish and date the successive phases and influences in its evolution. The contributions of all periods to the significance of a site should be respected. Interpretation should also take into account all groups that have contributed to the historical and cultural significance of the site"[20].

Recognition of "cultural diversity" means respecting the contribution of every historical period, culture and religion, including those who are a minority in a country. However, there have been interventions for the "interpretation" and rehabilitation or rebuilding of historical monuments, which could alter the meaning or even ignore part of the monument's history.

5.1 Natural and Cultural Heritage of the Ohrid Region, FYROM (World Heritage List - 1980, Cultural + Natural Heritage)

The town of Ohrid and its historic-cultural region are located in a lakeside setting of exceptional beauty, while its architecture represents one of the best preserved and most complete ensembles of traditional urban architecture in the Balkans. By the last decades of 20th century Ohrid largely maintained the character of a typical traditional Balkan town, with important Byzantine monuments and local mansions preserved. Unfortunately, in recent times the historical urban space seems to have surrendered to an unusual "exploitation" and "transformation" which undermine those key values, just for which the city of Ohrid joined the World Heritage List.

St. Sophia cathedral church in Ohrid is the most sacred monument by the Slavs in the Balkans. It was originally an early Christian basilica, to which an exceptional exonarthex was added (beg. 14th century), while its dome collapsed in the period of Ottoman domination. Under the UNESCO initiative its southern wall was reinforced, while its magnificent wall paintings (11th century) were restored. In the sanctuary area a *minbar* was maintained, since the church was converted into a Moslem mosque, composed of parts of the original Byzantine pulpit. It was, therefore, an essential element of the diachronic history of the religious monument. After interventions of the recent years, the *minbar* was removed and replaced by a completely new marble altar screen, while a new marble floor covered the historic tiled floor.

In addition, in 893 St. Clement had raised his historic three-aisled church next to Ohrid Lake, which, thanks to his missionary work, became "the first University of the Slavs in the Balkans". Here was also the tomb of the saint. The church later acquired an addition to a cross-inscribed church. After the Ottoman conquest the ensemble was destroyed and an Imaret mosque was erected in its place. The imaret was preserved until 2000, when it was decided to be demolished and an imaginary reconstruction of St Clements's historical ensemble – a selective imitation of Byzantine churches of the wider region - to be raised in its place (Fig. 9).

---

[20] ICOMOS Charter on the Interpretation and Presentation of Cultural Heritage Sites ("Ename Charter"), 2008, Principle 3: Context and setting, par. 1-2-3.

Fig. 9. The imaginary New St Clement in Ohrid, 2000

Unfortunately, all these acts are nothing less than a falsification of the historical reality which is not possible to be understandable and recognizable by ordinary pilgrims.

5.2   Reconstruction of Mother of God of Kazan Cathedral, Republic of Tatarstan, Russian Federation (World Heritage List - 2000)

The cathedral used to host *Our Lady of Kazan*, a holy icon of the highest stature within the Russian Orthodox Church, representing the Virgin Mary as the protector and patroness of the city of Kazan, and a palladium of all of Russia. It had been erected at the place of finding the icon (in 1579) by the project of the famous architect of Moscow Ivan Starov from 1798 to 1808. To the beginning of the 20th century the historical monastery of Our Lady of Kazan turned into a big architectural complex of buildings constructed at different times. Unfortunately, a major part of the convent buildings was demolished in Soviet time. The main cathedral of the monastery was pulled down, while a six-tiered 55 m high bell-tower of the 18th century was lost.

The cathedral is closely related to the UNESCO World Heritage Site, the Historic and Architectural Complex of the Kazan Kremlin, due to their long-term bonds of spirituality and shared history. The reconstruction of the cathedral (Fig. 10) and the rehabilitation of the Virgin monastery can be thought an example of *integrated conservation,* while fulfilling the criterion of the *Authenticity of Connection*, since can be

Fig. 10. Recreating the cathedral of the Kazan Icon of the Mother of God

faithfully and precisely recreated as an expression of continuity with their pre-destruction social, environmental, and cultural conditions, by:

- Re-establishing the "dialogue" between the triumphant framework of the 19th century buildings and the medieval buildings of Kremlin; recreating the city shaping dominant of the cathedral in the city historical panorama while preserving the values of the existing dominants and thus, contributing to strengthening *the authenticity and integrity of the built environment;*
- Invigorating the long-term bonds of spirituality and shared history with Kremlin and thus *increasing its Outstanding Universal Value* as World Heritage List monument;
- Realizing an important step for the revival of spiritual and religious life of Kazan and Russia as a whole. The cathedral serves as a symbol of historical continuity of generations, revival of traditions, multi-confessional and multi-ethnic relations of people. Thus, by becoming an Orthodox religious and cultural center of the world, the rebirth of the *"genus loci"* and the historic mission of the Virgin Monastery will be achieved.
- Contributing to the continuity of *sustainable development* of the territory.

However, rehabilitation should address significant issues, like: supporting the new church construction among the remains of the original church; ensuring the conservation, presentation and visiting of the existing historical ruins in the basement (crypt); ensuring functionality of the monastery site during major religious ceremonies, which should gather hundreds of thousands of pilgrims from all Russia.

The reconstruction of the cathedral can be a challenge, on behalf of the international community, to make sure that the doctrinal texts and principles are truly universal and capable of being applied in the local contexts.

5.3   Kizhi Pogost, Russian Federation (World Heritage List - 1990)

The architectural ensemble of the Kizhi Pogost is located in the southern part of the small Kizhi Island, in the Archipelago of Lake Onega. Nowadays it is the only ensemble with two 18th-century multi-domed wooden churches preserved in Russia: the Church of the Transfiguration and the Church of the Intercession and an octagonal wooden bell tower built in 1862 and considerably reconstructed in 1874 (Fig. 11). The Church of the Transfiguration is a monument with exceptional architectural and structural features. It has no parallel in either Russian or global wooden architecture. It was used during the summer. The church, whose central cupola culminates at 37 m, is a masterpiece of a multi-storey, multi-cupola, and single-block structure. Inside, under the so-called 'heaven' - a superb vault shaped like a truncated pyramid - there is a gilded wood iconostasis holding 102 icons from the 17th and 18th centuries. The Church of the Intercession, the winter church is a simpler structure. The eight cupolas encircle the 27 m high central onion dome, which covers the central parallelepiped space, and gives it a more static appearance. The 30 meters-high bell tower is of the traditional "octagon on cube" type with a high cube. The belfry crowns the structure. It has nine posts supporting the tent roof with an onion dome covered with shingles.

Fig. 11.   Kizhi Pogost under restoration process – the dismantled 6th tier of the Transfiguration church

The Kizhi Pogost is a unique monument of Russian wooden architecture, a universally recognized masterpiece of world architecture. It is noted for the harmony of its dimensions and shapes, and the artistic unity of its structures, built at different times.

The architectural beauty of the ensemble is emphasized by the expressive landscape, which can be considered as a national landscape.

Authenticity. The Kizhi Pogost represents an important step in the establishment of Orthodoxy in the Russian North. The Kizhi Pogost is an illustration of a carpenter pushing a technique to its furthest limits. The traditional building techniques and the structural and decorative elements that have been used in Russian architecture for centuries are brilliantly and perfectly implemented in the ensemble structures. The structures have not been significantly reconstructed and have preserved a substantial part of the original elements and material. To maintain the conditions of authenticity, restoration criteria and guidelines are crucial to address the treatment of elements from different periods, of witness marks, among other issues.

Protection and management requirements. The establishment of the buffer zone for the Kizhi Pogost represents a crucial step in preserving the visual integrity of the historic landscape and ensuring the integrity of the property and its setting. Much attention is paid to establishing effective partnerships between authorities, businesses and communities, to the strategic protection of this historical landscape, to the promotion of the Kizhi Pogost as a cultural and historical destination.

The restoration of the Church of the Transfiguration, has now entered its final stages as the top tiers have been completely dismantled and transited to the restoration complex. The lessons learned in the process will be critical for addressing further specific conservation challenges. Restoration and conservation of the church interior has also been carried out on the decorated rafters of the heaven ceiling, as well as the royal doors. The very high quality of workmanship is guiding the principles of the current project for restoration, and although the timber of the churches is in a very bad condition, there is a permanent consideration as to how to minimize new interventions.

The provision of tourist services will be combined with the development of the rest of the traditional economy - cultivation of land, development of traditional handicrafts, etc., in view to create the conditions for the long-term settlement of the inhabitants of the surrounding islands, based on mixed economic sources. The State Party focus on the establishment of sufficient protection measures both within and outside the buffer zone, including land-use and non-constructive zones legislation, as well as to strictly apply fluvial regulation in order to prevent any impact on the Outstanding Universal Value of the property[21].

6   Conclusions

The increased consciousness of public authorities and local communities, expressed in the constant care for the protection, preservation and adaptation of historical religious monuments, along with innovative practices -worthy of praise- characterize our era. However, selected examples of World Heritage Christian religious monuments

---

[21] Haugen A M, Prepis A (2018) Report on the ICOMOS Advisory Mission to Kizhi Pogost, Republic of Karelia (Russian Federation, 13-17/3/2018), Paris, text pp 31 + annexes pp100 + 64 photos.

discussed above pose a serious concern whether doctrinal international texts and principles are always capable to be applied in practice and in the local context.

The Context of the Religious Heritage of the Past in the Modern World
Historical religious monuments should be treated within the frames of an integrated conservation process, which enables the continuation of the urban cultural heritage, being kept in an appropriate human setting, and particularly their assignment towards a continual process of "creation" of a new synthesis, combining their elements of authenticity and integrity, together with sustainable development of both the immediate environment of the historical monument and its wider setting.

In the process of a new urban planning project focusing on the World Heritage religious monuments all relevant stakeholders should be connected and collaborating: national – local - religious authorities, communities, investors, NGO's etc., while rising awareness of the local community on the benefits and constraints, of positive and celebratory measures, sharing an overall vision for the integrated development of the site.

Under the pressing needs of modern society or of political decisions supporting cultural, religious and ethnic identities the rehabilitation of living religious ensembles may lead to possible alterations or to jeopardize the authenticity of the monuments, even of exceptional historical significant.

Sustainable Development Problems
A significant number of living religious ensembles face growing problems from the large number of visitors which is constantly showing upward trends. There are cases that communities (monastic or civil) are attracted by the significant benefits of the tourism and retreat to increasing pressures resulting in deterioration of the historical environment and abuse of religious way of life of the sanctuaries. A management plan is always absolutely necessary to reduce the negative impact of tourism and to increase its positive results.

Restoration, Rehabilitation and Interpretation Problems
The question about the limits of reconstruction is one of the main issues, particularly in relation to the notions of authenticity and integrity: irreversible interventions or major reconstructions can lead to a loss of authenticity, to contribute to irreversibly compromising the justification for the criterion for which the property was inscribed on World Heritage List.

If the question of material reality is fundamental, it is also the symbolical meaning of the monument equally important, as well. UNESCO has pointed out that the intangible component is as important as the physical, when it has to do with the main issue concerning the Outstanding Universal Value of a monument.

Recognition of "cultural diversity" means respecting the contribution of every historical period, culture and religion, including those who are a minority in a country. The religious architectural heritage of the past should be considered an important factor for the identity of local people, which can substantially contribute to the homogenization of multicultural societies and the achievement of social peace.

On the Greatest Challenge in the Management of Living Religious Heritage: Linking the Authenticity of Heritage and the Authenticity of Tourist Experiences to the Authenticity of Religious Tradition

Ioannis Poulios[(✉)]

Hellenic Open University, Patra, Greece
jannispoulios@hotmail.com

Abstract. The greatest challenge in the operation and management of living religious heritage is considered to be the reconciliation of heritage protection, tourism development and maintenance of religious function. The key concept associated to these differing uses of heritage is the same, 'authenticity'; yet, this concept is sensed and applied in differing ways: authenticity of heritage, authenticity of tourist experiences, and authenticity of religious Tradition.

The paper explores the three different concepts of authenticity on a theoretical level. Subsequently, three case studies are discussed: the monastic site of Meteora in Greece, in which the concepts of authenticity are separated from each other; the project of the conservation and restoration of the Tomb of Christ in Jerusalem – a central theme of the present Conference –, in which the authenticity of heritage is linked to the authenticity of religious Tradition; and the Toplou Monastery in Crete, Greece, in which the authenticity of tourist experiences is linked to the authenticity of religious Tradition.

The theoretical part uses material from disciplines associated to the three concepts of authenticity: heritage conservation, business/tourism management, and theology. Regarding the case studies: Meteora is based on my Ph.D. at University College London and on subsequent research; the conservation project of the Tomb of Christ on the exhibition guide of 'The Tomb of Christ: the Monument and the Project' at the Byzantine and Christian Museum and on my personal visit to the exhibition; and Toplou Monastery on my personal visit to the Monastery.

The ultimate aim is to embrace heritage protection and tourism development within the maintenance of the religious function of heritage. To this end, the attempt is not to draw a direct link between heritage protection and tourism development – as is normally the case – but an indirect one, through the maintenance of religious function. Also, the religious communities are promoted as the community group with the highest responsibility in the operation and management of their sites: good practices are highlighted, as well as practices that should be better avoided.

Keywords: Living religious heritage · Authenticity · Heritage management · Tourism · Tradition of the Church · Meteora · Tomb of Christ · Toplou Monastery

© Springer Nature Switzerland AG 2019
A. Moropoulou et al. (Eds.): TMM_CH 2018, CCIS 962, pp. 262–271, 2019.
https://doi.org/10.1007/978-3-030-12960-6_17

1   Introduction

The greatest challenge in the operation and management of living religious heritage is considered to be the reconciliation of heritage protection, tourism development and maintenance of religious function [21, 23]. The key concept associated to these three differing uses of heritage is the same, authenticity; yet, this concept is sensed and applied in differing ways: authenticity of heritage, authenticity of tourist experiences, and authenticity of religious Tradition.

2   The Three Concepts of 'Authenticity': Authenticity of Heritage, Authenticity of Tourist Experiences, and Authenticity of Religious Tradition

2.1   Authenticity of Heritage

Authenticity emerged as the key concept of heritage conservation internationally, with the adoption of the Venice Charter in 1964 and especially the UNESCO Convention Concerning the Protection of the World Cultural and Natural Heritage (the World Heritage Convention) in 1972 and the accompanying Operational Guidelines. In the context of the World Heritage Convention, authenticity may be seen as an 'effort to ensure that those values are credibly or genuinely expressed by the attributes that carry those values' [20].

Authenticity is essentially a product of Western European cultural history [8, 12, 22], and is rooted in a feeling of dissatisfaction with the present caused by the rapid change and mobility experienced by the Western world in the last centuries. In this rapidly changing reality, the past affords a comfortable and controllable context, and is thus seen in a nostalgic way. In this context, the discipline of heritage conservation has as its fundamental objective the preservation of physical heritage of the past (with an emphasis on material remains) from loss and depletion in the present [16]. A notion of discontinuity is thus imposed between the monuments, considered to belong to the past, and the people and the social and cultural processes of the present/future [9]. The role of the protection of the physical heritage is assigned to specially trained conservation professionals, i.e. archaeologists and conservators, the so-called 'experts', while local communities and religious communities are given a clearly secondary role, if any at all.

Despite attempts to expand the concept of authenticity, such as the adoption of the Nara Document on Authenticity in 1994 and the Nara+20 Document in 2004, authenticity is still attached to the discontinuity between the monuments of the past and the people of the present, the preservation of the material/fabric of the monuments and the power of the conservation professionals at the expense of the local and religious communities [17].

The attitude of conservation professionals against the local communities and the religious communities can be demonstrated in the cases of the World Heritage Sites of the Great Zimbabwe [14] and Angkor in Cambodia [13].

2.2 Authenticity of Tourist Experiences

'Experience' is a key concept in the tourism and the entertainment industry [6], and recently in the cultural and creative industries as well [18]. 'Experience', differentiated from 'service', is a personal, particularly strong connection, based on emotions and imprinted in memory, that the company develops with its customers [6, 19]. Thanks to the experience, the loyalty of the customer to the company is enhanced and thus the customer becomes a 'friend' of the company. Experience comprises various services that contain personal, innovative elements in the points of contact of the company with the customer and that are connected to each other in a unified context, with unified objectives [25].

Authenticity has emerged as a key component of the 'experience model'. The 'experience', as described above, targets all customers. At the same time, however, individual, 'authentic' experiences are designed for different customer segments, so that the customers develop an even more personal and stronger, an intimate connection with the company [7].

A characteristic example of a cultural organisation that, as explicitly stated by its director, applies the experience model, centred on authenticity, is the Cerritos Public Library in the State of California, US. Cerritos Library does not simply offer books, book services and programming ('service'), but also 'quiet areas for study and contemplation as well as lively areas where the imagination could run wild' ('experience') (Waynn Pearson cited in [11]; see also [3]). At the same time, individual, 'authentic' experiences are designed for different customer/user segments: for example, the 'Study Room of the Old World', which is decorated with old furniture and a fireplace, targets the older users, while the 'Children's Library', which is equipped with statues of dinosaurs and an aquarium, is for the children [10]. Another example of a cultural organisation that introduces elements of the experience model, centred on authenticity, is the Acropolis Museum in Athens, Greece [18]. A strongest element of experience is the visual connection between the Parthenon sculptures (exhibited at the upper level of the Museum) with the Parthenon Temple on the Acropolis Site. At the same time, individual, 'authentic' experiences are designed for different customer/visitor segments, such as the educational programmes on goddess Athena addressed to the children.

2.3 Authenticity of Religious Tradition

The concept of authenticity in the context of the Orthodox Church is linked to that of Tradition. Tradition means any teaching or practice that has been transmitted from generation to generation throughout the life of the Church; it is 'the very life of the Holy Trinity as it has been revealed by Christ Himself and testified by the Holy Spirit' [1, 5]. Church is considered a community of saints operating on the basis of Tradition: saints are the authentic, the real Christians, 'the living examples of authenticity', they 'become Tradition themselves' and are 'sons of God by the grace' [4, 24].

Tradition defines the Church as a whole, including the Holy Scripture, the writings of the Holy Fathers, the decisions of Ecumenical and local Councils, the administration, the liturgical life, and the art of the Church in all its expressions such as architecture, sculpture, painting, poetry and music.

The worship, i.e. the Holy Liturgy, is the most significant aspect, the core of the Tradition, since it unifies the faithful with Christ [15]. It is the Holy Liturgy that gives meaning to all the other elements of the Tradition of the Church. All the aforementioned elements of the Tradition, including art, are purely functional, acquiring their existence and meaning serving the worship of God [24].

In this context, the primary aim of the religious communities is to maintain the function of their churches and monasteries as places of worship. The arrangement and use of space in churches and monasteries is centred on the central church building, where worship is conducted.

3 Separating the Authenticity of Heritage and the Authenticity of Tourist Experiences from the Authenticity of Religion Tradition: A Case Study of Meteora, Greece

Meteora, a World Heritage site, comprises monasteries built on top of high rocks – the term 'Meteora' means 'floating in the area'.

Meteora can be seen as an example in which the concepts of authenticity became separated from each other over the course of time [16]. Specifically, since the construction of the monasteries in various periods between the 11th and the 15th century until approximately World War II, the artistic appreciation of the monastery architecture and art was inevitably attached to their monastic function, and the visitors were attracted to the site for exclusively religious purposes, as pilgrims.

Similarly, the 1960s was a period of few visitors in the site, before the establishment of an organised tourist system. In this period, Meteora functioned primarily as a monastic site. The local community was involved in the ritual life of the site, comprising the congregation of the monasteries, which means that at that time the interest in the authenticity of tourist experiences was embraced within the authenticity of religious Tradition. The monastic communities and the local community, with the support of the official Church, attempted to protect the material of the site, something that indicates that the interest in the authenticity of heritage was embraced within the authenticity of religious Tradition.

The situation changed in the 1970s and the early 1980s, with an increase in the number of visitors in the site and the development of state-sponsored organised tourism that served primarily non-religious purposes. In this period, the monastic communities were primarily concerned about the financial gains derived from tourism and did not actively encourage the visitors to participate in the ritual life of the site, while the local community started to be less involved in the ritual life of the site, as the congregation of the monasteries, and increasingly involved in tourism. This means that the authenticity of tourist experiences started to evolve separately from the authenticity of religious

Tradition. Also, at that time, the State became increasingly concerned about the protection of the material of the site mainly as a means to serve and promote tourism, with the consent of the monastic communities that did not have to pay for the protection needs. This shows that the authenticity of heritage started to evolve separately from the authenticity of religious Tradition, and in connection to the authenticity of tourist experiences.

The mid-1980s, especially the mid-1990s, to present is the period of the development of mass tourism industry on the site. Mass tourism has had huge implications for the site and for the broader region. The monastic communities became even more actively concerned about the financial benefits derived from it. Elements of the local community became clearly interested in the tourism industry, ceasing to constitute the congregation of the monasteries. Also, at that time the State established the heritage significance of the site at an international level, by promoting the site for World Heritage inscription, and linked the inscription to the promotion of tourism at the area. The World Heritage inscription process was carried out without the involvement of the monastic communities. Therefore, the site remained a monastic one, operating on the basis of the authenticity of religious Tradition, but the authenticity of heritage and the authenticity of tourist experiences were developed and established clearly separately from the authenticity of religious Tradition, with the acquiescence and even the encouragement of the monastic communities. Today, the operation of the site has become formalised as follows, responding mostly to tourism needs: the monasteries are mainly occupied by the visitors from ca 9 in the morning to ca 5 in the afternoon (possibly with a small break); outside these hours the monasteries are exclusively used by the monastic communities. Furthermore, it is important to stress that the entrance fees go to the monastic communities and not to the state/the Ministry of Culture; the monastic communities also make money through pilgrim donations and through their museum shops.

Tourism has become the decisive factor for the operation and use of the site, often at the expense of the monastic function and the heritage protection. Specifically, the monastic communities find it hard to conduct worship (the core of the Tradition of the Church: see above) in the monastery space that is occupied by tourists, i.e. in the *katholicon* and the refectory, and thus feel the need to construct new space for the conduct of worship, separated from tourists. Yet, such construction works are illegal given the limitations imposed by the national and World Heritage status of the site. However, the monastic communities proceed with construction works in an authorised way (i.e. without the agreement of the state authorities) and through their own financial resources gained through tourism. A most characteristic example to this end is the unauthorised five-storey wing/building and the unauthorised two-storey building in the Roussanou Monastery [16]. Still, despite such construction works at the site, the monastic communities often find it hard to conduct worship because of tourism, and thus feel the need to leave the site other monastic areas.

4 Linking the Authenticity of Heritage to the Authenticity of Religious Tradition: A Case Study of the Restoration and Conservation of the Tomb of Christ in Jerusalem

In the project of the restoration and conservation of the Tomb of Christ in the World Heritage city of Jerusalem, there was a consistent attempt to link the authenticity of heritage to the authenticity of religious Tradition, as demonstrated in: (a) the management of the project; (b) the technical operation of the project; and (c) the communication of the project to the general public e.g. through the exhibition entitled 'The Tomb of Christ: the Monument and the Project' hosted first by the National Geographic Museum in Washington DC and then by the Byzantine and Christian Museum in Athens.

In terms of management, the project was conducted upon the initiative, under the supervision and to some extent with the financial contribution of the three Christian communities, the Guardians of the Holy Tomb, who are considered to be the bearers of the authenticity of the religious Tradition in connection to the Holy Tomb: i.e. the Greek-Orthodox Patriarchate of Jerusalem, the Franciscan Order in the Holy Land, and the Armenian Patriarchate of Jerusalem. The National Technical University of Athens Interdisciplinary Team, i.e. those responsible for the authenticity of heritage of the Tomb, run the project in close cooperation with the three Christian communities. Representatives, namely the leaders, of the three Christian communities were physically present at key moments of the project, as for instance at the opening of the Tomb of Christ for the first time after five hundred years.

In terms of the technical operation of the project, the National Technical University of Athens Interdisciplinary Team, through their intervention on the physical heritage (the material) of the monument, gave emphasis on, and actually served, the religious significance and function of the Tomb of Christ. Consequently, the authenticity of heritage was highlighted in connection to, and embraced within, the authenticity of religious Tradition. It is worth noting to this end that, as noted in the associated exhibition in the Byzantine and Christian Museum (see also below), the National Technical University of Athens Interdisciplinary Team chose not to imprint their names in any way on the monument itself, so that they do not divert the focus from the religious significance of the Tomb (author's personal remark on the Byzantine and Christian Museum exhibition).

The exhibition, curated and organised by the National Technical University of Athens Interdisciplinary Team in collaboration with the National Geographic Museum and the Byzantine and Christian Museum, was made possible thanks to the blessings of the three Christian communities. The exhibition directly linked the restoration and conservation project (the authenticity of heritage) to the religious significance and function of the Tomb of Christ (the authenticity of religious Tradition). This was achieved in a variety of ways. First, the exhibition made strong statements on the religious significance and function of the Tomb of Christ highlighting the authenticity of religious Tradition. Characteristic examples are as follows:

'THE HISTORY OF THE CHURCH OF RESURRECTION begins with the Crucifixion and Resurrection of Jesus Christ' ([2], p. 8).

'A monument of the utmost importance to Christianity' (Charis Mouzakis, member of the Interdisciplinary Team, cited in [2], p. 11).

'When, after five centuries, on 26 October 2016, we opened the Tomb of Christ and National Geographic transmitted the news and the image to the world, over two billion people kneeled with us, in spirit, before it. The Tomb of Christ is alive for all humanity' ([2], p. 12).

Second, the exhibition emphasised the celebration of the Resurrection that takes place every Easter Sunday in the Church of Resurrection in Jerusalem (where the Tomb of Christ is located), with the miraculous (with or without quotation marks) transmission of the Holy Light – which is considered a most remarkable, as well as visible, sign of the living presence of the Holy Spirit (the authenticity of the religious Tradition) throughout the history of the Church to the present. The celebration of the Resurrection, with the transmission of the Holy Light, is portrayed in the video presentations of the exhibition (author's personal remark on the Byzantine and Christian Museum exhibition). Furthermore, as it was characteristically noted,

'The Holy Light, which the Patriarch of Jerusalem transmits on Holy Saturday, illuminates their coexistence, which, more than two centuries later, was expressed through their [the Christian communities'] common agreement regarding the rehabilitation project of the Holy Aedicule of the Holy Sepulchre' ([2], p. 7).

Third, the exhibition makes a clear statement that the findings that arose throughout the project and especially after the opening of the Tomb confirm the accounts of the New Testament on the Crucifixion and the Resurrection of Christ, in the context of the authenticity of the religious Tradition. A strong example to this end:

'The project has finished. Research continues. Historians, Archeologists, theologians, sociologists from around the world, based on the data of the project, will have a lot to say in the future about the values and the history of the Holy Aedicule and the Tomb of Christ. We have highlighted its values and we have scientifically confirmed its history' ([2], p. 17).

Fourth, the exhibition under discussion was connected with another temporary exhibition run at the same period in the Byzantine and Christian Museum, on an international competition of contemporary icons on the theme of the Resurrection of Christ, crafted in a variety of countries such as Latvia, Ukraine, Romania, the Northern Republic of Macedonia/Skopje, Poland, Russia, Serbia and Greece. In fact, the exhibition on the icons of the Resurrection was displayed at the entrance that led the visitors to the exhibition under discussion, serving in a way as an introduction to the exhibition (author's personal remark on the Byzantine and Christian Museum exhibitions). Through the connection of the two exhibitions, the relevance of the significance of the Resurrection of Christ to the contemporary world was highlighted.

5 Linking the Authenticity of Tourist Experiences to the Authenticity of Religious Tradition: A Case Study of the Toplou Monastery in Crete, Greece

Toplou Monastery in Crete offers individual, 'authentic' experiences, centred on the significance and function of the Monastery, designed for different customer segments. Indicative examples of activities to this end, developed upon the initiative and under the supervision of the Toplou monastic community, are the following:

The space is arranged in such a way that a variety of modern-day visitor facilities are provided that are centred around the *katholicon* of the monastery: an ecclesiastical museum, which was developed in cooperation with the local Antiquities Service; a small café; an olive oil factory; a winery; and a sales point for the products of the monastery, mostly olive oil and wine. The Monastery also offers the visitors the opportunity to stay as guests and participate in the monastic life.

The Monastery is active in the promotion and exportation of olive oil to foreign markets all over Europe, with tailor-made promotion activities e.g. to the German-speaking market (pers. com. Elena Paschinger, tourism communications specialist and blogger).

The Monastery – in cooperation with local partners such as the local Bishopric and the Technological Education Institute of Siteia on Eastern Crete, and taking advantage of nearby mature tourist destinations such as Vai Beach, the town of Siteia and Siteia Geopark – links religious tourism to other types of tourism such as culinary tourism, and is working towards the development of cultural routes.

6 Conclusion

For the reconciliation of heritage protection, tourism development and maintenance of religious function, i.e. for the linking of the three authenticities, it is most important to examine the way each of the authenticities evolve and the relationship among them, as well as the approach of the religious communities to this evolving relationship. Specifically:

As the case studies of the Meteora and the Toplou Monasteries demonstrate, the most decisive factor in the operation and management of living religious heritage is usually tourism development. As a general rule, tourism does not emerge as a result of the activity or the interests of the religious communities of the sites, but is the result of broader, global changes supported by government authorities. In the majority of living religious heritage sites, however, the religious communities tend to accept tourism. It is, therefore, important to study at which scale the religious communities accept tourism in relation to the other two authenticities. To this end, Meteora monastic communities, on the one hand, gave emphasis on tourism development, and did not encourage the participation of the visitors in the conduct of worship at their monasteries (i.e. did not embrace the interest in tourist experiences within the authenticity of the religious Tradition) and in the long term found it difficult also for themselves to conduct worship (i.e. to continue their own connection to the authenticity of the religious Tradition), seeking alternative, new space to construct within the site of Meteora – and even seeking space outside the site. Toplou monastic community, on the other hand, encouraged the participation of the visitors in the conduct of worship at their monastery (i.e. embraced visitors within the authenticity of the religious Tradition).

The Toplou Monastery and the project on the restoration and conservation of the Tomb of Christ give some suggestions of broader applicability on the approach of the religious communities towards the evolving relationship of the three authenticities. First, in any heritage protection or tourism development project, the primary aim should be the maintenance of the religious function of the site through the conduct of worship (i.e. the maintenance of the authenticity of religious Tradition). Thus, the attempt should not be to draw a direct link between heritage protection and tourism development– as is normally the case–, but an indirect one, i.e. both heritage protection and tourism development through the maintenance of religious function. Second, the management leadership of any heritage protection or tourism development project should be in the hands of, and under the continual supervision of, the religious communities. Third, partnerships for the implementation of the project, as well as the communication of the project to the broader public, should serve the aforementioned primary aim (i.e. the emphasis on the authenticity of the Tradition) and should be built under the aforementioned management scheme (i.e. under the leadership of the religious communities).

References

1. Bebis, G.: Tradition in the Orthodox Church (2014). http://www.goarch.org/ourfaith/ourfaith7116. Accessed 7 Jan 2014
2. Byzantine and Christian Museum 2017: *The Tomb of Christ: The Monument and the Project* (Digital Exhibition of Advanced Technology). Exhibition guide of the homonymous exhibition. https://tmm-ch2018.com/files/booklet.EN.pdf. Accessed 24 Aug 2018
3. Cerritos Library 2015: About the Library Today. http://menu.ci.cerritos.ca.us/cl_about.htm. Accessed 19 Apr 2012
4. Damianos (Archbishop of Sina, Fara and Raitho): Ορθοδοξία και Παράδοση/The Orthodox Church and the Tradition. In: Vlachos, I. (ed.) *Η Αποκάλυψη του Θεού* («κατά τας των αγίων θεοπνεύστους θεολογίας και το της Εκκλησίας ευσεβές φρόνημα»)/*the Revelation of God*, pp. 161–166. Holy Monastery of the Birth of the Virgin, Leivadia (1987)
5. Florovsky, G.: Γρηγόριος ο Παλαμάς και η Πατερική Παράδοση/St Gregory Palamas and the Tradition of the Fathers. In: Christou, P. (ed.) Πανηγυρικός τόμος εορτασμού της εξακοσιοστής επετείου του θανάτου του Αγίου Γρηγορίου του Παλαμά Αρχιεπισκόπου Θεσσαλονίκης: 1359–1959, 240–254, p. 241. M. Triantafyllou, Thessaloniki (1960)
6. Gilmore, J., Pine, J.: The Experience Economy: Work is Theatre & Every Business a Stage. Harvard University Press, Boston (1999)
7. Gilmore, J., Pine, J.: Authenticity: What Consumers Really Want. Harvard University Press, Boston (2007)
8. Jokilehto, J.: Authenticity: a general framework for the concept. In: Larsen, K.E. (ed.) Nara Conference on Authenticity in Relation to the World Heritage Convention, Proceedings, 17–34, Nara, Japan, 1–6 November 1994, pp. 18–29. UNESCO World Heritage Centre, Paris (1995)

9. Jones, S.: 'They made it a living thing didn't they...': the growth of things and the fossilization of heritage. In: Layton, R., Shennan, S., Stone, P. (eds.) A Future for Archaeology: The Past in the Present, 107–126, p. 122. UCL Press, London (2006)

10. Kourgiannidis, L.: Σχεδιάζοντας «Εμπειρίες» στις Βιβλιοθήκες: το Παράδειγμα της Δημόσιας Βιβλιοθήκης Cerritos, Η.Π.Α./Designing 'Experiences' in Libraries: a case study of the Cerritos Public Library, USA. Dissertation Hellenic Open University, Patra (2011)

11. Library Journal 2003. Selling the Learning Experience. http://lj.libraryjournal.com/2003/03/people/movers-shakers-2003/waynn-pearson-movers-shakers-2003/. Accessed 20 Apr 2012

12. Lowenthal, D.: Changing criteria for authenticity. In: Larsen, K.E. (ed.) Nara Conference on Authenticity in Relation to the World Heritage Convention: Proceedings, 121–136, Nara, Japan, 1–6 November 1994, pp. 125–127. UNESCO World Heritage Centre, Paris (1995)

13. Miura, K.: Conservation of a 'living heritage site'- a contradiction in terms? A case study of Angkor World Heritage Site. Conserv. Manage. Archaeol. Sites 7(1), 3–18 (2005)

14. Ndoro, W.: Your Monument Our Shrine: The Preservation of Great Zimbabwe. Department of Archaeology and Ancient History, Uppsala University, Uppsala (2001)

15. Papadopoulos, S.: Ο χρόνος/The Time. ETBA (Greek Bank of Industrial Development/Ελληνική Τράπεζα Βιομηχανικής Αναπτύξεως), Simonopetra, 30–45, pp. 44–45. ETBA, Athens (1991)

16. Poulios, I.: The Past in the Present: A Living Heritage Approach – Meteora, Greece. Ubiquity Press, London (2014). Open Access http://dx.doi.org/10.5334/bak

17. Poulios, I.: Gazing at the 'blue ocean', and tapping into the mental models of conservation: reflections on the Nara+20 document. Heritage Society 8(2), 158–177 (2015)

18. Poulios, I., Nastou, D., Kourgiannidis, E.: Bridging the distance between heritage conservation and business management: heritage as a customer 'experience' – a case study of the Acropolis Museum in Athens. In: Karachalis, N., Poulios, I. (eds.) Athens, Modern Capital and Historic City: Challenges for Heritage Management at Times of Crisis, PHAROS (Netherlands Institute of Athens), vol. XXI.1, pp. 97–122 (2015)

19. Schmitt, B.H.: Experimental Marketing: How to Get Customers to Sense, Feel, Think, Act, and Relate to Your Company and Brands, New York, p. 25 (1999)

20. Stovel, H.: The world heritage convention and the convention for intangible cultural heritage: implications for protection of living heritage at the local level. In: The Japan Foundation, Utaki in Okinawa and Sacred Spaces in Asia: Community Development and Cultural Heritage, 23–28 March 2004, pp. 129–135. The Japan Foundation, Tokyo (2004)

21. Stovel, H., Stanley-Price, N., Killick, R. (eds.) Conservation of Living Religious Heritage. Papers from the ICCROM 2003 Forum on Living Religious Heritage: Conserving the Sacred, p. 131. ICCROM, Rome (2005)

22. Titchen, S.M.: On the construction of 'outstanding universal value'. some comments on the implementation of the 1972 UNESCO world heritage convention. Conserv. Manage. Archaeol. Sites 1(4), 235–242 (1996)

23. Ucko, P.J.: Foreword. In: Carmichael, D.L., Hubert, J., Reeves, B., Schanche, A. (eds.) Sacred Sites, Sacred Places, pp. xiii–xxiii. Routledge, London (1994)

24. Vlachos, I.: Η Αποκάλυψη του Θεού («κατά τας των αγίων θεοπνεύστους θεολογίας και το της Εκκλησίας ευσεβές φρόνημα»)/The Revelation of God, pp. 167–168. Holy Monastery of the Birth of the Virgin, Leivadia (1987)

25. Voss, C., Zomerdijk, L.: Innovation in experiential services - an empirical view. In: Innovation in Services, pp. 97–134. DTI, London (2007). http://www.dti.gov.uk/files/file39965.pdf. Accessed 10 June 2012

Reuse, Circular Economy and Social Participation as a Leverage for the Sustainable Preservation and Management of Historic Cities

·

Towards a New Heritage Financing Tool for Sustainable Development

Bonnie Burnham[(⊠)] [iD]

Cultural Heritage Investment Alliance, New York, NY, USA
bonnie@bburnham.us

Abstract. The man-made world is undergoing dramatic transformations through globalization and the migration of rural populations to urban areas. All our cities, towns and human settlements will be impacted by this change.

Despite the acknowledgement of the issue and the mandate to conserve cultural and natural heritage as point 11.4 of the UN Sustainable Development Goals, governmental and philanthropic support have not been significant enough to implement a global conservation strategy on the scale needed. Local heritage organizations, whether public sector or non-profit, are chronically short of funds and have few prospects of gaining support from outside sources. In order to be successful on a larger scale, heritage preservation must be coupled with planning and development to support community growth and the mission of sustainable development.

This paper explores how a cultural heritage investment fund could work globally to attract funds from lenders and investors, and re-invest those funds in community revitalization with heritage as its centerpiece. Three urban case histories illustrate how financing can catalyze the mutually reinforcing benefits of heritage preservation and the goals of social inclusion, environmental conservation, and the creation of circular economies.

Keywords: Investment · Heritage · Revolving fund

1 A Changing Perception of Heritage

"Let us, while waiting for new monuments, preserve the ancient monuments." *Victor Hugo*

Concern for the survival and safeguarding of heritage became a part of the public ethos in the 19th century in the Western world, with the growth of museums and the heritage advocacy of John Ruskin and William Morris in England, Alois Riegl in Austria, and Victor Hugo in France. Since that time, the built heritage has been seen as a pillar of memory and identify, a tangible intangible, whose greatest value is its very existence, a part of the commons shared by all. In economic terms, heritage, symbolizing humanity's collective accomplishment, has been viewed as a fiscal liability to be borne by the State, incapable of generating economic output and requiring maintenance.

This notion has begun to change in recent decades as the economic outputs from activities associated with heritage places have become more widely recognized, and the multiple values associated with heritage – tangible and intangible, and including use

© Springer Nature Switzerland AG 2019
A. Moropoulou et al. (Eds.): TMM_CH 2018, CCIS 962, pp. 275–288, 2019.
https://doi.org/10.1007/978-3-030-12960-6_18

value – have been quantified. A large body of analysis is dedicated to defining and measuring the multiple values of heritage – existence value, use value, and intangible value [1–3].

Today most communities value their heritage as an economic generator as well as a social anchor that reinforces identity, contributes to community wellbeing, and makes a given environment more attractive and competitive as a place to live. Heritage sites generate knowledge, social opportunities, vibrant creative cultures, and positive environmental investments associated with recreation, visitation and enjoyment. All of these activities and outputs, or externalities, are economic generators. The challenge has been to "internalize the externalities," to capture the resources generated through the process of safeguarding heritage in order to make it sustainable.

A new use imperative was added to the functions of heritage with the adoption of the 2015 UN Sustainable Development Goals. Within goal 11: Make Cities and Human Settlements Inclusive, Safe, Resilient and Sustainable, Goal 11.4 mandates governments to "strengthen efforts to protect and safeguard the world's cultural and natural heritage." [4] But there is no specific strategy associated with this goal. In fact, the inclusion of heritage within the UN SDGs also adds another layer of responsibility for the use of heritage sites, buildings and properties to produce social benefits as well as to serve the interests of environmental sustainability. With the 2015 SDGs, heritage becomes a key strategic anchor for the sustainability of the planet and its people. With this responsibility comes a new opportunity to work holistically, across a range of values and goals, rather than vertically, through governmental financing mechanisms supplemented occasionally by philanthropy – to safeguard and utilize heritage as an asset for development at the level of communities large and small (Fig. 1).

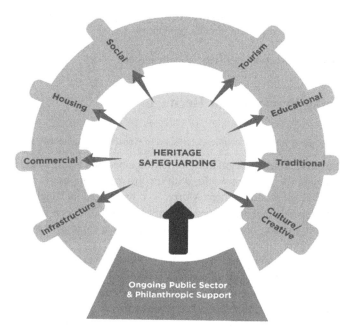

Fig. 1. Activities associated with heritage use generate revenue, but this revenue is generally captured by upstream and downstream actors, and rarely give created value back to heritage [5].

2   The City Problem: A Crisis of Competing Objectives and Clashing Dynamics

"Rapid economic and institutional transformation subjects the built environment to varying degrees of risk.... National development policies focused on economic issues do not adequately support conservation objectives and may even clash with them, while the dynamics of real estate markets reinforce disparities in valuation between the old and the new. They create situations in which the value of land in accessible sites is depressed by the condition or present use of historic buildings on the land."

Mona Serageldin in Values and Heritage Conservation [6]

Half of the world's people live in cities today, and by 2050 three-quarters of the world's population will be urban. All our cities, towns, and human settlements are being impacted by this change. Against this backdrop, the traditional scale, technology and use of buildings are disappearing at an unprecedented pace. This represents not only a tragic loss of human legacy and diversity, but also a squandering of embodied energy, materials, and resources that could be productive contributors to sustainable development. To reverse this trend, an ecosystem consistent with the multiple values and potential uses associated with heritage must be constructed. With heritage resources at the center, this ecosystem is capable of contributing to the goal of safeguarding assets while contributing to a circular economy (Fig. 2).

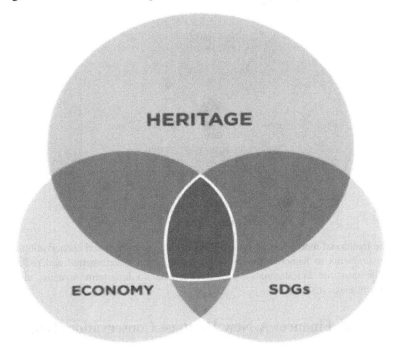

Fig. 2. The highlighted area of intersection, between designated heritage sites and properties, local economy, and the achievement of sustainable development goals, represents the ecosystem where planning and development offers opportunities to engage public and private finance in a collaborative effort.

3   Heritage Financing

The traditional model of heritage financing brings together sizeable investments by public authorities and private financing provided through interested philanthropists and corporate sponsors in a typical ratio of two to one. Decisions concerning which heritage properties to conserve in are made by public authorities on the basis of significance, public use potential, donor interest, and budget (Fig. 3).

With new demands on governments to respond in crisis mode to the needs for services to support a growing world population, and to the impacts of increasing catastrophes provoked by climate change and resource depletion, the pool of public funding for cultural amenities is shrinking. Preserving the human legacy without a new financial model is an impossible task – the more so in emerging and rapidly growing economies where the multiple values to be realized through conservation are in conflict with the short-term economic return of new development.

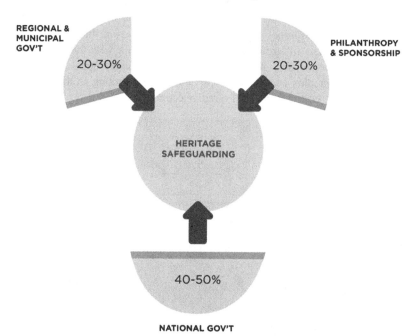

Fig. 3. The traditional model for the financing of heritage safeguard and conservation relies on national governments to launch a process of engaging local governmental and philanthropic partners. The one-time investment of capital generates no long-term revenue, although it produces positive externalities.

4   Innovation Finance: A New Heritage Conservation Tool

The use and value of assets is determined by a supply and demand chain. Public strategies grow out of an understanding of this, and incentives and tax systems are created to control and balance the supply and demand for goods and experiences. The economy of heritage management is unstable and unpredictable. Despite high demand it requires subsidy, and the availability of resources is difficult to predict. The goal of innovation finance for heritage is to transform this unstable economy into a sustainable one. To achieve this goal will require a change of perception from seeing heritage as only having iconic, embodied value to perceiving it as the core of an ecosystem benefitting the whole society. There is a range of possibilities to develop financial products to implement this change. These methodologies have not been applied on a large scale. However, there are enough well established local and regional precedents to justify the concept of creating a system that:

- broadens the uses of heritage within a community framework,
- leverages complementary development,
- associates all stakeholders in the expected benefits that will accrue over a long period of time, and
- internalizes the financial effects created by the change agents to achieve sustainability.

A range of non-profit organizations working at a local or national scale offer innovation finance for heritage sites and for environmental reclamation, which demonstrate the viability of this tool. None has the capability or ambition, however, to work at a global scale (Fig. 4).

Precedent Funds & Entities

Strategic European Loan Fund

Revolving fund focused on properties restituted to Eastern European Jewish communities.

Restauratiefonds.

National Restoration Fund

Set of revolving funds that provide first mortgages and other types of financing to historic buildings in the Netherlands.

THE NEW YORK LANDMARKS CONSERVANCY

Historic Properties Fund

Makes loans to assist in the restoration of landmarked properties in New York City, with expertise in residences and community facilities.

 The Architectural Heritage Fund

Provides grants and loans from revolving fund to projects in planning, acquisition, and pre-development stages in the United Kingdom.

 Ginkgo

Leveraged impact investment fund that supports brownfield redevelopment of formerly publicly-owned land in France and Belgium.

 National Trust Community Investment Corporation

National Trust Historic Real Estate Fund

Makes subordinate and equity investments in high-impact historic rehabilitation projects in the United States.

Fig. 4. Numerous local and national organizations operate funds that provide pre-development planning, capacity building, and loan funding for heritage properties. These organizations demonstrate the viability of a reimbursable financing model to catalyze investments from the public and private sector.

5   The Cultural Heritage Investment Alliance

This paper describes the creation of an international financing vehicle with the purpose of catalyzing the opportunities to engage a range of investment vehicles in projects to develop heritage buildings, districts, areas and cities as sustainable ecosystems to support present and future communities.

The Cultural Heritage Investment Alliance (CHIA) is a newly established organization that seeks to invest in historic assets to generate economic and social benefits for communities worldwide. It is in formation as a US-based non-profit organization, led by an experienced team of specialists in architecture and planning, cultural affairs, finance and heritage conservation. Initial capitalization and project mobilization are underway. Once capitalized, CHIA will establish subsidiary operating companies as needed to implement projects around the globe.

An initial capitalization of $60 million, including $20 million in grants for pre-development planning, is the goal for the first 10-year operational phase of CHIA. This capital will be invested in approximately 40 high-impact projects, which will be completed during the blueprint phase. Under conservative assumptions, CHIA's finance fund could repay investors their $40 million of principal at the end of ten years while making current interest payments.

CHIA offers investors the opportunity not only to help save the world's treasured places, but strengthen community cohesion, but also create new livelihoods and catalyze economic growth and broadly sustainable development.

6   Financial Products

Four financial products will be offered during the initial phase of CHIA operations:

6.1   Bridge Loans

A loan to provide cash flow and accelerate activity in anticipation of pre-committed take-out by a public and/or philanthropic source. These types of investments are good matches for typical grant-funded cultural heritage restoration projects focused on individual buildings of high significance. The principal impact is the public cultural use and enjoyment of a resource that has previously been neglected. Bridge loans can also be used to address emergency situations in the aftermath of a catastrophe, to prevent further deterioration while public funding commitments are consolidated.

6.2   Term Loans

A term or "semi-permanent" (3–10 year repayment period) self-amortizing loan that enables development and rehabilitation projects to move forward. CHIA will be one of the few lenders in this space, making it a highly differentiated and sought-after product. These loans would seek to integrate historic assets into development projects and catalyze community engage, job creation, sustainable tourism, and local economic growth.

6.3   Pre-development Financing

A single or multi-phase commitment to a sponsor/partner organization to plan for work and prepare for implementation. These may be structured as a "recoverable grant" type of investment, in which the loan is essentially a convertible note with no restriction on expiration or liquidation. This financing will be coupled with technical assistance to bring projects to point where they can are eligible for term loan financing, and will be intended to attract other financial partners.

6.4   Country-Dedicated Loan Facilities

A flexible facility intended to serve multiple projects in a single geography. This could be a single debt facility that is drawn on by a project sponsor for multiple projects, or a fund that has multiple users. This vehicle would potentially address capital control and legal infrastructure issues in emerging markets (Fig. 5).

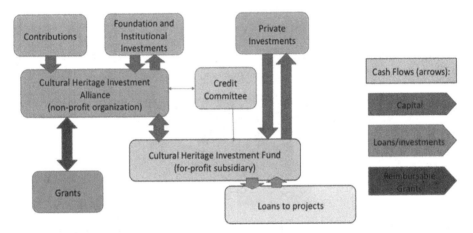

Fig. 5. Financed through investment, loans and contributions from a range of sources, CHIA will reinvest its resources in aspects of projects that are not traditionally supported with government heritage funding, such as affordable housing, affordable housing; the creation of maker spaces that attract innovative entrepreneurs, creative industries and community use; education and training facilities; and visitor amenities, that have a capacity to produce a modest return on investment.

6.5   Other Models

There are other ways in which innovation financing can be introduced into this marketplace, including crowd funding, social impact bonds issued by municipalities, and direct equity investment. These categories of funding are compatible with the debt financing proposed above, and may be added to CHIA's financing offerings in the future.

The chart below shows the use of innovation financing to create a partnership or trust that will in turn capture more of the resources generated as a result of safeguarding heritage. This "internalization of externalities" will create a capital stream to provide for long-term sustainability and leverage up to 10 times the direct contribution of public and philanthropic donors (Fig. 6).

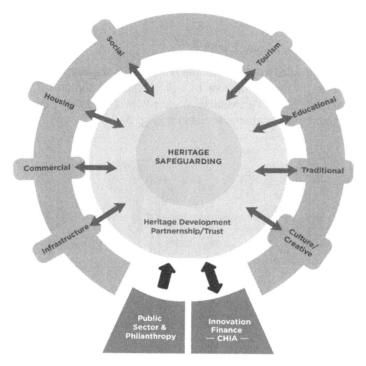

Fig. 6. Catalytic capital to complement public and philanthropic funding will support revenue-generating activities that help heritage sites to become and remain economically sustainable.

7   Case Studies

Three Examples of how heritage projects with an integrated vision could be orchestrated through a collaboration between government agencies, municipalities, investors, concessionaires, and property owners, catalyzed with innovation financing and orchestrated by an independent project agency that ensured the internalization of revenues adequate to achieve sustainability (Fig. 7).

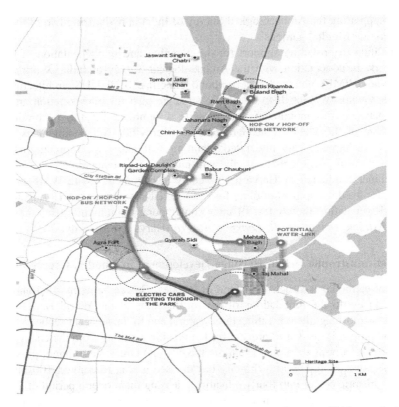

Fig. 7. Agra Behond the Taj Mahal intervention area. Photo Courtesy Skidmore Owings & Merrill, New York

7.1   Agra Beyond the Taj Mahal

In December 2017 the World Bank and the State of Uttar Pradesh signed an agreement for a $40 million loan, with a 19 year maturity and a $57 million overall cost. Per the World Bank website, "The development objective of the Uttar Pradesh Pro-Poor Tourism Development Project for India is to increase tourism-related benefits for local communities in targeted destinations…. In Agra, the project Agra Beyond the Taj Mahal will focus on retelling the story and history of the city, its more than 150 sites and monuments and its rich living heritage by stimulating "Agra beyond the Taj"—a move away from a Taj Mahal-centric tourism model—to retain visitors and increase their spending in the city. It targets the locations that are already seeing notable tourist footfall, bridging its major attractions, which are today visited in isolation, such as the Taj Mahal and the Agra Fort, while promoting nearby lesser visited attractions, such as the traditional Kachhpura village in front of Mehtab Bagh's Mughal garden.

To ensure a destination-level approach, the project will finance the preparation of a tourism development plan for Agra, leverage and partner with the private sector, as well as other key agencies working in the city, such as the World Monuments Fund, which is supporting the Archaeological Survey of India in revitalizing two of the city's four remaining Mughal gardens.

In addition to providing support for the planning and implementation of conservation work, in cooperation with the Archaeological Survey of India, World Monuments Fund (WMF) obtained the pro-bono expertise of the US architecture firm Skidmore, Owings & Merrill to develop a strategic plan for transportation and landscaping between the monuments, to provide a template for the restoration and improvement of the Shah Jehan Garden, and to provide sanitary amenities for the village of Kachhpura. This integrated planning will allow local procurement and implementation of projects that comply with international standards.

A seminar conducted by Harvard University in conjunction with WMF produced scenarios for the development of other former Mughal sites occupying 15 km along the Yamuna River within the context of their contemporary community use, and a publication, *Extreme Urbanism III, Looking at Agra* (Harvard University, 2015) [7].

7.2 Post-catastrophe Urban Heritage Development in Rhodes, Greece

A series of earthquakes struck the island of Rhodes, Greece between 2014 and 2017, causing damage to the foundations of the fortifications of the World Heritage City, which are its most significant architectural feature, and to the adjacent seabed. A major restoration project is needed to prevent further erosion resulting from the changed tidal patterns. In the current economic situation, the city and Greek state cannot provide the capital to undertake this work. The city has devised a plan to convert 100 buildings within the historic center into tourism facilities, leasing them over a period of 15 years to concessionaires, and using the funds raised to cover the cost of the fortification repairs. The project aims to bring high-quality, low impact tourism into the historic city, which has been increasingly abandoned by residents. There are opportunities for new cultural facilities, high-end tourism amenities, and the revival of craft traditions that have died out in recent years as Rhodes has become a destination for beach and cruise ship tourism. Planning for the historic center opens the opportunity for a circular city strategy that would help to revitalize the life of the city, draw permanent residents, and help bring about a recovery from decline that balances the uses of the city's heritage fabric and leverage the investment in restoration (Fig. 8).

7.3 Rehabilitation of the Tripoli International Fairground, Lebanon

The Tripoli International Fairground is a 250 acre self-contained site between the important medieval historic center of the Silk Road town of Tripoli, on Lebanon's north coast, and the city's Aegean port of El-Mina. Commissioned in the 1960s and designed by the renowned Brazilian modernist architect Oscar Niemeyer with a vision to become a symbol of Lebanon as a progressive, modernizing country and site of a World's Fair, it was nearly completed when the 1976 eruption of civil war forced the project to be abandoned. Following the end of the war, the resources were never

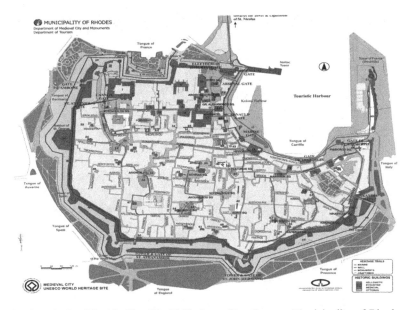

Fig. 8. Rhodes Historic City World Heritage Site. Source: Municipality of Rhodes.

available to focus on the use of the fairground, and it has remained abandoned, although intact, since that time. Listed on the World Monuments Watch in 2006, and periodically the focus of media attention, the Fairground is used today for occasional events, but fails to cover its own small maintenance budget. A major development initiative focused on the nearby port and tax-free zone has brought greater attention and interest to the Fairground, and the UNESCO office in Beirut is developing a strategy for conservation planning and management, and inclusion of the International Fairground on World Heritage Tentative List for Lebanon. This interest comes not a moment too soon, as years of deferred maintenance have resulted in a partial collapse of one of the site's 15 structures. An emergency assessment has produced the diagnosis that damage caused by neglect is widespread. Half the site is undeveloped, and could be developed for appropriate public use in a city that is densely inhabited. Each building has a potential to generate revenue, controlled under the Fairground Board. The original purpose of providing a place for presentation of goods, services, traditional arts and crafts, and Lebanese culture has the potential to be realized if pressure for development does not undermine this opportunity. The site has the potential for the full panoply of uses to be realized, creating a dynamic social, cultural, educational, community and touristic interplay (Fig. 9).

Fig. 9. Tripoli, Lebanon international fairground. Designed by Oscar Neimeyer. (Photo: Courtesy World Monuments Fund).

8 Obstacles to Realizing an Integrated Circular Heritage Strategy

For these and myriad other heritage development opportunities to be realized, a range of issues need to be overcome which are presently an impediment to realizing a vision of circularity and a financial strategy that internalizes the external outputs of development. The barriers are institutional, financial, social, technical, and commercial, and include:

8.1 Institutional

A linear mindset and lack of integration between cultural authorities and other public agencies; complicated regulatory structures; and few examples of integrated public action that has been successful.

8.2 Financial

High transaction costs and long development timetables that require a long-term commitment without solid guarantees of return on investment; the need for upfront financing, countered by a range of risks; the lack of metrics to demonstrate financial viability.

8.3 Social

A lack of awareness or sense of urgency that heritage is facing risks; resistance to change; the exclusion of heritage revitalization from most cities' circular planning strategies.

8.4 Technical

A lack of integration between heritage and environmental conservation goals; a lack of information exchange; few metrics to measure the circularity and benign environmental impacts of heritage conservation.

8.5 Commercial

Lack of experience in and legal framework for public entities to engage the private sector in a range of partnerships to support conservation, whether through subsidies, licensing, concessions or other means.

9 A Way Forward

> "Culture is key to what makes cities attractive, creative and sustainable. History shows that culture is at the heart of urban development, evidenced through cultural landmarks, heritage and traditions. Without culture, cities as vibrant life- spaces do not exist; they are merely concrete and steel constructions, prone to social degradation and fracture. It is culture that makes the difference. How can culture be integrated into urban strategies to ensure their sustainability?
>
> *UNESCO in Culture Urban Future"* [8]

These obstacles can be overcome by the creation of a marketplace to bring together diverse actors to integrate their activities, achieve powerful impacts, and broadcast these examples through professional networks to a wider audience. To achieve this will require institutional, market, and community cooperation, and leadership by a catalytic entity. The Cultural Heritage Investment Alliance aspires to provide that forum.

References

1. Avrami, E., della Torre, M., Mason, R.: Values and Heritage Conservation, Research Report. Getty Conservation Institute (2000)
2. Licciardi, G., Amirahmasebi, R.: The Economics of Uniqueness: Investing in Historic Cores and Cultural Heritage Assets for Sustainable Development. World Bank (2012)
3. Zancheti, S.M., Simila, K.: Measuring heritage conservation performance. In: Proceedings of the 6th International Seminar on Urban Conservation, Recife, Brazil, March 2011. ICCROM (2014)

4. https://sustainabledevelopment.un.org/sdg11
5. Cominelli, F., Greffe, X., Carbonaro, G.: Designing Investment Funds for Unesco Sirtes: Demand, Governance and Structure. Université de Paris-Sorbonne 1 (2014). courtesy of the World Heritage Centre
6. Serageldin, M.: Preserving the historic fabric in the context of fast-paced change. In: Avrami, E., Dellal Torre, M., Mason, R. Values and Heritage Conservation, Research Report. Getty Conservation Institute, p. 51 (2000)
7. Mehrotra, R., Diwadkar, V., Moratilla, J.M.: Planning for Conservation, Looking at Agra (Extreme Urbanism 3). Harvard Graduate School of Design (2015)
8. Culture Urban Future, Global Report on Culture for Sustainable Urban Development. UNESCO, p. 3 (2016)

Public Built Cultural Heritage Management:
The Public-Private Partnership (P3)

Cristina Boniotti[(✉)]

Department of Architecture, Built Environment and Construction Engineering,
Politecnico di Milano, Piazza Leonardo da Vinci, 32, Milan, Italy
cristina.boniotti@polimi.it

Abstract. The topic of the paper relates to the role of conservation and valorization in the management process of built cultural heritage, more specifically in the case of publicly owned, complex properties.

Although well-established opinions state that the conservation of cultural heritage as common goods basically pertains to the purview of the public sector, the participation of private resources and adoption of new business models may pose an opportunity for the public administration to intercept funds that have originally not been intended for heritage.

Partnership is an organizational issue that implies some degree of cooperation between different partners. The interest towards partnership schemes is the product of the multiple interactions they are capable of creating and the variety of operational instruments employed to implement them. The importance lent to the relationships established in public-private partnerships warrants that their theoretical models, their organization, and some considerations concerning normative aspects undergo careful scrutiny. Since public-private partnership has already been adopted in the past and in diverse contexts, as for instance in infrastructure development, the paper does not focus on innovating this alternative way of funding, but rather on describing and analyzing this emerging phenomenon of transition between public and private organizations in the cultural heritage field, as it has not been widely adopted.

Keywords: Public built cultural heritage · Management ·
Public-private partnership

1 Public-Private Interactions and Negotiations

Cultural heritage bears the potential to beget some sound economic value that, by the way, does not at all conflict with its aesthetic worth and can be measured by means of numerical indicators both from a tangible and intangible perspective. Boasting an economic value that is subject to most commonly accepted rules of production, exchange, pricing, and cost-benefit analysis, cultural heritage is likely to strongly influence a country's overall outlook, which implies that a lot of direct and indirect benefits can be achieved for the benefit of communities.

Public properties are a significant cultural heritage and a wide range of real estate, characterized by an important architectural value as well as a degree of social and economic potential. They are required to be adequately conserved and valorized.

© Springer Nature Switzerland AG 2019
A. Moropoulou et al. (Eds.): TMM_CH 2018, CCIS 962, pp. 289–298, 2019.
https://doi.org/10.1007/978-3-030-12960-6_19

The concept of valorization adopted herein is consistent with the definition of *valorizzazione* illustrated in Article 6 of the Italian Cultural Heritage and Landscape Code, as "[...] the exertion of the functions and regulation of the activities aimed at promoting cultural heritage knowledge and ensuring an asset's optimal conditions of use and public fruition, also by people with disabilities, in order to promote the development of culture. It also includes the promotion and support of cultural heritage conservation. In reference to the landscape, the valorization also includes the requalification of the compromised or degraded buildings and areas, under protection, or the construction of new, coherent, integrated landscape values". The term at issue stems from the Late Latin word *valor, valoris*, which means "value, worth". While a financial asset yields some cash flow, a heritage asset provides a flow of various other values and benefits. As to the latter aspects, we ought to mention David Throsby's theory stating that the value embodied in or generated by heritage assets comprises both economic and cultural values consisting of aesthetic, symbolic, spiritual, social, historic, authenticity, and scientific values [1–4]. By the same token, valorization refers to the ways by which we ensure that the flow of values is constant over time. Please note that the concept of value is a dynamic one in that people's perceptions vary over time [5].

In light of the difficulties many Governments typically faces in conserving and managing their rich public cultural heritage, which often lingers in a condition of neglect, the objective is to identify a set of additional tools capable of providing adequate financial resources as well as skills. The study aims at fostering a preliminary reflection on possible sector-specific models for public built cultural heritage management that have not been well defined yet, especially so in reference to one of the institutional options, namely the adoption of public-private agreements.

The adoption of an up-to-date approach capable of projecting the definition of cultural heritage beyond its traditional boundaries could pose new opportunities. It foresees the converging of investments from different sectors into cultural heritage by means of negotiation dialogues fostering the use of non-heritage funding originating from other domains such as the labor market, regional development, and creative industries, to achieve heritage and non heritage-related goals [6]. This alternative model of cultural heritage enhancement implies that a trade-off be pursued between different parties and the subsequent introduction of the concept of trading zone, which is a form of dialogue and cooperation at once, despite the divergences existing amongst various sectors [7, 8]. This collaboration makes resources become available for conservation and valorization activities, thus boosting the exchange of good practices and abilities, and creating new networks.

Although well-established opinions state that the conservation of cultural heritage as common goods basically pertains to the purview of the public sector, the participation of private resources may pose an opportunity for the public administration to intercept new funding channels. During a workshop organized by the Directorate-General for Research and Innovation of the European Commission, even the Horizon 2020 Expert Group on Cultural Heritage highlighted that the public sector should refocus its own approach by incentivizing and encouraging the private sector to get involved and invest in cultural heritage through new financial instruments such as tax breaks, differentiated VAT brackets, well-designed grants, loan programs, and public-private partnership (P3) schemes [9].

Consistent with the previously illustrated models, partnership is an organizational issue that implies some degree of cooperation between different partners. The interest towards partnership schemes is consistent with the multiplicity of interactions they create and the variety of operational instruments whereby they are implemented.

2 Public-Private Partnership

2.1 Definition and Characteristics

P3 consists in a form of long-term cooperation between public and private entities aimed at absolving public duties such as design, construction, management, maintenance of public works or public services – in it, resources and risks are shared on the basis of each party's skills and contribution [10]. It amounts to a transaction between governmental and non-governmental actors and represents a coordination between different organizations aimed at manufacturing goods or services for the benefit of the community as a whole [11, 12]. Under regular conditions, groups of stakeholders from different sectors will interact, thus combining their resources, negotiating profits, and pursuing economic benefits as well as publicly relevant aims.

In the cultural heritage field on the other hand, P3s might well lead to the creation of a financing model based on the competition between private investors fixing buildings and heritage assets, delivering cultural and environmental services within areas of historic relevance and enjoying a share of revenue consistent with the risk they have born. Indeed, risk analysis plays an important role in value-for-money achievement.

Truth be told, the term "public-private partnership" itself still comprises a very wide spectrum of financing arrangements and legal models in which private organizations take upon themselves activities naturally own to public bodies [11]. No wonder this term's significance is currently subject to heavy debate and being criticized to the point that P3 is seen by some as little more than a politically handy language game [13].

Anyway, some features can be considered as being actual and constant throughout all types of arrangements:

- Resource and risk pooling, in order to face complex, uncertain, or risky situations.
- Knowledge and responsibility pooling, as a form of co-production between actors who pursue different logics.

In fact, one important aspect is that the partners involved are not only interested in return on investment, but also in profiting from each actor-organization's peculiarities. The gathering of resources of differing natures and combination of different cultures and interests create synergies, pose some value added and a potential, and define innovative policies and solutions. Together with the sharing of financial resources, the sharing of ideas fosters the involvement of weaker partners, which are usually marginalized in situations mainly focused on economic aspects [14].

Besides, private resources can bring advantages in finance and fruition owing to their:

- Concurrency of public (social use) and private interest (profit).
- Efficiency, based on one entity's having to comply with the ways and means conducive to implementing one's project if the expected cash flow is to be attained.
- Effectiveness, in that a public activity turned into a business is evaluated by employing the profitability and competitiveness criteria.
- Transparency, as the involvement of actors required to work under conditions of close interdependence provides additional cross-checks.

In this way, public administrators have the inherent capacity of involving more and more economic operators in the heritage game, whereas private actors may well provide their managerial, commercial and innovative skills to the effect of obtaining some economic return [10].

Besides, tripartite partnerships might provide comprehensive interventions involving, e.g., infrastructures, public and private spaces, etc., continuity, and coordinated actions, which happen to be the basic conditions for urban heritage conservation and renewal actions to become successful [15].

More specifically, heritage-oriented P3s boast a number of peculiarities compared to infrastructure and public works development-oriented P3s:

- They are long-term collaborations between public and private entities aimed at conserving and valorizing one or more pieces/instances of built cultural heritage.
- They are a collection of private investments allocated for the purpose of built heritage conservation and valorization, provided the asset becomes public again once all goals have been accomplished, which makes it a right of use for a definite amount of time rather than a form of sale.
- Resources and risks are shared proportionally and based on each partner's own field of expertise in order to face complex, uncertain, or risky situations.
- It is an administrative model that allows for raising funds for the pursuit of profit-oriented as well as non-profit oriented goals. Profitability should be considered not only in terms of use-based values (balance sheet), but also non use-based values (long-term advantages such as the fostering of positive externalities, territorial capital, etc.).
- It is generally aimed at mending the gap between cost and value, which consists in the difference between the amount of money invested for a project and the one the marketplace is willing to pay for it [16]. Considering that the cost for a restoration is very often higher than its value, the P3 approach allows to manage the asset while obtaining a return on investment thanks the cash flow being generated.

2.2 Partners Roles, Responsibilities and Risks

Since P3 requires a pooling of diverse resources and some cooperation between actors that differ in terms of interests and are subject to different legal and economic limitations, it is thus necessary to clearly define the roles, responsibilities, and risks falling to each one of the stakeholders involved.

This implies sharing the so-called three Rs: resources, responsibilities and risks [17]. In every operation, the balance between roles and responsibilities and that between the public and private sectors changes. For example, in situations characterized by some restrictions in a player's access to equity and funds, the public sector should always take the upper hand [16]. Therefore, each P3 scheme needs its own specifications.

In order to establish a successful P3, costs, benefits, and risks must be properly shared by all partners involved. In fact, every member should be entrusted with the tasks and responsibilities they are best at [18].

Thanks to the P3, public partners become capable of involving in the process a greater number of economic operators, thus improving knowledge and proficiency in their organizations as a response to society's demand for conservation. Even governments should play a more active role. Based on every situation's specific circumstances, the public administration must exercise its function of protection by overseeing the process through different documents and contracts such as those establishing subsidiary companies, groups, and appointing project managers [11]. Public entities are expected to contribute to the transaction by providing financial and policy incentives, regulatory relief, long-term protection of cultural heritage, and the property itself. Moreover, they shall ensure that the context undergoes adequate regeneration, provide infrastructure and services to the asset, etc. [16].

On the other hand, private partners are best equipped for providing expertise, know-how, their high-degree planning and trade skills, and special sense for revenue. They are expected to contribute to the transaction by means of financial capital, debt negotiation and securing, skills in real estate development and/or construction, and be accountable for the design, conservation, valorization, long-term tenancy or management (long-term ownership, long-term management, marketing of space), etc. [16].

More specifically, certainty as to a P3 initiative's economic sustainability commands that a proper risk assessment be performed. The main categories to be included in a risk analysis are as follows:

- Political risk: arises from political incidents and/or catastrophic events.
- Environmental risk: natural disasters (earthquakes, tides, hurricanes, etc.), malicious disasters (instances of arson, attacks, etc.). Coverage of this risk is mandatory.
- Unexpected-event risk: strikes and riots. Coverage of this risk is mandatory.
- Financial risk: a risk related to the balance between incoming cash flows and outflows, inflation, and interest rates. It needs to be shared and mitigated.
- Development risk: a pre-completion project risk related to general operation's times, costs, and preset quality thresholds. It needs to be shared and mitigated.
- Design risk: a pre-completion project risk related to the observance of design times, costs, and preset quality thresholds. It needs to be shared and mitigated.
- Restoration risk: a pre-completion project risk related to the observance of restoration times, costs, and preset quality thresholds. It needs to be shared and mitigated.
- End user demand-related and/or revenue risk: a post-completion project risk related to revenues and the consequent debt payment. It needs to be shared and mitigated.
- Operating and maintenance-related risk: a post-completion project risk related to preventive and planned conservation costs, and valorization costs. It needs to be shared and mitigated [16, 19–22].

As stressed above, each type of risk has either to be covered by the party (be it the Government or a contractor) best suited to manage it or shared [23]. The warranties and clauses directed at hedging, mitigating, and sharing risks are what is referred to as a "security package".

2.3 Diversified Instruments

Every single partnership project calls for the use of a definite contractual instrument.

Several P3 instruments were introduced in the last few decades in Italy as well as in Europe. These tools, through which a partnership is applied, are quite diversified: in some cases they can be classified as forms of contract, while in others as subsidiary companies [11]. In fact, the idea of P3 often refers to a wide range of management alternatives and instruments, each of which aimed at different public objectives. It is therefore a flexible model subject to different interpretations and taking the shape of various procedures [24]. Applications may vary also based on the fields they belong to and each national legislation. However, all these forms of joint participation between public and private bodies are aimed at the same goals, namely the use of financial, managerial, and creative private resources needed to realize or manage public-service activities or facilities [25].

Apart from concession of construction, concession of management and project financing, namely the most widespread models, the involvement of private sector institutions may be attained also by employing several additional mechanisms.

According to Lawrence Martin, P3 has many meanings, and various tools can be used to help governments achieve their purpose, in this case the preservation of built cultural heritage. Although this term may in itself not convey the same meaning to everyone, we should nonetheless focus on any given specific tool's actual financing arrangements. Based on their features, P3s are generally classified as:

- Design-Build (DB).
- Design-Build-Maintain (DBM).
- Design-Build-Finance (DBF).
- Design-Build-Finance-Maintain (DBFM).
- Design-Build-Finance-Operate (DBFO).
- Design-Build-Finance-Operate-Maintain (DBFOM).

which entails ever-increasing rates of private-sector responsibility [21]. In fact, the rise of private sector involvement is proportional to the increase in private sector risk. In the context of the cultural heritage field the "Build" is replaced by the "Conserve" component [16].

3 Re-Examination of Public Policies: Shared Identity and Collective Behavior

Therefore, the potential of partnership warrants that public policies and administration be re-examined more thoroughly.

As stated by Susan Macdonald: "Most governments face significant challenges in their efforts to conserve and manage their cultural heritage assets and few have the necessary resources (money and skills) required to fully achieve their conservation goals. In many places government has been seen as the primary guardian of the nation's heritage, but increasing pressure to fulfill other public demands, requires community commitment and private engagement in order to help governments retain these assets for future generations" [26].

For future reference, the effort to involve actors not usually interested in such endeavors will have to be consolidated by conveying the importance of culture in the local economy and promoting collaboration among actors commonly not used to cooperating with other entities [27].

The exchange of ideas and know-how stemming from various fields and disciplines might well lend some precious contribution to local economies and foster the development of more strategic networks aimed at yet novel and more ambitious goals. Hence, the process is not only of a technical nature, but shall consist of special care for relations and coordination between diverse sectors such as the manufacturing, construction, environment, and health industries and actors such as the Superintendence for Architectural and Landscape Heritage, construction companies, public institutions, citizenships, universities, etc. The latter approach will likely be much more sustainable than its alternatives.

A high-level cooperation network and a well-established private patronage are the expression of how social and economic subjects may be mobilized within the governmental process, mainly so on a local scale [28]. The complexity of the networks being established, itself an indicator of the diversity of the actors involved, and their density, which accounts for the existence of direct links arising from them, are the paramount indicators of governance quality [29]. Besides, private investments create a positive impact at a social and economic level by producing beneficial local and regional externalities [30].

New governance models will have to take into account private actors as decision makers and participants in the management of common goods together with public entities. Considering the budget constraints harrowing most European Governments, the P3 approach may well ensure private funding and know-how in all-inclusive and long-term projects and further provide some useful guidance in informing future initiatives. The forthcoming strategies aimed at built heritage management are trying to enhance the understanding and awareness of the full potential of cultural heritage. The spread of these approaches might well serve as a basis for future projects, develop effective policies for heritage on the territory, strengthen the relationship with private sectors, and involve other kinds of market-based partners and stakeholders into the process.

According to the Hangzhou declaration, this model is a possible way of pursuing sustainable development in that all four pillars of sustainability are jointly adopted in a holistic approach, thus building relations across the cultural, social, economic, and environmental levels: "The great and unexplored potential of public-private partnerships can provide alternative and sustainable models for cooperation in support of culture. This will require the development, at national level, of appropriate legal, fiscal, institutional, policy and administrative enabling environments, to foster global and innovative funding and cooperation mechanisms at both the national and international levels" [31].

Acknowledgments. The author conveys her sincere gratitude to Prof. Stefano Della Torre from the Department of Architecture, Built Environment and Construction Engineering of the Politecnico di Milano for supervising the current research activity and to Prof. Caroline Cheong from the Department of History of the University of Central Florida for her invaluable comments and suggestions.

References

1. Throsby, D.: Cultural capital. J. Cult. Econ. **23**, 3–12 (1999)
2. Throsby, D.: Cultural capital and sustainability concepts in the economics of cultural heritage. In: de la Torre, M. (ed.) Assessing the Values of Cultural Heritage, pp. 101–117. The Getty Conservation Institute, Los Angeles (2002)
3. Throsby, D.: Heritage economics: a conceptual framework. In: Licciardi, G., Amirtahmasebi, R. (eds.) The Economics of Uniqueness. Investing in Historic City Cores and Cultural Heritage Assets for Sustainable Development, pp. 45–74. The World Bank, Washington, DC (2012)
4. De la Torre, M.: Assessing the Values of Cultural Heritage. The Getty Conservation Institute, Los Angeles (2002)
5. Rojas, E.: Governance in historic city core regeneration projects. In: Licciardi, G., Amirtahmasebi, R. (eds.) The Economics of Uniqueness. Investing in Historic City Cores and Cultural Heritage Assets for Sustainable Development, pp. 143–181. The World Bank, Washington, DC (2012)
6. CHCfE Consortium: Cultural Heritage Counts for Europe, full report (2015). http://www.encatc.org/culturalheritagecountsforeurope/outcomes/. Accessed 20 Aug 2018
7. Balducci, A., Mäntysalo, R. (eds.): Urban Planning as a Trading Zone. Springer, Dordrecht (2013). https://doi.org/10.1007/978-94-007-5854-4
8. Gustafsson, C.: The Halland Model. A Trading Zone for Building Conservation in Concert with Labour Market Policy and the Construction Industry, Aiming at Regional Sustainable Development. University of Gothenburg (2011)
9. European Commission, Directorate-General for Research and Innovation: Getting cultural heritage to work for Europe. Report of the Horizon 2020 Expert Group on Cultural Heritage (2015). https://ec.europa.eu/programmes/horizon2020/en/news/getting-cultural-heritage-work-europe. Accessed 20 Aug 2018
10. Cori, R., Paradisi, I.: Una ipotesi di lavoro: l'applicazione ai servizi del sistema dei beni culturali del Project Financing. In: Leon, A., Verdinelli De Cesare, P. (eds.), Qualità dei bandi per l'acquisto di servizi nel sistema dei beni culturali. I servizi del sistema dei beni culturali: come interpretarli, combinarli, innovarli, qualificarli, vol. 2, pp. 41–52 (2011). http://www.svilupporegioni.it/site/sr/home/argomenti-di-rilievo/semplificazione-amministra-tiva/scheda16002850.html. Accessed 20 Aug 2018
11. Codecasa, G., Di Piazza, F.: Governare il partenariato pubblico-privato. Strategie di governo e strumenti del management pubblico nei progetti di riqualificazione urbana. In: Codecasa, G. (ed.), Governare il partenariato pubblico e privato nei progetti urbani, pp. 1–20. Maggioli Editore, Santarcangelo di Romagna (2010)

12. Codecasa, G.: Ripensare i governi urbani. Spunti per un'agenda di ricerca. In: Codecasa, G. (ed.), Governare il partenariato pubblico e privato nei progetti urbani, pp. 151–198. Maggioli Editore, Santarcangelo di Romagna (2010)
13. Hodge, G., Greve, C.: Public-private partnerships: governance scheme or language game? Aust. J. Public Adm. **69**(S1), S8–S22 (2010)
14. Gualini, E.: Il progetto come arena contesa. Entitlement, responsabilità e riflessività della partnership in pratiche di governance urbana. In: Codecasa, G. (ed.) Governare il partenariato pubblico e privato nei progetti urbani, pp. 111–150. Maggioli Editore, Santarcangelo di Romagna (2010)
15. Fox, C., Brakarz, J., Cruz, A.: Tripartite Partnerships. Recognizing the Third Sector: Five case studies of urban revitalization in Latin America. Inter-American Development Bank, Steven Kennedy, SBK&A, Washington, DC (2005)
16. Rypkema, D., Cheong, C.: Public-Private Partnerships and Heritage: A Practitioner's Guide. Heritage Strategies International, Washington (2012)
17. Macdonald, S., Cheong, C.: The Role of Public-Private Partnerships and the Third Sector in Conserving Heritage Buildings, Sites, and Historic Urban Areas. The Getty Conservation Institute, Los Angeles (2014)
18. Rojas, E.: Urban Heritage Conservation in Latin America and the Caribbean. A Task for All Social Actors. Inter-American Development Bank, Sustainable Development Department, Technical Papers Series, Washington, DC (2002)
19. Bellintani, S., Ciaramella, A.: L'audit immobiliare. Manuale per l'analisi delle caratteristiche degli edifici e dei patrimoni immobiliari. Il Sole 24 ore, Milano (2008)
20. Baiardi, L.: La valorizzazione e il concetto di redditività degli immobili. In: Tronconi, O., Baiardi, L., Valutazione, valorizzazione e sviluppo immobiliare, pp. 61–101. Maggioli Editore, Santarcangelo di Romagna (2010)
21. Martin, L.: Public Procurement Practice. Public-Private Partnership (P3): Facilities and Infrastructure, Guidance issued by the National Institute of Governmental Purchasing to state and local government procurement officials (2016). http://www.nigp.org/docs/default-source/New-Site/public-private-partnership-(p3)-facilities-and-infrastructure. Accessed 20 Aug 2018
22. Merola, F.: Investimenti in asset reali: dall'immobiliare alle infrastrutture, lesson held at Luiss Business School, EREF Executive Program in Real Estate Finance, 10 November 2017
23. Lawther, W., Martin, L. (eds.): Private Financing of Public Transportation Infrastructure: Using Public-Private Partnerships. Lexington Books, Lanham (2015)
24. Benecchi, S., Savi, I.: Governare il partenariato. Il caso di Parma. "Parma prova e… sviluppa". In: Codecasa, G. (ed.), Governare il partenariato pubblico e privato nei progetti urbani, pp. 41–70. Maggioli Editore, Santarcangelo di Romagna (2010)
25. Moroni, S.: Risorse carenti, discrezionalità auspicabile, impreparazione diffusa? Appunti sugli strumenti partenariali. In: Codecasa, G. (ed.): Governare il partenariato pubblico e privato nei progetti urbani, pp. 101–110. Maggioli Editore, Santarcangelo di Romagna (2010)
26. Macdonald, S.: Leveraging heritage: public-private, and third-sector partnerships for the conservation of the historic urban environment. In: Le patrimoine, moteur de développement. Heritage, a driver of development, proceedings of the 17th ICOMOS general assembly symposium, Paris, ICOMOS, pp. 893–904 (2011). http://openarchive.icomos.org/1303/. Accessed 20 Aug 2018

27. Della Torre, S.: Shaping tools for built heritage conservation: from architectural design to program and management. Learning from "Distretti culturali". In: Van Balen, K., Vandesande, A. (eds.), Community Involvement in Heritage, pp. 93–101. Garant, Antwerp – Apeldoorn (2015)

28. Capello, R.: Built cultural heritage as an economic developer. Lesson held at the Politecnico di Milano, Mantua Campus, part of the Ph.D. Summer School "Trading values for preservation and planning in world heritage cities" by Prof. Stefano Della Torre, 5 September 2016 (2016)

29. Dente, B., Bobbio, L., Spada, A.: Government or governance of urban innovation? Plann. Rev. **41**(162), 41–52 (2005)

30. Murzyn-Kupisz, M.: The socio-economic impact of built heritage projects conducted by private investors. J. Cult. Heritage **14**(2), 156–162 (2013)

31. UNESCO United Nations Educational, Scientific and Cultural Organization: The Hangzhou Declaration: Placing Culture at the Heart of Sustainable Development Policies. Culture: key to sustainable development, Hangzhou International Congress China, 15–17 May 2013 (2013). http://www.unesco.org/new/fileadmin/MULTIMEDIA/HQ/CLT/images/FinalHangzhouDecl-aration20130517.pdf. Accessed 20 Aug 2018

A Programme for Sustainable Preservation of the Medieval City of Rhodes in the Circular Economy Based on the Renovation and Reuse of Listed Buildings

Antonia Moropoulou[1]([✉]), Nikolaos Moropoulos[2],
George Andriotakis[3], and Dimitrios Giannakopoulos[4]

[1] NTUA School of Chemical Engineering, Athens, Greece
amoropul@central.ntua.gr
[2] Marathon, Greece
[3] Rhodes, Greece
[4] Athens, Greece

Abstract. The Medieval City of Rhodes was listed in 1988 as a World Heritage Site and as such it must be preserved and developed in the context of an integrated large-scale plan that on one hand maintains its unique character, and on the other it blends the preservation of the Medieval City with the development of the modern town of Rhodes. This paper presents a programme for the sustainable preservation of the Medieval City of Rhodes in Greece. The programme is based on the renovation and reuse of listed buildings and will contribute to the further development of circular economy initiatives. It is self-funded, and its benefits are expected to be in the Medieval City fortifications, the moat, the walls, the infrastructure, provision of subsidized public housing and other relevant projects in the City of Rhodes. A total of 300 buildings comprise the programme's scope. They will be renovated and reused in two phases. The first phase comprises four projects and is envisioned to have a time horizon of 25 years and has in its scope the renovation and reuse of 150 of the 300 available buildings. Once renovated, the buildings will be put to four types of use: Tourism and Recreation, Private use of emblematic buildings by international and other organizations to promote extroversion, Commercial use to revitalize traditional arts and crafts and Public housing. A program to renovate and reuse these buildings has been formulated and is proposed to be implemented in the context of the decree by the Minister of Culture (October 2017) who has set up a committee prepare a framework contract to that effect between the Ministry of Culture and the Municipality of Rhodes. Three public interest partners have already been involved in the project: The Fund of Archaeological Proceeds, TAP, the Local Development Enterprise of Rhodes – DERMAE, and the National Technical University of Athens – NTUA. The remaining planning stage challenge for the Ministry of Culture and the Municipality of Rhodes is to determine the scheme under which the public interest partners can collaborate with private interest partners to generate and share value while at the same time preserving the buildings.

© Springer Nature Switzerland AG 2019
A. Moropoulou et al. (Eds.): TMM_CH 2018, CCIS 962, pp. 299–321, 2019.
https://doi.org/10.1007/978-3-030-12960-6_20

Keywords: Medieval · City · Rhodes · Circular · Economy · Renovation · Reuse

1 Introduction

The Medieval City of Rhodes is a UNESCO World Heritage City [1] as having cultural and historical significance that serves the collective interests of humanity and as such it must be preserved and developed in the context of an integrated large scale plan that on one hand maintains its unique character, and on the other it blends the development of the Medieval City with the development of the modern town of Rhodes.

This preservation must be sustainable and project the cultural, social, and historical features of the Medieval City of Rhodes, so that it justifies its characterization as a World Heritage City and meets the demanding relevant criteria of UNESCO.

The Medieval City of Rhodes, and more specifically fortifications, bastions, ramparts are undergoing intense decay pathology phenomena. Aggressive sea salt is sprays triggering salt decay and materials disaggregation in synergy with phenomena of structural pathology, due to severe earthquakes, structural failures (due to material losses) and dissolution of lime mortars, under the exercise of lateral thrusts in the contra Scarpa.

Innovation as a key driver towards circular economy is related to interdisciplinary knowledge-based decision making, digital driven preservation, integrated environmental impact assessment and preservation management for resilience enhancement and reconstruction.

The Municipality of Rhodes aspires to meet the challenge presented by the Medieval City of Rhodes and the multifaceted complexity of the relevant integrated programme and is working with the Ministry of Culture to formulate a framework agreement in this direction.

As mentioned in the UNESCO description of the features of the Medieval City of Rhodes: *"All the built-up elements dating before 1912 have become vulnerable because of the evolution in living conditions and they must be protected as much as the great religious, civil and military monuments, the churches, monasteries, mosques, baths, palaces, forts, gates and ramparts"* [1].

The built-up elements are the focus of the programme described in this paper. There are more than 300 listed heritage buildings in the Medieval City of Rhodes that are available today, can be renovated and reused.

The expected benefits of the program to the Greek State and the Municipality of Rhodes are the following:

- Sustainable preservation of the Medieval City fortifications, building structures in the moat and medieval walls by the sea (I).
- Maintenance of the Medieval City's infrastructure (II).
- Provision of subsidized public housing (III).

- Projects (IV) by:
  - Fund of Archaeological Proceeds (TAP)
  - Local Development Enterprise of Rhodes (DERMAE)
  - Municipality of Rhodes
- National Technical University of Athens - Laboratory for the preservation of cultural heritage monuments (NTUA AIEN)
- International Centre for the protection, preservation and sustainable development of Heritage Towns

The programme will contribute to the development of circular economy initiatives, where resources are used in a more sustainable way. The re-use of the listed buildings will bring benefits for both the environment and the economy, and at the same time provide housing needed by the low income or dispossessed citizens of Rhodes.

In the first section of the paper we discuss the international trends and guidelines on sustainable presentation and management of Historic Cities, followed by an overview of the relevant initiatives for the Medieval City of Rhodes. The third section presents the projects and the key features of the programme.

2 International Trends, Guidelines and Innovation for the Sustainable Preservation and Management of Historic Cities

The 2030 Agenda [2] for sustainable development adopted at 2015 by the United Nation - General Assembly declares among others sustainable development goals to transform our world "and to make cities and human settlements inclusive, safe, resilient and sustainable" (goal 11).

The Agenda aims to substantially increase the number of cities adopting and implementing integrated policies and plans by 2020. These plans will contribute towards inclusion, resource efficiency, mitigation and adaptation to climate change and risk reduction.

Due to unprecedented urban growth, uncontrolled development, exploding tourism and population shifts, compounded by climate change, natural hazards and renewed sectarian conflict, Cultural Heritage at risk constitutes a potential impoverishment of the heritage of all the peoples of the world. Cultural Heritage at risk is putting cities "identity" at risk jeopardizing future development and is also depriving, local communities of values that bind them together and provide the resilience to face the challenges ahead.

UNESCO enlists emblematic historic cities in the World Heritage List according to well defined criteria and conditions of integrity and authenticity. The list is maintained by UNESCO based on periodic monitoring, reporting and assessment which is responsibility of the partner states [1].

The World Heritage Convention aims at the identification, protection, conservation and transmission to future generations of cultural heritage of outstanding value, in synergy with sustainable development.

Effective management involves a cycle of short, medium and long-term actions to protect, conserve and present the nominated property. An integrated approach to planning and management is essential to guide the evolution of properties over time and to ensure maintenance of all aspects of their Outstanding Universal Value, including risk preparedness as well as management plans and training strategies.

Related legislations, policies and strategies should ensure the protection of Cultural Heritage and promote and encourage the active participation of the communities and stakeholders concerned with the property as necessary conditions to its sustainable protection, conservation, management and presentation. Public interest agencies should do so in close collaboration with property managers, the local agency with management authority and other partners, and stakeholders in property management.

An effective management system depends on the type, characteristics and needs of the nominated property and its cultural and natural context. Management systems may vary according to different cultural perspectives, the resources available and other factors. They may incorporate traditional practices, existing urban or regional planning instruments, and other planning control mechanisms, both formal and informal. Impact assessments for proposed interventions are essential for all World Heritage properties.

A new framework is needed to provide worldwide funding for cultural heritage at risk, that can bridge funding gaps and bring together the resources of the public and private sectors. No such framework exists today.

Networking between public agencies, local non-profit entities working to preserve the heritage, communities and private investors, is needed in order for heritage conservation to be expanded and support broader objectives and longer - term goals, assuring that private new investments will be mobilized by public, national and local government incentives.

Financing tools are required to support key areas where cultural heritage losses are escalating as a result of development pressures and hazards.

Finding new sources of revenue (other than government funding and charity) that support the heritage field more comprehensively requires the integration of development with heritage preservation, and experience in the creation of public value through public - private real estate initiatives.

The Medieval City of Rhodes provides the setting for a pilot programme which promotes, for the first time ever, the circular economy agenda in the sustainable preservation of Historic Cities.

Innovation as a key driver towards circular economy is related to interdisciplinary knowledge-based decision making, digital driven preservation, integrated environmental impact assessment and preservation management for resilience enhancement and reconstruction.

Innovative funding can permit the preservation of cultural heritage through its re-use and can bridge Private-Public Partnerships with Social Economy.

By generating revenue and value a sustainable preservation and management plan of a Historic City may contribute to the development of Heritage Driven Economy.

3 Initiatives for the Sustainable Preservation and Management of the Medieval City of Rhodes

The City of Rhodes has been founded by the Ancient Dorik Exapolis (Knidos, Alikarnassos, Lindos, Kamiros, Ialisos and Kos) 2400 years ago. The Medieval City of Rhodes has been built upon the Byzantine Colloquium in the Early Christian period (324-642 A.D) and has evolved through the Middle Byzantine period (642-1071 A.D) and the Late Byzantine period (1071-1453 A.D). The city has taken its current look by the knights of St. John of Jerusalem and built by the knights' order of St. John of Malta, the Hospitalles (Hospitaller period 1309-1522 A.D). Following the Post Byzantine period and the Turkish occupation (1523 1912 A.D), the Italian occupation (1912-1947 A.D) has left a strong imprint to the Medieval City through major restorations of the Grand Masters' Palace and the Fortifications.

In 1947 the Dodekanese island complex is incorporated into the Greek State and Rhodes contributes actively to the modern Greek history. Initiatives for the revitalization of the Medieval City were undertaken by the Deputy Ministry of Youth in the Ministry of Culture under the auspices of UNESCO. Along with the collaboration of the Ephorate of Antiquities of Dodecanese and the Municipality of Rhodes implemented major interventions in the Medieval Fortifications, the Moat, the Ramparts, the Bastions of Caretto and St. George as well as the installation of the "Melina Merkouri" theatre and the restoration of the "Virgin of the Castle" (15th Century Church). These interventions demonstrate the capacity of the Medieval City to be regenerated and culturally rehabilitated by raising the awareness of the local community and activating international participation.

In 1982 the Deputy Ministry of Youth [3] in the Ministry of Culture submitted the demand to UNESCO and the Ephorate of Antiquities of Dodecanese prepared and submitted all the documentations for enlisting the Medieval City of Rhodes in the World Heritage List. Following an ICOMOS evaluation [4], UNESCOs' World Heritage Committee enlisted the Medieval City of Rhodes in the World Heritage List (1988) [5].

As a follow up to the initiative, a programme agreement [6] has been signed between the Ministry of Culture and the Municipality of Rhodes for the preservation of the Medieval City and its cultural rehabilitation (1984-2004 A.D). Many conservation and restorations interventions where planned and implemented by the Common Technical Bureau of the Medieval City as well as by the Ministry of Culture and the Municipality of Rhodes, each and independent afterwards.

Today, thirty years after the enlisting of the Medieval City of Rhodes in the UNESCO World Heritage List, the public interest entities have to contribute to the development of a strategic plan for the sustainable preservation and management of the Medieval City of Rhodes and ensure that funds are made available. In addition, the necessary management arrangements have to be put in place in cooperation with local stakeholders.

To this end, on the 1st June 2015, Deputy Minister of Culture of Greece, Mr. N. Xydakis, under the auspices of Minister of Culture Mr. A. Baltas, on the occasion of the Marie Sklodowska-Curie program "ITN-DCH: Initial Training Network for Digital

Cultural Heritage" announced in Rhodes the establishment of an "International Centre for Protection, Conservation and Sustainable Development of Historic Cities".

On October 2017 the Minister of Culture Lydia Koniordou set up a committee to propose an updated program agreement to face the challenge of the sustainable preservation and management of the Medieval City of Rhodes [7].

The programme presented here was discussed and approved by the committee (Rhodes, 3 April 2018) [8] and was further discussed at the South Aegean Development Conference (Rhodes, 17 April 2018).

At the High-Level Conference [9] of the European Union for the year 2018 (Brussels, 22 March 2018), dedicated to Cultural Heritage preservation, the Medieval City of Rhodes programme was referred as a future development to extend usability of digital driven Cultural Heritage preservation to the scale of Historic Cities.

4 Knowledge Based Diagnosis and Strategic Preservation Planning

4.1 Threads in the Preservation of the Medieval City of Rhodes

The Medieval City of Rhodes, and more specifically fortifications, bastions, ramparts are undergoing intense decay and present pathology phenomena. Aggressive sea salt spray is triggering salt decay and materials degradation in synergy with phenomena of structural pathology, due to severe earthquakes, under the exercise of lateral thrusts in the contra Scarpa, as well as structural failures due to material losses and disintegration of the joint lime mortars (Figs. 1, 2 and 3).

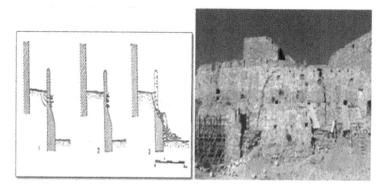

Fig. 1. Masonry collapses due to hard carbonated crusts [10]. Schematic representation by Poziopoulos [11].

The use of incompatible repair materials, like cement concrete as joint mortars by the De Vecchi restoration during the Italian occupation, or the use of incompatible substitution buildings stones by later interventions, trigger the acceleration of salt decay.

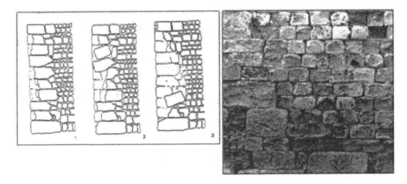

Fig. 2. Masonry collapses due to hard carbonated crusts [10]. Schematic representation by Poziopoulos [11].

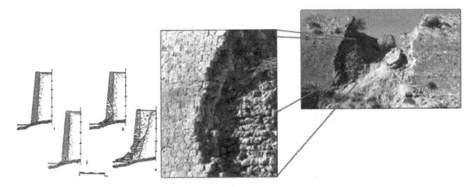

Fig. 3. Masonry collapses due to hard carbonated crusts [10]. Schematic representation by Poziopoulos [11].

The replacement of the decayed pore stones with new stones of smaller porosity, enhances the decay phenomena in the surrounding original material at the interfaces between the two stones as visualized by Digital Image Processing and validated by Fiber Optics Microscope (FOM) inspection and Scanning Electron Microscope (SEM) examination (Fig. 4).

The use of cement as a restoration binder material, as in the case of the Medieval City of Rhodes, enhances the alveolar weathering of the surrounding original material (pore stone) and leads to the creation of large craters, due to incompatibility of the two materials in terms of terms of dramatically thermohygric performance (water absorption and water-vapour permeability) attributed to their different microstructure as visualized by Digital Image Processing and validated by Fiber Optics Microscope (FOM) inspection and Scanning Electron Microscope (SEM) examination (Fig. 5).

Moreover, the synergy of the exerted environmental stresses is depicted upon the fortifications of the Medieval City by the environmental impact assessed (Fig. 6).

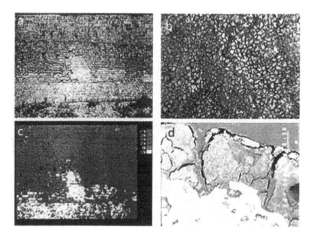

Fig. 4. (a) original photograph, (b) Fiber Optics Microscope inspection, (c) Digital Image Processing, (d) Scanning Electron Microscope (SEM) examination. Incompatible materials and conservation interventions- substitution by more compact porous stones of buildings stones in the fortifications-Contra Scarpa moat suffering by typical hard carbonate crust [12].

Fig. 5. (a) original photograph, (b) Fiber Optics Microscope inspection, (c) Digital Image Processing, (d) Scanning Electron Microscope (SEM) examination. Incompatible materials and conservation interventions - restoration with cement repair joint mortars of the sea-walls suffering by alveolar weathering [12].

Positive impact may be developed through proper environmental management as in the case of the Cleaning, the Cultural reuse of the moat, and the rehabilitation of the rain water drainage systems which are reversing negative impacts of planting, rising dump.

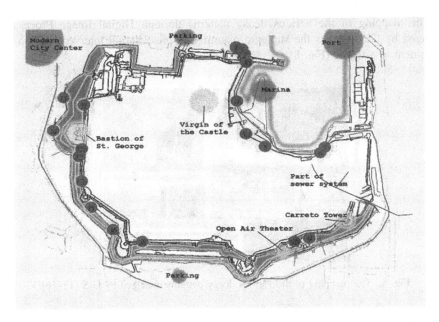

Fig. 6. GIS map of Environmental loads and uses to reverse the negative environmental impact. The map depicts in Brown color the cleaning of the moat, in Yellow color the Cultural reuse, in Red color the pollution sources, in Green color arrows the planting, in Dark grey line the drainage system. [13, 14] (Color figure online)

4.2 Interdisciplinary Knowledge Based Integrated Diagnosis and Strategic Preservation Planning

Directions, criteria and planning methodology for an integrated preservation plan have been elaborated by the NTUA research team [15, 16]. GIS management of NDTs' data permits the environmental impact assessment on the Medieval City of Rhodes and integrated diagnosis towards strategic preservation planning (Fig. 7).

Fig. 7. GIS management of NDTs' data environmental impact [13–16].

The mapping of the various decay patterns through Digital Image Processing managed by GIS permits the strategic planning of preservation interventions on the monument as a whole (Fig. 8).

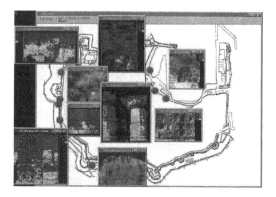

Fig. 8. The mapping of the various decay patterns managed by GIS [13–16].

The false colors are selected to render the intensity of the decay according to the patterns of decay as it is shown in the figure below (Fig. 9).

Gray Level	Color	Decay Type	Lithotype
Black	Black	Full Scale Cavities	
	Blue	Extended Alveolar Weathering	
	Violet	Extended Grain Detachment	Surfaces with grain detachment or highly porous and susceptible to decay
	Dark Brown	Gray / Black crust	Surfaces of medium porosity and exposure of a new front after the detachment of the exterior
	Light Brown	Degradation of the exterior	
	Black yellow	Hard carbonate crusts / biological crusts	
	Yellow	Hard carbonate crusts and/or salt efflorescence / sub-efflorescence	Solid (intact) small porosity surfaces of new construction recently quarried
White			

Fig. 9. False colors rendering decay patterns and intensity [17].

The above classification in order of increasing degree of alveolar decay is validated by the microstructural investigation by Mercury Intrusion Porosimetry and the Pore volume distribution for each degree of alveolar decay (Fig. 10).

Resilience Enhancement and Reconstruction of the fortification of the Historic City are planned for this layer [3, 15, 16].

The National Technical University of Athens has continuously monitored and assessed the preservation state of the Medieval fortification and ramparts on an interdisciplinary basis since early 80's to date and is capable to support scientifically a sustainable preservation plan (Fig. 11).

Fig. 10. Microstructural investigation (Mercury Intrusion Porosimetry) [18].

Fig. 11. Interdisciplinary scientific research by NTUA's team

5 The Programme

The programme comprises four projects, and a total of 300 buildings, each project including a subset of buildings grouped by the type of their reuse.

- Project A: Tourism and Recreation Services
- Project B: Private Use of emblematic buildings by international and other organizations to promote extroversion
- Project C: Commercial use to revitalize traditional arts and crafts
- Project D: Public housing

Projects A, B and C will be self-funded, whereas project D will be funded by projects A, B and C or by Public - Private Partnership.

These buildings belong to the Greek State [19].

Figure 12 shows a plan of the Medieval City of Rhodes, scale 1:10000. Some of the buildings that are available now and can be included in the programme are indicated by the red horizontal lines.

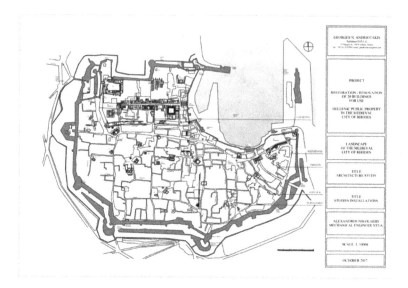

Fig. 12. Plan of the Medieval City of Rhodes (Color figure online)

A program to renovate and reuse these buildings has been formulated and is proposed to be implemented in the context of the decree by the Minister of Culture (October 2017) who has set up a committee prepare a framework contract to that effect between the Ministry of Culture and the Municipality of Rhodes.

The first phase [20] of the program is envisioned to have a time horizon of 25 years and has in its scope the renovation and reuse of 150 of the 300 available buildings. The second phase will be formulated based on the results of the first phase and is estimated to require another 25 years for its implementation. During the second phase the remaining 150 buildings will be renovated and put to various uses. This paper presents the first phase of the programme only.

5.1   Project A

Overview

The buildings of project A will be leased for a period of 25 years to a private enterprise which offers tourism and recreation services. The leasee will renovate the buildings and, once in operation will pay an annual leasing fee to the Municipality of Rhodes.

Local architects and engineers will prepare the architectural and engineering designs and carry out the renovation of the buildings, following the specifications [21] that will be put in place by the National Technical University of Athens (NTUA). NTUA will be one of the partners in the programme's framework contract.

The maintenance cost of the buildings, both materials and labour, for the leasing period will be borne by the leasee.

Project A is structured into three subprojects, A1, A2, A3, each one of them including 20 buildings and leading to a separate leasing contract. The data presented in this section are based on a sample of 20 buildings that are readily available.

The architectural designs and cost estimates prepared by the authors substantiate the following data:

- Number of buildings: 20
- Number of beds: 180–200
- Renovation cost estimate: 3 million Euros
- Daily revenue by bed: 70 Euro
- Days in operation per year: 300

The design for the floor plans for the ground, 1st and 2nd floors of the building at Tlipolemou Street are shown on Fig. 13.

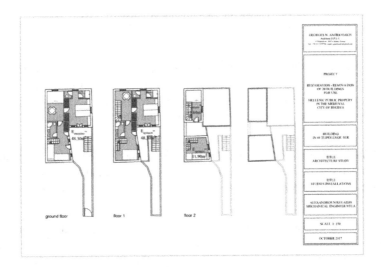

Fig. 13. Floor plans of the building at Tlipolemou Street

The design for the façade of the building at Lachitos Street are shown on Fig. 14.

Revenue

Based on the data presented above, the annual gross revenue of the leasee are estimated at 3,024,000 Euros.

If the Municipality of Rhodes will receive 10% of the leasee's gross operating revenue as leasing fee, its annual revenue will amount to 302,400 Euros. This is an example and is presented in this paper for illustration purposes. The actual percentage will be negotiated and agreed between the Municipality of Rhodes and the leasee.

Fig. 14. Facade of the building at Lachitos Street

Timing

Subproject A1 can start immediately following the signing of an agreement between the Municipality of Rhodes and the Greek State, in the context of the programme's overall framework agreement. All 20 buildings are available, as already mentioned.

Subprojects A2 and A3 can start based on the buildings' availability.

Duration

It is estimated that the procedures for the award of the "design, build and operate" leasing contract, the architectural and engineering designs, and the renovation itself will require a period of 18 months.

Benefits

The Municipality of Rhodes will implement projects to preserve the Medieval City fortifications, building structures in the moat and medieval walls by the sea (I). The projects will be funded by long term loans that the Municipality secure with the revenue streams generated by project A. It is estimated that over a period of 10 years an amount of 5 million Euros will be used to fund the projects. This amount will be obtained from long term loans secured against project A's revenue streams.

In addition, the Municipality will fund the provision of public housing to its citizens for a period of 10 years, at a cost of 2.5 million Euros, to be obtained from long term loans secured against project A's revenue streams.

Funding

The amount of 7.5 million Euros required to fund the benefit projects stated above, will be obtained by long term loans that the Municipality of Rhodes will secure against the revenue streams of project A.

As shown in Table 1, a 2.5 million Euro long term loan with repayment in 10 years and annual interest rate of 5.00% has fixed annual repayments of 319,000 Euros.

Table 1. Loan repayment schedule (indicative)

Annual interest rate	Payment in year	Number of monthly payments	Principal amount	Fixed monthly payment	Total amount to be paid back	Total interest	Fixed annual payment
3.00%	10	120	2,500,000	−24,140	−2,896,822	396,822	−289,682
4.00%	10	120	2,500,000	−25,311	−3,037,354	537,354	−303,735
5.00%	10	120	2,500,000	−26,516	−3,181,965	681,965	−318,197
3.00%	20	240	2,500,000	−13,865	−3,327,586	827,586	−166,379
4.00%	20	240	2,500,000	−15,150	−3,635,882	1,135,882	−181,794
5.00%	20	240	2,500,000	−16,499	−3,959,734	1,459,734	−197,987
3.00%	25	300	2,500,000	−11,855	−3,556,585	1,056,585	−142,263
4.00%	25	300	2,500,000	−13,196	−3,958,776	1,458,776	−158,351
5.00%	25	300	2,500,000	−14,615	−4,384,425	1,884,425	−175,377

It can therefore be safely assumed that the Municipality of Rhodes can get long term loans up to 7.5 million Euros secured against the annual revenue stream of all three subprojects, amounting to 907,200 Euros.

5.2   Project B

Overview
Project B buildings are complex and have special statute and will be leased for a period of 25 years to a party which will use it privately, in a way that is compatible with the continuous build-up of the Medieval City of Rhodes as a blend of the old with the leading edge of today [22, 23]. Examples are international organisations, diplomatic offices and consulates of foreign countries, private enterprises active in technology, media, advertising, and so on.

There will be a total of 20 buildings in project B and the costs will be borne by the leasee.

Revenue
It is estimated that on the average the annual revenue of the Municipality of Rhodes by building will amount to 30,000 Euros, bringing the project B total annual revenue stream to 600,000 Euros.

Timing
Following a promotion activity to attract entities [24, 25] to invest in it, the project can start in the second half year since the start of the programme, which coincides with the start of project A (subproject A1).

Duration
It is estimated that the procedures for the award of the "design, build and operate" leasing contract, the architectural and engineering designs, and the renovation itself will require a period of 18 months.

Benefits

The revenue of project B will fund the following projects:

- Studies for special projects and cultural events (Fund of Archaeological Proceeds, TAP)
- Upgrade of museums' infrastructure (Fund of Archaeological Proceeds, TAP)
- Structuring, management, strengthening, and modernisation of projects and activities in the context of the framework agreement for the Medieval City of Rhodes (Cultural Municipal Enterprise and Local Development Enterprise of Rhodes - DERMAE)
- Maintenance of the Medieval City's infrastructure (Municipality of Rhodes)
- Research and Training in the context of the framework agreement (Laboratory for the preservation of cultural heritage monuments – NTUA AIEN)
- Set up and operation of the International Centre for the protection, preservation and sustainable development of Heritage Towns

5.3   Project C

Overview

The buildings in project C will be used as the main or branch office or outlet of a business entity and will be leased for a period of 25 years. Examples are retail shops and small businesses. The objective is to revitalise traditional crafts and innovative start-ups in various sectors of the economy.

There will be 20 buildings in project C.

The cost of renovation of each building will be borne by the leasee.

Revenue

It is estimated that on the average the annual revenue of the Municipality of Rhodes by building will amount to 15,000 Euros, bringing the project C total annual revenue stream to 300,000 Euros.

The revenue will materialise 12 months following the completion of the building's renovation.

Timing

Following a promotion activity to attract entities [26, 27] to invest in it, the project can start in the second year since the start of the programme, which coincides with the start of project A (subproject A1).

Duration

It is estimated that the procedures for the award of the "design, build and operate" leasing contract, the architectural and engineering designs, and the renovation itself will require a period of 18 months.

Benefits

Maintenance of the Medieval City's infrastructure performed by the Municipality of Rhodes (II).

5.4   Project D

Overview

Project D buildings will be offered to financially challenged citizens of Rhodes for a period of 10 years minimum in return of a token rent.

There will be 50 buildings in project D.

The cost of renovating a building in project D is estimated to be 50,000 Euros. Therefore, the total renovation cost of project D is 2,500,000 Euros.

Project D will be funded by a long-term loan of an equal amount that the Municipality of Rhodes will secure against the revenue stream of one of the three project A subprojects (see project A).

Timing

Following a promotion activity to attract entities to invest in it, the project can start in the second half of the second year since the start of the programme [28], which coincides with the start of project A (subproject A1).

Duration

It is estimated that the procedures for the award of the "design and build" contract, the architectural and engineering designs, and the renovation itself will require a period of 18 months.

Benefits

The Municipality of Rhodes has a comprehensive programme for the support of its financially challenged citizens, many of whom reside in the Medieval City of Rhodes.

Project D will supplement the benefits offered by this programme by making 50 renovated buildings available to citizens for a token rent.

5.5   The Programme Partners

The programme agreement is between the Ministry of Culture and the Municipality of Rhodes. As third parties to participate the following public interest entities will be engaged in the programme design and implementation:

- Fund of Archaeological Proceeds, TAP
- Local Development Enterprise of Rhodes – DERMAE
- National Technical University of Athens – NTUA

The responsibilities of each partner will be clarified in the context of the framework agreement between the Ministry of Culture and the Municipality of Rhodes.

In the sections where we presented the projects of the programme, we alluded to the involvement of private interest partners. This was done for illustration purposes. The potential engagement of private interest partners needs to be clarified and decided by programme owners, i.e. the Ministry of Culture and the Municipality of Rhodes.

International best practice indicates that there are significant benefits to be gained by the fruitful collaboration of public and private interest partners, if there is a supporting legal framework and the public entities can manage the relevant agreements.

6   Programme Implementation

6.1   Indicative Time Plan

The programme will commence with project A, subproject A1. The buildings will start generating revenue for the Municipality of Rhodes 18 months after the start of sub-project A1.

Project B will start on the 7th month following the start of the programme. The buildings will be operational 18 months later and the first revenue stream will flow at the beginning of the 19th month following the start of the programme.

Project C will start at the beginning of the second year following the start of the programme. The buildings will be operational 18 months later and the first revenue stream will flow at the beginning of the 42nd month following the start of the programme.

Finally, project D will commence at the beginning of the third year since the start of the programme. The buildings will be available to be inhabited by citizens at the beginning of the 43rd month after the start of the programme.

Figure 15 presents a high-level view of the programme's time plan. Time is shown as columns, each column representing half a year.

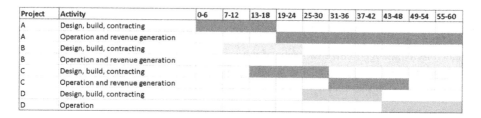

Fig. 15. Program time plan (indicative)

There are two main groups of activities shown for each project. Design build and contracting [29, 30], followed by operation and revenue generation (only for projects A, B and C).

6.2   Funding and Revenue

Projects A, B and C are self-funded, as all costs, one-off and recurring, will be borne by the leasees.

Project D will be funded by a long-term loan to be secured against the revenue stream of one of the three project A subprojects.

Table 2 shows the estimated annual revenue by project.

Table 2. Summary of phase I

Project	Number of buildings	Municipality annual revenue	Project start (month)
A	60	907,200	1
B	20	600,000	7
C	20	300,000	13
D	50	0	25
Total	150		

6.3 Benefits

The expected benefits of the programme the following:

- Sustainable preservation of the Medieval City fortifications, building structures in the moat and medieval walls by the sea (I)
- Maintenance of the Medieval City's infrastructure (II)
- Provision of subsidized public housing (III)
- Various projects (IV)

Projects in category I will be funded by long term loans of 10 million Euro (5 million in the first decade and another 5 million in the second decade), secured against the revenue streams of two project A subprojects.

Projects in category II will be funded by the revenue streams of project C.

Projects in category III will be funded by long term loans of 5 million Euro (2.5 million in the first decade and another 2.5 million in the second decade), secured against the revenue streams of one project A subproject.

Finally, projects in category IV will be funded by the revenue streams of project B.

All three groups of benefits of the program will materialize during the first phase of the programme.

6.4 Action Plan - Iterative

The recommended next steps are the following:

- Review of relevant international best practices.
- Review of the legal framework and finalization of the framework agreement between the Ministry of Culture and the Municipality of Rhodes.
- The programme owners decide on the involvement of private interest partners in the programme.
- Preparation of the eligibility files of the buildings to be selected. Each file will include all necessary legal, city planning, architectural, and other technical documents.

7 Conclusions

The Medieval City of Rhodes is a UNESCO World Heritage City, having cultural and historical significance that serves the collective interests of humanity. As such it must be preserved and developed in the context of an integrated large-scale framework programme that on one hand maintains its unique character, and on the other it blends the preservation of the Medieval City with the development of the modern city of Rhodes.

This preservation must be sustainable, express and promote the cultural, social, and historical features of the Medieval City, so that it justifies its characterization as a World Heritage City, meeting the demanding relevant criteria of UNESCO.

The Municipality of Rhodes aspires to meet the challenge presented by the Medieval City of Rhodes and the multifaceted complexity of the relevant integrated programme and is already working with the Ministry of Culture and the NTUA to formulate a framework agreement in this direction, featuring the renovation and reuse of listed buildings of public property.

The Medieval City of Rhodes provides the setting for an innovative programme which develops and promotes, for first time ever, the circular economy in the context of Historic Cities' sustainable preservation.

The programme is based on the renovation and reuse of listed buildings and will contribute to the further development of circular economy initiatives [31]. It is self-funded, and its benefits are expected to be in the Medieval City fortifications, the moat, the walls, the infrastructure, provision of subsidized public housing and other relevant projects in the City of Rhodes. A total of 300 buildings comprise the programme's scope. They will be renovated and reused in two phases. The first phase comprises four projects and is envisioned to have a time horizon of 25 years and has in its scope the renovation and reuse of 150 of the 300 available buildings [32]. Once renovated, the buildings will be put to four types of use: Tourism and Recreation, Private use of emblematic buildings by international and other organizations to promote extroversion, Commercial use to revitalize traditional arts and crafts and Public housing. A program to renovate and reuse these buildings has been formulated and is proposed to be implemented in the context of the decree by the Minister of Culture (October 2017) who has set up a committee prepare a framework contract to that effect between the Ministry of Culture and the Municipality of Rhodes. Three public interest partners have already been involved in the project: The Fund of Archaeological Proceeds, TAP, the Local Development Enterprise of Rhodes – DERMAE, and the National Technical University of Athens – NTUA. The remaining planning stage challenge for the Ministry of Culture and the Municipality of Rhodes is to determine the scheme under which the public interest partners can collaborate with private interest partners to generate and share value while at the same time preserving and reusing the buildings.

Innovation as a key driver towards circular economy is related to interdisciplinary knowledge-based decision making, "digital" driven preservation, integrated environmental impact assessment and preservation management for resilience enhancement and reconstruction.

At the High-Level Conference of the European Union [9] for the year 2018 (Brussels, 22 March 2018), Cultural Heritage preservation – to which the conference was dedicated - was referred to as a future development to extend the application and scope of "digital" driven Cultural Heritage preservation to the scale of Historic Cities. Innovative funding can enable the preservation of cultural heritage through its re-use and can bridge Private-Public Partnerships with Social Economy while promoting Circular Economy.

Hence, the sustainable preservation and management programme of the Medieval City of Rhodes has the potential to implement the concept of Heritage Driven Economy by generating revenue whilst contributing to the preservation of the Medieval City, to the social cohesion of the City of Rhodes, and to the development of Circular Economy initiatives [20].

References

1. Unesco WHC 17/01/2012: Operational Guidelines for the implementation of the World Heritage Convention, July 2017
2. Agenda: 70/1 Transforming our world: the 2013 Agenda for sustainable Development (2015). United General Assembly 15-16301(E)
3. Moropoulou, A., Andriotakis, G.: Cultural Heritage Preservation. The European case of the Medieval City of Rhodes. Conference of Peripheral Maritime Regions of the EEC, Porto (Endogenous development strategies - Charte Europeenne Littoral), 27–31 September 1982
4. ICOMOS: World Heritage List No 493. Rhodes, December 1987
5. UNESCO: Convention concerning the protection of the world Cultural and Natural Heritage, World Heritage Committee, Twelfth Session. Brazil, December 1988
6. Program Agreement, Ministry of Culture, Fund of Archaeological Proceeds and Municipality of Rhodes, Athens (1985)
7. Michailidou, M.: Ephoriate of Dodecanese- Ministry of Culture "History of the Medieval City of Rhodes" Documents of the Committee, August 2017
8. Moropoulou, A., Politis, E., Chatzidiakos, F., Korka, E., Michailidou, M.: Second meeting of the committee for the Study and Preparation of the New Common Agreement for the Medieval City of Rhodes. Documents of the Committee, Rhodes, April 2018
9. Moropoulou, A.: Digital Solutions and Multi-Layer Innovations in the rehabilitation of the Holy Aedicule of the Holy Sepulchre in Jerusalem. In: Innovation and Cultural Heritage, High level Horizon 2020 Conference of the European Year of Cultural Heritage Brussels (2018)
10. Moropoulou, A.: Decay phenomena and mechanisms. Interdisciplinary NTUA Master Course – Protection of Monuments
11. Poziopoulos, A.: History and conservation problems of the Medieval City of Rhodes. In: Proceeding Scientific Meeting 27–29 November 1986 UNESCO – Hellenic Ministry of Culture – Municipality of Rhodes, p. 375 (1992). ISBN 9602141315
12. Moropoulou, A., Koui, M., Tsiourva, Th., Kourteli, Ch., Papasotiriou, D.: Macro- and micro nondestructive tests for environmental impact assessment on architectural surfaces. In: Vandiver, P.B., Druzik, J.R., Merkel, J.F., Stewart, J. (eds.) Materials Issues in Art and Archaeology V, vol. 462, pp. 343–349. Materials Research Society, Pittsburgh (1997)

13. Moropoulou, A., et al.: Techniques and methodology for the preservation and environmental management of historic complexes - the case of the Medieval City of Rhodes. In: Moropoulou, A., Zezza, F., Kollias, E., Papachristodoulou, I. (eds.) Proceedings of 4th International Symposium on the Conservation of Monuments in the Mediterranean Basin, vol. 4, pp. 603–634. Technical Chamber of Greece, Rhodes (1997)

14. Moropoulou, A., Koui, M., Avdelidis, N.P., Kourteli, Ch.: Preservation planning as a tool for a sustainable historic city. In: Brebbia, C.A., Ferrante, A., Rodriguez, M., Terra, B. (eds.) Advances in Architecture 9, Urban Regeneration and Sustainability: The Sustainable City, pp. 327–336. Wessex Institute of Technology (2000)

15. Moropoulou, A., Kourteli, Ch., Achilleopoulos, N.: Environmental management and preservation of the medieval fortifications of the City of Rhodes. In: International Conference Secular Medieval Architecture in the Balkans, 1300–1500, and its Preservation, AIMOS - Society for the Study of Medieval Architecture in the Balkans and its Preservation, and Organization for the Cultural Capital of Europe - Thessaloniki 97, 3–5 November (1997). Book of abstracts

16. Moropoulou, A.: Criteria and planning methods of regional integrated development programs, with emphasis to the ecological and cultural tourism, based on the utilisation, protection and preservation of the environment and the historic complexes. In: International Conference on Archaeology and Environment in Dodecanese: Research and Cultural Tourism, Rhodes, 1–4 November 2000. Book of abstracts

17. Moropoulou, A., Koui, M., Theoulakis, P., Kourteli, Ch., Zezza, F.: Digital image processing for the environmental impact assessment on architectural surfaces. J. Environ. Chem. Technol. **1**, 23–32 (1995)

18. Theoulakis, P., Moropoulou, A.: Microstructural and mechanical parameters determining the susceptibility of porous building stones to salt decay. Constr. Build. Mater. **11**(1), 65–71 (1997)

19. Michailidou, M.: Ephoriate of Dodecanese- Ministry of Culture "Historic buildings and Monuments owned by the Public" Documents of the Committee, January 2018

20. Moropoulou, A.: Innovative sustainable preservation and management of Historic Cities towards Cultural Driven Economy: The Medieval City of Rhodes as a pilot program. In: CHCD 2018 Symposium, Beijing (2018)

21. Moropoulou, A., Delegou, E.T.: Innovative technologies and strategic planning methodology for assessing and decision-making concerning preservation and management of historic cities. In: Moropoulou, A., Kollias, E., Papatheodorou, G. (eds.) Proceedings of the 7th International Symposium of the Organization of World Heritage Cities, Rhodes (2003). CD-ROM Proc

22. Moropoulou, A., Kouloumbi, N., Haralampopoulos, G., Konstanti, A., Michailidis, P.: Criteria and methodology for the evaluation of conservation interventions on treated porous stone susceptible to salt decay. Prog. Org. Coat. **48**(2-4), 259–270 (2003)

23. Avdelidis, N.P., Moropoulou, A., Stavrakas, D.: Detection and quantification of discontinuities in building materials using transient thermal NDT techniques: modeling and experimental work. Mater. Eval. **64**(5), 489–491 (2006)

24. Moropoulou, A., Koui, M., Kourteli, Ch., Theoulakis, P., Avdelidis, N.P.: Integrated methodology for measuring and monitoring salt decay in the Medieval City of Rhodes porous stone. J. Mediterr. Archaeol. Archaeom. **1**(1), 37–68 (2001)

25. Moropoulou, A., Haralampopoulos, G., Tsiourva, T., Auger, F., Birginie, J.M.: Artificial weathering and non-destructive tests for the performance evaluation of consolidation materials applied on porous stones. Mater. Struct. **36**, 210–217 (2003)

26. Moropoulou, A., Koui, M., Kourteli, Ch., Avdelidis, N.P., Achilleopoulos, N.: GIS management of NDT results for the spatial estimation of environmental risks to historic monuments. In: Brebbia, C.A. (ed.) Management Information Systems, pp. 207–215. Wessex Institute of Technology (2000)

27. Moropoulou, A., Koui, M., Avdelidis, N.P.: Innovative strategies for the preservation of historic cities by ND monitoring techniques and GIS management of data regarding environmental impact on historic materials and structures. In: Proceedings of 9th International Symposium Congress on Deterioration and Conservation of Stone, Venice, pp. 119–127 (2000)

28. Moropoulou, A., Theoulakis, P., Tsiourva, Th., Karoglou, M., Koui, N.: Innovative strategic planning of conservation materials and interventions at the Medieval Fortifications of Rhodes. In: Proceedings International Symposium 15 years of Restoration in the Medieval Town of Rhodes, Rhodes (2001)

29. Moropoulou, A., Theoulakis, P., Dellas, K., Dellas, G., Tzakou, A.: Environmental management of historic complexes or cities: a new conservation strategy. In: Thiel, M. J. (ed.) Conservation of Stone and Other Materials, RILEM-UNESCO, vol. 2, pp. 845–852. E&FN SPON, Chapman & Hall, Paris (1993)

30. Kioussi, A., Karoglou, M., Bakolas, A., Labropoulos, K., Moropoulou, A.: Documentation protocols to generate risk indicators regarding degradation processes for cultural heritage risk evaluation. In: Grussenmeyer, P. (ed.) XXIV International CIPA Symposium, 2–6 September 2013, Strasbourg, France, International Archives of the Photogrammetry, Remote Sensing and Spatial Information Sciences ISPRS, vol. XL-5/W2, pp. 379–384 (2013)

31. Moropoulou, A., Konstanti, A., Aggelakopoulou, E., Kokkinos, Ch.: Historic Cities as open labs of research and post graduate education. In: 7th International Symposium of the Organization of World Heritage Cities, Rhodes (2003)

32. Moropoulou, A., et al.: Study of mortars in the Medieval City of Rhodes. In: Thiel, M.J. (ed.) Conservation of Stone and Other Materials, RILEM-UNESCO, vol. 1, pp. 394–401. E&FN SPON, Chapman & Hall, Paris (1993)

33. Author, F.: Article title. Journal 2(5), 99–110 (2016)

Inception – Inclusive Cultural Heritage in Europe through 3D Semantic Modeling

Advanced 3D Survey and Modelling for Enhancement and Conservation of Cultural Heritage: The INCEPTION Project

Roberto Di Giulio⬛, Federica Maietti(✉)⬛, and Emanuele Piaia⬛

Department of Architecture, University of Ferrara,
Via Ghiara 36, 44121 Ferrara, Italy
{dgr,federica.maietti,emanuele.piaia}@unife.it

Abstract. Digital documentation technologies combined with innovative analytical techniques and digital tools can be an effective strategy supporting multidisciplinary documentation and modelling aimed at conservation, enhancement and preservation of Cultural Heritage. New technologies and digital devices should play an important innovative role to understand, access, enhance and preserve Cultural Heritage.

In this framework, the ongoing project INCEPTION *Inclusive Cultural Heritage in Europe through 3D semantic modelling* – funded by the European Commission within the Programme Horizon 2020 – proposes a workflow aimed at the achievements of efficient 3D digitization methods, post-processing tools for an enriched semantic modelling, web-based solutions and applications to ensure a wide access to experts and non-experts. The implementation of data collection processes and the development of semantically enriched 3D models is an effective way to enhance the dialogue between ICT technologies, different Cultural Heritage users and different disciplines, both social and technical.

Moreover, INCEPTION deals with complex heritage architectures and sites; in order to manage this complexity, a common protocol for data capturing has been developed, considering the uniqueness of each heritage site, accuracy and reliability, additional data and semantic proprieties to be recorded for different heritage applications.

Keywords: Heritage digital documentation · Advanced data capturing · Semantic modelling · Inclusive accessibility · Enhancement · Conservation

1 Introduction

Cultural Heritage conservation is more and more linked to the opportunities of documentation, condition assessment, monitoring and predictive analysis by means of nondestructive procedures. Protection, conservation and sustainable maintenance of Cultural Heritage is one of the Europe's priority to preserve our common assets to future generation, to foster social cohesion, to reinforce a sense of belonging to a common European space and to support job creation and economic growth.

A. Moropoulou et al. (Eds.): TMM_CH 2018, CCIS 962, pp. 325–335, 2019.
https://doi.org/10.1007/978-3-030-12960-6_21

Heritage management is a strong interdisciplinary field and many actors are involved in the complex process that, from the documentation up to the restoration, leads to the preservation, enhancement and sustainable exploitation of assets.

In this framework, the European Commission supports more and more several challenges related to different fields of Cultural Heritage toward a European common identity, transnational dialogue and understanding.

Moreover, the increasingly widespread use of different diagnostic methodologies and three-dimensional survey devices opens new scenarios within the non-destructive assessment and characterization of structures and materials, toward an overall and holistic digital documentation of cultural heritage [1].

Fig. 1. INCEPTION "corporate" image. 3D digitization methods, post-processing tools for an enriched semantic modelling, web-based solutions and applications to ensure a wide access to experts and non-experts are the core of the project.

As stated during the Innovation & Cultural Heritage Conference, held on 20 March 2018 in Brussels [2], the future European research on cultural heritage needs a holistic and critical research agenda and an inclusive interdisciplinary approach. One of the key questions is how to best use the opportunities provided by digitalization in the valorization of cultural heritage. Several presentations, based on ongoing Horizon 2020 funded projects, showed the manifold advances of digitalization for conservation, presentation and consumption of heritage. "Digitalization can be an effective instrument of democratization of cultural heritage as it opens new forms of access, e.g. by allowing entrance to previously closed heritage places or museum collections, or by allowing memberships in heritage communities without physical presence in a locality. […] More research is needed concerning the use of digital heritage by the different social, cultural and professional groups, which takes into consideration the social effects of virtual realities and the visibility of actorship in the processes related to cultural heritage practices" [3].

Advanced 3D documentation identifying different layers of data to be recorded for heritage knowledge, enhancement and conservation is one of the main outcomes of the European Project "INCEPTION - Inclusive Cultural Heritage in Europe through 3D semantic modelling" (Fig. 1), funded by the European Commission under the Programme Horizon 2020 and started in 2015. The project proposes the enhancement of efficiency in 3D data capturing procedures and devices, especially dealing with complex heritage "spaces": cultural heritage sites, historical architectures, archaeological sites characterized by non-conventional features, location and geometries [4] (Fig. 2).

Fig. 2. Integrated digital documentation of complex heritage buildings. The example of Palazzo del Podestà in Mantua (DIAPReM Centre).

The overall project workflow is developed starting from requirements, integrated data capturing and holistic heritage documentation, the semantic enrichment via 3D modelling in H-BIM environment, and the models deployment and valorization through the INCEPTION platform. Indeed, the main innovations under INCEPTION will be delivered through an open standard platform to collect, implement and share digital models. Platform interface and functionalities will allow users to download and upload models, work with H-BIM models with different level of details, enrich contents and information linked to geometric models in an interoperable way and explore a wide range of data and contents.

The project has the ambition to strongly support the development of a pan-European approach to data usage for decision making related to conservation and preventive interventions and for supporting site management and sustainable exploitation of assets by integrating new methods for condition assessment survey of Cultural Heritage based on predictive analysis (diagnostic, conservative, morphometric) and non-destructive procedures.

2   The INCEPTION Project Overall Vision

INCEPTION aims to develop new tools for the interoperability and the inclusive sharing of three-dimensional heritage models towards new forms of access and awareness of the European cultural heritage. In the middle of the fourth and last year of activity, the project is working on models deployment and widespread sharing, starting from a methodological and technical advancement in 3D data capturing and holistic digital documentation [5].

The project consists of five main actions:

- the definition of a common framework for catalogue methodology, by mapping the stakeholder' knowledge demands;
- the development of an advanced integrated 3D data capturing;
- the identification of the Cultural Heritage buildings semantic ontology and data structure for information catalogue and modelling in Heritage-BIM environment, based on Open BIM standard;
- the implementation of the INCEPTION Semantic Web Platform. Deployment and valorization of 3D heritage models, to enable the sharing and enrichment of the information and interpretation of the models by users.

2.1   Project Workflow

The overall project workflow is developed starting from requirements (according to specific users and needs, what kind of data and information can be collected and managed by a 3D model), the integrated data capturing and holistic heritage documentation. Semantic modelling for Cultural Heritage buildings in H-BIM environment and the development of the INCEPTION platform for deployment and valorization of enriched 3D models will allow accomplishing the main objectives of accessing, understanding and strengthening European Cultural Heritage [6].

According to the abovementioned workflow, the H-BIM modelling procedure starts with a depth analysis of user needs (considering both expert and non-expert users). The identification of the Cultural Heritage buildings semantic ontology and data structure for information catalogue will allow the integration of semantic attributes with hierarchically and mutually aggregated 3D digital geometric models for management of heritage information. The development of the INCEPTION platform is the key-targeted achievement of the overall project, in order to accomplish the main objectives of accessing, understanding and strengthening European cultural heritage by means of enriched 3D models. The main aim of the platform is to innovate 3D models "forever", "for everybody", "from everywhere", by developing, collecting and sharing interoperable 3D semantic models (Fig. 3).

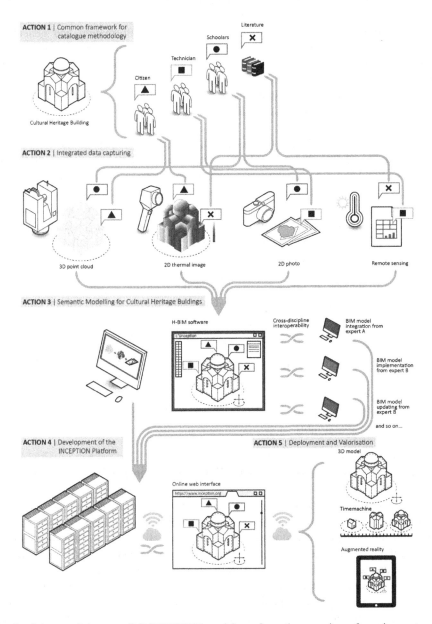

Fig. 3. Schema of the overall INCEPTION workflow, from the mapping of requirements and holistic documentation up to the semantic modelling and model sharing through the INCEPTION platform.

In the cloud-based platform, the main input is a BIM model of a Heritage building or site. All the BIM models are uploaded in the platform as IFC (Industry Foundation Class) standard files, in order to allow the access regardless of the software used to generate the BIM model. The IFC file is then processed by means of several server-side custom Windows services, extracting all the semantic information and generating Resource Description Framework (RDF) triples, according to the INCEPTION H-BIM ontology, serialized as Turtle (TTL) files. All these triples are stored in a semantic triple store, accessed via HTTP through a dedicated Apache Fuseki SPARQL server. The platform provides the user with the possibility of enriching the models with new semantic metadata, new data and attachments related to the whole CH site or to a specific geometrical element.

Therefore, semantic H-BIM models uploaded on the platform will allow users to interact with the models; thus, not only "to access" and "to understand" the models, but also to be stimulated "to give the users' perceptual responses to the models" through multi-layered information and documentation systems eligible for spatial and multi-criteria queries in a virtual 3D environment [7]. The end-users will be able to access information utilizing a standard browser, and they will be able to query the database using keywords and an easy search method.

This could be particularly relevant to foster the use of 3D models for diagnostic procedures and as a very efficient assessment and decision-making tool. Digital data can be used to manage dimensional analysis of the buildings or sites [8], typological analysis, technological analysis and material assessment to assess the current condition (typical materials and structures and their damage) and state of conservation, technical solutions for preservation and overall maintenance planning.

The INCEPTION platform offers the opportunity to manage contents such as single elements of a building to visualize different historic phases, stored data of previous interventions and links with external databases to increase the quality of reliable information. The possibility to share and make available documents linked to the geometric 3D model and the capacity to download 3D models with different scales of detail are very relevant tools supporting diagnostic procedures [9].

3   Integrated 3D Survey and Holistic Documentation

Methods and processes for data collection are continuously developing and today are characterized by an effective interdisciplinary. Skills on 3D laser scanner survey, diagnostic procedures and historical researches, as well as environmental condition assessment or management of metric and dimensional data support the vision of integrated digital documentation for cultural heritage assessment [10]. The INCEP-TION's approach is focused on the 3D data capturing and modelling of heritage buildings and sites, through digital representation of the shape, appearance and con-servation condition, in addition to a set of semantic information able to enrich research and deployment applications. The project is developing new research avenues in the field of heritage 3D data acquisition and modelling in order to guide the process of digitization of cultural heritage and innovation strategies to the three-dimensional modelling.

Advancement of 3D data capturing under INCEPTION starts from a primary concept: the 3D survey of complex heritage architectural spaces needs a common protocol for data capturing considering the uniqueness of each site, quality indicators, time-consumption, cost-effectiveness, data accuracy and reliability, additional data and semantic proprieties to be recorded for heritage applications, adaptability to different sites with different historical phases.

The architectural space becomes the foundations, the common core and the "connection" for the creation of a protocol for optimizing the 3D documentation of cultural heritage. The methodology set as a priority the unconventional features/ geometries, unique and complex within Heritage, avoiding the "segmentation" of data acquired and facilitating data access and use through an inclusive approach [11].

The integration of digital data and the possibilities of re-use digital resources is an important challenge for protection and conservation of the historic buildings as well as for an efficient management in the long term [12] (Fig. 4).

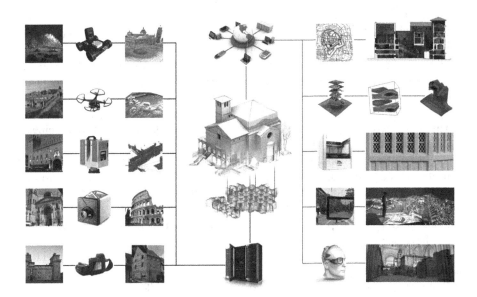

Fig. 4. From the left: integrated data capturing through the data acquisition protocol, advancement in 3D models processing in BIM environment and model sharing.

The data acquisition protocol has been developed within a more general methodological procedure of heritage documentation: integrated digital documentation under INCEPTION is carried out through a holistic approach. Through the 3D digitization of a monument, various aspects of recording concerning diagnostics (building material and decay patterns), construction and architectural documentation, new digital content management may be achieved. In this context, interdisciplinary approach means that the advantages and the output of every method applied to a monument, can lead to the

3D modelling approach, improving an inclusive understanding of European cultural identity and diversity [13].

3.1   The INCEPTION Data Acquisition Protocol

In order to face the main challenges related to 3D survey of complex architectures and to start solving the issue of the large amount of captured data and time-consuming processes in the production of 3D digital models, an Optimized Data Acquisition Protocol (DAP) has been set up. The purpose is to guide the processes of digitization of cultural heritage, respecting needs, requirements and specificities of cultural assets.

The DAP provides a workflow for a consistent development of survey procedures for tangible cultural heritage and define a common background for the use of H-BIM across multiple building types and for a wide range of technical users. Furthermore, this protocol will be useful for any agency, organization or other institution that may be interested in utilizing survey procedures aimed at 3D H-BIM semantic models creation and their implementation for the INCEPTION platform.

The DAP is intended to ensure uniformity in 3D digital survey for all the buildings that will be part of INCEPTION platform, considering a wide range of 3D data capturing instruments [14] because of multiple users and different techniques related to specific disciplines. 3D survey instruments and techniques continue to evolve, and this protocol will continue to be reviewed and updated to reflect advances in industry technology, methodology and trends; in every case, the protocol application will ensure data homogenization between surveys tailored to different requirements.

The survey workflow is split into eight main steps that define specific requirements and their related activity indicators: Scan Plan, Health and safety, Resolution Requirements, Registration mode, Control network, Quality control, Data control and verification, Data storage and archive. Each step of the workflow becomes a measuring system to verify the requirements of the survey, and the ability of finding the right answer define the level quality [15]. Every single step becomes an activity indicator that contributes to get a specific evaluation ranking. Four incremental evaluation categories have been defined (B, A, A+, A++).

"B" is the minimum evaluation category to be compliant with the INCEPTION Platform. It's intended to be used for very simple buildings or for the creation of low-detailed BIM model for digital reconstruction aimed at VR, AR and visualization purposes. In this case, the metric value of the model is less important than the morphological value.

"A" is the evaluation category suitable for documentation purposes where the metric and morphological values are equivalent in term of impact on the survey that needs to be preliminary scheduled and designed. The registration process of 3D captured data cannot be based only on morphological method but it should be improved by a topographic control network or GPS data.

"A+" is the most suitable for preservation purposes because only the surveys compliant with this category could be a useful tool for restoration projects that need extremely correct metric data. From these surveys, BIM models as well as 2D CAD drawings until 1:20 scale are available. The project phase gets more importance than previous categories in order to schedule and manage the survey campaign and choose the right technical instruments to perform the data capturing. The management and the correction of metric errors are based on topographic techniques, in particular for what could concern the registration of different scan. The documentation phase will be developed organizing the information into Metadata and Paradata. Elements of quality control are integrated into the process.

"A++" is suitable for very complex buildings where the capturing process need to be documented and traced in order to get the maximum control on data or when monitoring process developed in a non-continuous time span take place. This category could be useful even if different teams of technicians work together, simultaneously or in sequence, with different capturing instruments and different accuracies. The A++ category allows analyzing how a survey has been performed in every single phase: moreover, this capability allows integrating a survey in different times.

4 Conclusions

"Heritage information – the activity and products of recording, documenting, and managing the information of cultural heritage places – should be not only an integral part of every conservation project but also an activity that continues long after the intervention is completed. It is the basis for the monitoring, management, and routine maintenance of a site and provides a way to transmit knowledge about heritage places to future generations" [16].

The possibility to achieve interoperable models able to enrich the interdisciplinary knowledge of European cultural identity is one of the main outcomes of INCEPTION. The implementation of data collection processes and the development of semantically enriched 3D models is an effective way to enhance the dialogue between ICT technologies, different Cultural Heritage experts, users and different disciplines, both social and technical.

Collecting information to preventive conservation, monitoring and maintenance of European cultural historical heritage is undeniably connected with the use of methods, tools and techniques that in recent years have become technologically advanced and widespread. The technological evolution of survey systems represents an important innovation to manage three-dimensional databases [17]. The integration of other non-destructive procedures as thermal imaging, intensity value, integrated sensors, spectrophotometry, sonic surveys, etc. allows the collection and integration of other important digital data. These databases are geometric archives that can be used for research goals and for preservation, restoration and protection of cultural heritage.

Acknowledgments. The project is under development by a consortium of fourteen partners from ten European countries led by the Department of Architecture of the University of Ferrara. Academic partners of the Consortium, in addition to the Department of Architecture of the University of Ferrara, include the University of Ljubljana (Slovenia), the National Technical University of Athens (Greece), the Cyprus University of Technology (Cyprus), the University of Zagreb (Croatia), the research centers Consorzio Futuro in Ricerca (Italy) and Cartif (Spain).

The clustering of small medium enterprises includes DEMO Consultants BV (The Netherlands), 3L Architects (Germany), Nemoris (Italy), RDF (Bulgaria), 13BIS Consulting (France), Z + F (Germany), Vision and Business Consultants (Greece).

The INCEPTION project has been applied under the Work Programme *Europe in a changing world – inclusive, innovative and reflective Societies* (Call - Reflective Societies: Cultural Heritage and European Identities, Reflective-7-2014, Advanced 3D modelling for accessing and understanding European cultural assets).

This research project has received funding from the European Union's H2020 Framework Programme for research and innovation under Grant agreement no. 665220.

References

1. Maietti, F., Piaia, E., Brunoro, S.: Diagnostic integrated procedures aimed at monitoring, enhancement and conservation of cultural heritage sites. In: MALTA International Conference–Europe and the Mediterranean–Towards a Sustainable Built Environment, SBE 2016, pp. 309–316. Gutenberg Press, Malta (2016)

2. The high-level Horizon 2020 conference "Innovation and Cultural Heritage" was organized by the European Commission Directorate General for Research and Innovation, in close cooperation with Directorates General for Education and Culture and for Communications Networks, Content and Technology

3. Innovation and Cultural Heritage: Conference Report. ISBN 978-92-79-81847-9, https://ec.europa.eu/info/sites/info/files/conferences/ki-02-18-531-en-n.pdf. Accessed 23 Aug 2018

4. Di Giulio, R., Maietti, F., Piaia, E.: 3D documentation and semantic aware representation of Cultural Heritage: the INCEPTION project. In: Proceedings of the 14th Eurographics Workshop on Graphics and Cultural Heritage, pp. 195–198. Eurographics Association (2016)

5. Ioannides, M., et al.: Towards monuments' holistic digital documentation: the Saint Neophytos Enkleistriotis case study. In: Ioannides, M., et al. (eds.) EuroMed 2016. LNCS, vol. 10058, pp. 442–473. Springer, Cham (2016). https://doi.org/10.1007/978-3-319-48496-9_36

6. Maietti, F., Di Giulio, R., Balzani, M., Piaia, E., Medici, M., Ferrari, F.: Digital memory and integrated data capturing: innovations for an inclusive Cultural Heritage in Europe through 3D semantic modelling. In: Ioannides, M., Magnenat-Thalmann, N., Papagiannakis, G. (eds.) Mixed Reality and Gamification for Cultural Heritage, pp. 225–244. Springer, Cham (2017). https://doi.org/10.1007/978-3-319-49607-8_8

7. Ioannides, M., et al.: Online 4D reconstruction using multi-images available under Open Access. ISPRS Ann. Photogram. Remote Sens. Sapt. Inf. Sci. **II-5 W1**, 169–174 (2013)

8. Kyriakaki, G., et al.: 4D reconstruction of tangible cultural heritage objects from web-retrieved images. Int. J. Heritage Digit. Era 3(2), 431–451 (2014)

9. Yastikli, N.: Documentation of Cultural Heritage using digital photogrammetry and laser scanning. J. Cult. Heritage **8**(4), 423–427 (2007)

10. Doulamis, A., et al.: 5D modelling: an efficient approach for creating spatiotemporal predictive 3D maps of large-scale cultural resources. ISPRS Ann. Photogram. Remote Sens. Spat. Inf. Sci. **II-5/W3**, 61–68 (2015)

11. Di Giulio, R., Maietti, F., Piaia, E., Medici, M., Ferrari, F., Turillazzi, B.: Integrated data capturing requirements for 3D semantic modelling of cultural heritage: the INCEPTION protocol. Int. Arch. Photogramm. Remote Sens. Spat. Inf. Sci. **XLII-2/W3**, 251–257 (2017)

12. Verykokou, S., Doulamis, A., Athanasiou, G., Ioannidis, C., Amditis, A.: Multi-scale 3D modelling of damaged cultural sites: use cases and image-based workflows. In: Ioannides, M., et al. (eds.) EuroMed 2016. LNCS, vol. 10058, pp. 50–62. Springer, Cham (2016). https://doi.org/10.1007/978-3-319-48496-9_5

13. Maietti, F., Di Giulio, R., Piaia, E., Medici, M., Ferrari, F.: Enhancing Heritage fruition through 3D semantic modelling and digital tools: the INCEPTION project. In: IOP Conference Series: Materials Science and Engineering, vol. 364, p. 012089 (2018)

14. Stylianidis, E., Remondino, F. (eds.): 3D Recording, Documentation and Management of Cultural Heritage. Whittles Publishing, Dunbeath (2016)

15. Balzani, M., Maietti, F.: The architectural space in an inclusive protocol for the 3D integrated acquisition of Cultural Heritage for documentation, diagnosis, representation, enhancement and conservation. In: Bertocci, S., Bini, M. (eds.) The Reasons of Drawings. Thought, Shape and Model in the Complexity Management. Proceedings of the XXXVIII Convegno Internazionale dei Docenti della Rappresentazione - XIII Congresso Unione Italiana Disegno, pp. 1039–1044. Gangemi Editore, Roma (2016)

16. Letellier, R., Schmid, W., LeBlanc, F.: Recording, Documentation, and Information Management for the Conservation of Heritage Places. Guiding Principles. Getty Conservation Institute, J. Paul Getty Trust, Los Angeles (2007)

17. Maietti, F., Piaia, E., Turillazzi, B.: Digital documentation: sustainable strategies for Cultural Heritage assessment and inspection. In: MALTA International Conference–Europe and the Mediterranean–Towards a Sustainable Built Environment, SBE 2016, pp. 303–308. Gutenberg Press, Malta (2016)

INCEPTION: Web Cutting-Edge Technologies Meet Cultural Heritage

Ernesto Iadanza[1(✉)] ⓘ, Peter Bonsma[2], Iveta Bonsma[2],
Anna Elisabetta Ziri[3] ⓘ, Federica Maietti[4] ⓘ, Marco Medici[4] ⓘ,
Federico Ferrari[4] ⓘ, and Pedro Martín Lerones[5]

[1] Consorzio Futuro in Ricerca, Ferrara, Italy
ernesto.iadanza@unifi.it
[2] RDF Ltd., Sofia, Bulgaria
{peter.bonsma,iveta.bonsma}@rdf.bg
[3] Nemoris srl., Bologna, Italy
annaelisabetta.ziri@nemoris.it
[4] Department of Architecture, University of Ferrara, Ferrara, Italy
{federica.maietti,marco.medici,
federico.ferrari}@unife.it
[5] Fundación CARTIF, Boecillo, Valladolid, Spain
pedler@cartif.es

Abstract. INCEPTION project is a research and innovation project funded by the European Commission to realize "innovation in 3D modelling of cultural heritage through an inclusive approach for time-dynamic 3D reconstruction of artefacts, built and social environments. It enriches the European identity through understanding of how European cultural heritage continuously evolves over long periods of time".

In this paper are described some state of the art technologies adopted in developing the cloud web platform that is the core of the whole project.

After a detailed comparison of the features of the INCEPTION platform, compared with 27 other existing web sites, some of the most interesting solutions, based on the match between BIM (Building Information Modeling), Cloud and Semantic Web approach, are described.

This EU project is a clear example of cutting-edge technologies applied to the European Cultural Heritage.

Keywords: Heritage documentation · H-BIM · Semantic web · Cloud · BIM

1 Introduction

The main scope of the INCEPTION project, a research and innovation project funded by the European Commission under H2020-EU - Reflective societies - cultural heritage and European identity is, according to the official website, to realize "innovation in 3D modelling of cultural heritage through an inclusive approach for time-dynamic 3D reconstruction of artefacts, built and social environments. It enriches the European

A. Moropoulou et al. (Eds.): TMM_CH 2018, CCIS 962, pp. 336–346, 2019.
https://doi.org/10.1007/978-3-030-12960-6_22

identity through understanding of how European cultural heritage continuously evolves over long periods of time" [1].

The development of a specific cloud based INCEPTION platform is the key-targeted achievement of the overall project, in order to accomplish the main objectives of accessing, understanding and strengthening European Cultural Heritage (CH) by means of enriched 3D models.

The platform meets the main aim of realising innovation in 3D models "forever", "for everybody", "from everywhere", by developing, collecting and sharing inter-operable 3D semantic models.

It is designed to be used both by the CH site managers/owners and by the end users. While the first will feed the platform with 3D models and semantic information about their cultural sites, the end users will access the platform to navigate inside the European CH world. They will experience both on-site and off-site tools for a complete immersion in the site, also through Virtual Reality (VR) and Augmented Reality (AR) tools. The platform provides all the available semantic information linked to the whole 3D model or to the single geometrical elements.

The whole INCEPTION project is based on the bond between state-of-the-art architectural modeling technologies (BIM, Building Information Modeling) and the latest cutting-edge web technologies. The platform is grounded on semantic web technologies and makes extensive use of WebGL and RESTful APIs.

2 State of the Art

The features included in the INCEPTION platform has been compared to 27 of the major web-based platforms that allow downloading and exchanging 3D models.

While INCEPTION is a semantic BIM platform for CH buildings, we have analysed a wide variety of available web platforms to analyse the specific characteristics of each in relation to the many design constraints that we have set for the user-experience and technical requirements of the INCEPTION platform (such as enjoyment, navigation, interaction, etc.).

The innovative features and application potential of the INCEPTION platform are evident in relation to the major platforms seen as the state of the art. A comparative multi-variable analysis of different platforms is shown in Tables 1 and 2 below. A more detailed comparison can be found in the project deliverables, but cannot be fully disclosed here because of confidentiality constraints.

The comparison shows that the platforms related to Cultural Heritage are few and focused on specific project of documentation/enhancement. They are not highly visited and do not have many 3D models available because they are linked to specific artefacts related to a museum or cataloguing items in an archaeological site. The quality, however, is high because almost always there is a curatorship. The problem (regarding accesses) of these platforms is that their main objective is documentation, cataloguing, and sometimes enhancement.

The most active platforms (in terms of population and visits) are those with a generic vocation and strongly linked to the sharing of 3D models.

Table 1. List of platforms that have some points in common with the Inception Platform

	Application sector		
	Application	Genre	Link
Inception Platform	Architecture, Documentation, Animation, Gaming, Visualization	Architecture, Heritage Preservation	
3D Warehouse	Animation, Gaming, 3D Printing	Architecture, Cars, Furniture, Fanart	3dwarehouse. sketchup.com
3dexport	3D Printing, Gaming, Animation and Graphic Design	DIY, Jewelry, Decoration	it.3dexport.com
3DModelFree	Animation and Graphic Design	Interior Design, Architecture	www. 3dmodelfree.com
3DSky	Animation and Graphic Design	Interior Design, Architectural Visualization	3dsky.org
Archive3d	Animation and Graphic Design	Interior Design, Architectural Visualization	archive3d.net
Autodesk Online Gallery	3D Printing, Engineering, Architectural Visualization	All	gallery.autodesk. com
BIM. archiproducts	Architecture, Animation, Graphic Design	Furniture	bim.archiproducts. com
BIMobject	Architecture, Animation, Graphic Design	Furniture, Architecture	bimobject.com
Blend Swap	Animation, Graphic Design, 3D Printing	All	www.blendswap. com
Cgtrader	Gaming, Animation, Graphic Design	All	www.cgtrader.com
Clara.io	Animation, Graphic Design, and 3D Printing	All	clara.io
CYARK	Documentation	Cultural Heritage	www.cyark.org
GB3D Type Fossils	Documentation	Fossils, Cultural Heritage	www.3d-fossils. ac.uk
Grabcad	Engineering and 3D Printing	Tools, Equipment	grabcad.com
Library Smartbim	Architecture, Animation, Graphic Design	Furniture, Architecture	library.smartbim. com
Myminifactory	3D Printing	DIY, Jewellery, Decoration, Heritage Preservation	all3dp.com
National bim library	Architecture, Animation, Graphic Design	Furniture, Architecture	www. nationalbimlibrary. com

(continued)

Table 1. (*continued*)

	Application sector		
	Application	Genre	Link
Ornament3d	Documentation	Cultural Heritage	ornament3d.org
Sketchfab	3D Printing, Animation, Gaming	Fanart, Architecture, Education, Heritage Preservation	sketchfab.com
Smithsonian X3D	Animation, Graphic Design, 3D Printing	Heritage Preservation	3d.si.edu
Syncronia	Architecture, Animation, Graphic Design	Furniture	www.syncronia.com
Thingiverse	3D Printing	DIY	www.thingiverse.com
ThreeDScans	3D Printing, Animation, Graphic Design	Heritage Preservation	threedscans.com
Turbosquid	Gaming, Architectural Visualization, Graphic Design	All	www.turbosquid.com
Unity Asset Store	Gaming, Architectural Visualization	Universal	www.assetstore.unity3d.com
Zamaniproject	Documentation	Cultural Heritage	zamaniproject.org

The INCEPTION platform will be placed on a more generalist level in the Cultural Heritage (all CH heritage) with strong skills in technical/managerial/maintenance/conservation and development, as well as enhancement through the new possibilities offered by the Augmented and Virtual reality. The INCEPTION platform will not look like a closed documentation or cataloguing system, but a space for interchange of information for the dialogue between edutainment and the AEC (Architecture Engineering Construction) engineers, between first-class college students, scholars and tour operators.

Furthermore, the Semantic Web structure allows the platform to be interlinked with external CH available linked data and to be gradually enhanced by specific flexible data structures in the form of project specific ontologies.

None of the scanned platforms currently implements semantics about specific content on individual 3D models nor allows (structured or not) access to metadata or paradigms (if any). None of the scanned platforms implements true multi-user access with information, functions, or different data access modes depending on the different typologies of users.

There are projects that are more similar to the model proposed by INCEPTION, but we wanted to analyze platforms that had a market placement in relation to the number of uploaded and accessed models and their relevance to the AEC market: significantly, the great graphics engine of the Sketchfab platform, also used by other platforms. INCEPTION has chosen the Sketchfab platform for dissemination in this first phase of the project before implementing its platform, in relation to 3D models that can be implemented on the platform and/or products in case studies.

Table 2. Comparison of features available in several platforms

	Features of 3D models					Search		Visualization Tool						Accessibility and social		
	3D Model	3D Bim Model	3D cloud of point data	Metadata	Semantic Data	Search	Semantic Search	3D View	Interactive 3D View	V/R for Pc	V/R for mobile	Editing View	Editing Model	Inclusive GUI	User Feedback	Social media interaction
Inception Platform	x	x	x	x	x	x	x	x	x	x			x	x	x	x
3D Warehouse	x	x				x	x	x	x						x	
3dexport	x					x								x	x	x
3DModelFree	x					x										
3DSky	x					x								x	x	x
Archive3d	x					x	x									
Autodesk Online Gallery	x					x	x							x	x	
BIM.archiproducts	x	x				x	x							x	x	x
BIMobject	x	x				x										
Blend Swap	x					x		x	x	x	x	x	x		x	x
Cgtrader	x					x									x	x
Clara.io	x					x		x	x		x	x		x	x	x
CYARK	x		x					x	x							
GB3D Type Fossils	x			x		x	x									
Grabcad	x					x		x						x	x	x
Library Smartbim	x	x				x	x									
Myminifactory	x					x	x								x	x
National bim library	x	x				x	x									
Ornament3d	x					x		x								
Sketchfab	x		x			x		x	x	x	x	x	x		x	x
Smithsonian X3D	x					x		x	x	x		x		x		
Syncronia	x	x		x		x								x	x	x
Thingiverse	x					x	x							x	x	x
ThreeDScans	x					x		x						x		
Turbosquid	x					x	x							x		
Unity Asset Store	x					x	x								x	
Zamaniproject	x					x								x		

3   The INCEPTION Platform

In the cloud based INCEPTION Platform architecture (see Fig. 1), the main input is a BIM model of a CH site. The models can refer to several categories, such as museums, archaeological sites, historical sites and heritage buildings.

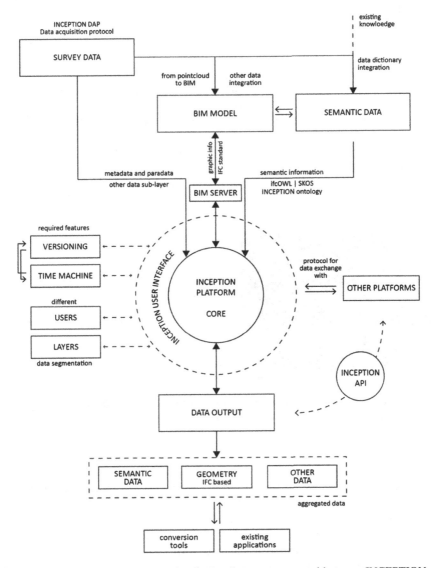

Fig. 1. Links and data flow between the platform's components and between INCEPTION and external tools

These models, that embed geometrical and semantic data, are created following a custom and well codified Data Acquisition Protocol (DAP), designed during the earlier phases of the INCEPTION EU Project. Describing this protocol is out of this paper's scope. All the BIM models are introduced in the platform as IFC (Industry Foundation Class) standard files. The adoption of this standard guarantees that the platform can be accessed regardless of the software used to generate the BIM model (e.g. Revit, ArchiCAD, etc.).

The IFC file is then processed by means of several server-side custom Windows services, that extract all the semantic information (both geometries and metadata) and generate Resource Description Framework (RDF) triples, according to the INCEPTION H-BIM ontology, serialized as Turtle (TTL) files.

All these triples are stored in a semantic triple store, accessed via HTTP through a dedicated Apache Fuseki SPARQL server.

The platform provides the user with the possibility of enriching the models with new semantic metadata. Indeed, the web client allows you to enrich the models with new data (e.g. a date, a value, some textual remarks, see Fig. 2) as well as with some attachments (e.g. pictures, thermographic images, 3D models of specific details, videos, etc.), related to the whole CH site or to a specific geometrical element.

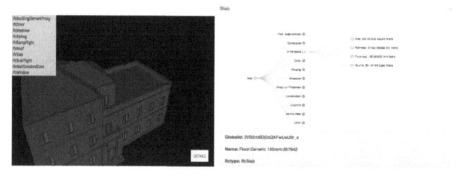

Fig. 2. Every element of the building can be used to perform a live SPARQL query that returns all the details for that element, according to the HBIM ontology. Each value can be updated via web, thanks to the SPARQL 1.1 Update functionalities.

In Fig. 3 it is shown a Collada (.DAE) file uploaded as attachment to a whole BIM model. According to its definition from Wikipedia, "Collada (acronym for COLLAborative Design Activity) is a is an interchange file format for interactive 3D applications. It is managed by the nonprofit technology consortium, the Khronos Group, and has been adopted by ISO as a publicly available specification, ISO/PAS 17506. COLLADA defines an open standard XML schema for exchanging digital assets among various graphics software applications that might otherwise store their assets in incompatible file formats. COLLADA documents that describe digital assets are XML files, usually identified with a .DAE filename extension" [2].

The use of Collada files, together with IFC files can be very useful: although this format lacks in the formal definition of entities, if compared to IFC, it provides an easy way to incorporate photorealistic textures. Moreover, it is possible to find some libraries, based on WebGL and Three.js, to visualize and manipulate these files through a common HTML5-enabled web browser. Most web browsers, today, support HTML5 both on desktop and on mobile devices. In INCEPTION we made an extensive customization of the above libraries to optimize the visualization of CH sites and exhibits [3].

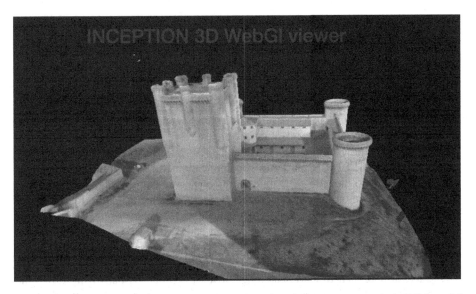

Fig. 3. An example of Collada (.DAE) file uploaded as attachment to a whole BIM model. Collada files can be useful since they can easily incorporate photorealistic textures.

4 Semantic Web Approach

The INCEPTION Platform interacts with its RESTful APIs by means of the above mentioned SPARQL 1.1 Protocol and RDF Query Language, which provides an SQL-like syntax and can be used to query the RDF triple store. From version 1.1 SPARQL offers full CRUD (Create, Read, Update and Delete) functionalities (e.g. SELECT, UPDATE) as well as a useful FILTER expression. It has been constructed to manipulate semantic web triples and data types and aggregations e.g. numbers, strings and URIs.

Clearly, the end user does not have to know how to perform a SPARQL query, since the platform interface itself converts graphical requests into queries. The following text represents exactly what is embedded in a SPARQL query to get all the WallStandardCase walls from a specific BIM mode, the query returns 121 wall elements:

```
PREFIX HBIM:   <http://www.inception-
project.org/HBIM.ttl#>
PREFIX rdf: <http://www.w3.org/1999/02/22-rdf-syntax-ns#>
    SELECT ?subject
    WHERE {

                ?subject rdf:type HBIM:WallStandardCase
;

    }
    LIMIT 1000
```

The semantic web approach of INCEPTION implies that many of these functions will be accessed by third party developers in back end (through web services) to integrate INCEPTION in their linked data systems.

Nevertheless, the access to UPLOAD and DOWNLOAD to the storage is fully managed by the INCEPTION platform web applications, that integrate functionalities for linking the files to the whole 3D model or to a specific element (e.g. a column or a door), uploading them on the cloud storage, creating the semantic triples to access and link them to the geometries, previewing, downloading, searching etc.

The INCEPTION platform web-applications can also act as a stimulus towards third parties to insert the INCEPTION H-BIM standard in their products, by raising awareness to the INCEPTION Platform capabilities.

One useful approach to start searching for a BIM model is narrowing down the geographic area in which to search. Users are nowadays accustomed to include management of geographic information in their daily activities on the web. The BIM model will be provided with geographic coordinates, thus enabling the use of common open-source GIS standards (such as CityGML and InfraGML) to query the system selecting areas of interest. A nice review on state-of-the-art integration between BIM and GIS is given by Liu et al. [4].

Functions like "search for all the models in Greece" or "select all the models in a 10 km area from this position" as well as mapping models on a GIS map will provide an intuitive and very productive way to start a search. To obtain this functionality, every 3D model uploaded is provided with latitude and longitude (easily obtained directly from the INCEPTION web application, leveraging on a Google Maps API) [5].

The coordinates are then transformed in semantic metadata using the following standard ontology http://www.w3.org/2003/01/geo/wgs84_pos, as shown in the following table:

```
HBIM:/HAMH-Museum/metadata#Building_1

<http://www.w3.org/2003/01/geo/wgs84_pos#lat>

"37.3517659"^^xsd:float


HBIM:/HAMH-Museum/metadata#Building_1

<http://www.w3.org/2003/01/geo/wgs84_pos#long>

"23.46703560000003"^^xsd:float
```

5 Cloud Storage

A cloud storage is dedicated to hosting all the files that complement data for the buildings, such as CAD files, historical documents, images and all the other type of files that could improve the level of information assigned to the object. These files are stored in the INCEPTION cloud storage and managed and organized through dedicated APIs, that also allow users to selectively access this data.

The user will have access to this archive through the INCEPTION platform, by exploiting custom APIs purposely developed for a broad interaction with external components. This open-source object-relational mapping (ORM) system allows thinking to the data model as a set of objects. All the entities created can be exposed as web services, therefore promoting the access to the INCEPTION platform to other solution providers.

6 Conclusion

In this paper we described the INCEPTION platform designed to exploit the concept of the semantic web. We have developed some web APIs to provide the INCEPTION H-BIM Interoperable Platform with a REST interface to access the BIM models allowing users to operate on 3D information as well as its related semantics. This core functionality is dedicated to the end-users as well as to developers and solution providers for interacting with external mobile devices and applications. Mobile phone applications, to be used on site, will access the INCEPTION H-BIM Interoperable Platform through this REST web-service, stimulating a flowering of web based mobile applications from third parties.

The web application, core of the whole platform, allows users to search 3D models using specific keywords contained in the semantic information. It displays a list of results in a textual and graphical form, giving the user the capability of clicking on single BIM objects and get access to specific (semantic) information and to correlate files and other linked data. This web application also gives third parties a live sample of the INCEPTION Platform capabilities, stimulating the insertion of the INCEPTION H-BIM standard in their products.

A cloud based file storage is dedicated to hosting all the files related to the CH projects, in several formats.

The end-user tools that are under development will strongly interact with the described features (APIs, reference queries and reference apps). These tools (in particular a condition assessment/asset management tool, an on-site AR mobile app and an off-site VR mobile app) will have both general and application-specific implications, as well as implications on the performances of the end-user tools.

References

1. INCEPTION EU Project. www.inception-project.eu
2. COLLADA FORMAT. https://en.wikipedia.org/wiki/COLLADA
3. WebGL Collada Library. https://threejs.org/examples/webgl_loader_collada.html
4. Liu, X., Wang, X., Wright, G., Cheng, J.C.P., Li, X., Liu, R.: A state-of-the-art review on the integration of Building Information Modeling (BIM) and Geographic Information System (GIS). ISPRS Int. J. Geo-Inf. **6**, 53 (2017)
5. Google MAPS Platform Documentation. https://developers.google.com/maps/documentation/

Cultural Heritage Sites Holistic Documentation Through Semantic Web Technologies

Anna Elisabetta Ziri[1(✉)] , Peter Bonsma[2], Iveta Bonsma[2],
Ernesto Iadanza[3] , Federica Maietti[4] , Marco Medici[4] ,
Federico Ferrari[4] , and Pedro Martín Lerones[5]

[1] R&D, Nemoris srl., Bologna, Italy
annaelisabetta.ziri@nemoris.it
[2] RDF Ltd., Sofia, Bulgaria
{peter.bonsma,iveta.bonsma}@rdf.bg
[3] Consorzio Futuro in Ricerca, Ferrara, Italy
ernesto.iadanza@unifi.it
[4] Department of Architecture, University of Ferrara, Ferrara, Italy
{federica.maietti,marco.medici,
federico.ferrari}@unife.it
[5] Fundación CARTIF, Boecillo, Valladolid, Spain
pedler@cartif.es

Abstract. One of the goals of the INCEPTION project, funded by the EC within the Programme H2020, is to explore and enhance H-BIM knowledge management in the sector of Cultural Heritage (CH), taking into account the richness of the interdisciplinary documentation associated with the different significances of the assets.

Architectural and historical documents, structural analysis, building material characterisation and other sourced documentation for a H-BIM virtual model of a built space can be of different formats, scopes and significances but their organisation, sharing and findability are crucial for the holistic e-documentation that can be enriched from interdisciplinary users and reused for different purposes.

Conceptual frameworks of a document management system interoperable with H-BIM should guarantee an easy way to organise digital documents from variable sources and associate them to specific parts of the buildings. This to avoid overwhelming information and to allow the user to explore the virtual model with the best granularity.

Within the INCEPTION project, a layered and interoperable H-BIM ontology has been developed to gather and store new information from the original BIM model and CH information, as well as to associate the correct architectural element to each structural or decorative part of the building.

Then the H-BIM ontology has been extended implementing a specific addition to create an association between external documentation and the whole 3D model or individual elements specified in the H-BIM graph.

This paper explains how the Semantic Web technologies allow retrieving and filtering the holistic documentation by the user needs and using it in the 3D model exploration and analysis, improving the user experience and the findability of different knowledge sources.

© Springer Nature Switzerland AG 2019
A. Moropoulou et al. (Eds.): TMM_CH 2018, CCIS 962, pp. 347–358, 2019.
https://doi.org/10.1007/978-3-030-12960-6_23

Keywords: Heritage documentation · H-BIM · Semantic Web

1   Introduction

INCEPTION is a research and innovation project funded by the European Commission under the Programme Horizon 2020, call Reflective-7-2014, *Advanced 3D modelling for accessing and understanding European cultural assets.*

Its goal is to understand European cultural identity and diversity in Cultural Heritage sites by facilitating collaborations across disciplines, technologies and sectors.

To realise this scope, one of the main achievements is the INCEPTION platform, which explores, implements and enhances 3D models in BIM (Building Information Model) environment, exchanging data format according to existing state-of-the-art standards and using Semantic Web technology.

IFC is an open BIM standard format for 3D digital reconstructions increasingly used in the architectural field. Its application to Cultural Heritage, the so-called H-BIM (Heritage Building Information Model), is still quite challenging [1] and it is still in a definition phase due to the complexity of the domain and the different scopes and usages of the 3D model of a CH asset [2].

The INCEPTION Platform will allow such models to be easily accessible and reusable by researchers, scientists, experts and creative practitioners working in the cultural and heritage industries. Having such a broad audience is mandatory to promote collaboration across sectors and facilitate cross-disciplinary researches, dissemination, education and business opportunities.

The broad application of the INCEPTION objectives shall result in three major statements that approach the context of cultural heritage stakeholders from a complex, multifaceted set of formats, sources that together offer more, and better information than the sum of individual data (thus the definition of "holistic" documentation).

Therefore, interoperability is one of the primary targets to allow a platform to be available and usable by users with different scopes but the same need to get and share information.

The Semantic Web technologies, led by the W3C consortium and broadly accepted as the international standard to share machine-readable knowledge content, is undoubtedly the best solution to save and reuse the CH documentation related to a building but significant for different domains of interest [3].

To achieve all those different objectives, an H-BIM ontology has been developed to remap the architectural features from ifcOWL (an open standard for BIM modelling in a semantic web language) and integrate them with the architectural elements specific to the CH domain issued from the INCEPTION set of demonstration cases.

The knowledge related to the 3D model can then be enriched over time by different users, approaches and skills and from various sources and formats.

The difference between several use cases related to CH immovable assets (buildings, monuments, museums and archaeological sites) strongly suggests a functional and layered architecture for the INCEPTION platform and the related ontology.

Analysing a set of use cases related to the INCEPTION demonstration cases, one of the primary concerns was the possibility to relate the complementary documentation to specific parts of the model.

This documentation has different sources, scopes and interests for the user: it is often available as a set of files of different format and content, with changing relevance and interest depending on the scope of the model exploration.

The solutions must not overwhelm the user but give him the possibility to explore at the same time the model and its further deepening with a reasonable level of detail and granularity.

The INCEPTION platform is grounded on an ontology that allows to link and define metadata for those files, as well as the possibility to store, retrieve and, if the format allows it, visualise them in the 3D model analysis and exploration.

2 H-BIM Holistic Documentation

In the cultural heritage field, the management of the site documentation is a necessity enhanced by the inter-disciplinarity of the stakeholders [4].

BIM is a digital representation of physical and functional properties of a building and the answer in the Architecture, Engineering and Construction (AEC) industry to the previous lack of a management system to interlink 2D and 3D information.

The information, to which the 'I' of the BIM acronym refers, is not only related to the construction representation but even to other information necessary for building management during its lifecycle. BIM is also thought as virtual modelling of the building, in which any kind of physical element logically correlates with the other through a data storage (usually a database, but we can also think of semantic repositories).

In the field of Cultural Heritage, managing information related to 3D models is mainly focused on reverse architecture, starting from data acquisition at an intermediate point in a building's lifecycle. The chosen time-period could also be related to a specific condition of the asset in the past, retrieved by research and historical sources. This typical use case highlights the documentation complexity and layering, even more relevant according to the 3D model scope.

Thus, not only we can have different digital reconstruction related to the various historical time frames, but also the related documentation can have multiple layers in time and sources.

In addition, external documentation can be originated before the digital model is created, as a source for it (point clouds, photographs, historical pictures representing the building at a given time) or can be added as a further analysis or study research, after the model completion.

It is essential to understand that the way of retrieving and taking advantage of the model documentation is just as important as its classification. Obviously if having a valuable set of files correctly classified through acknowledgeable standards and referenced to the entire 3D model is a starting point, those data are not usable if they are not available in an exploitable format or if we are facing a huge, not manageable, amount of information. Any exploring instrument and native interlinked format have to give the user a path and an easy way to search, explore, utilise and enjoy it. As the term holistic [5] suggests, we must think the documentation and the virtual building model, including the enrichment data interconnected with the architectural structure, as a whole.

INCEPTION lists nine main demonstration cases in six different European countries and ranging from museums, palaces, castles, churches and archaeological sites. The diverse nature of the sites requires several different approaches towards users, depending on the intrinsic value of the CH area but also on the user background and scope. Analysing the existing documentation and needs of those demonstration cases was of great importance in prioritising the platform tools and architecture to upload the related documentation and allow using it proficiently.

The typology of the external documentation can be various: in some cases, existing ontologies or structured databases were available. However, in most cases, the available sources were already existing external files or digitised objects.

Of course, there are a lot of different formats and possibilities: for example externally linked data, high definition details, textures, websites, databases or already defined ontologies. For each of these specific needs, INCEPTION provides web services to allow 3D models interoperability. However, to simplify and gather the already existing use cases issued from the demonstration cases, the focus has been put on an ontology that allows linking files created outside the platform to specific sections of the 3D model. The ontology language also targets the aim of reusing existing data, one of the issues addressed by INCEPTION.

3   Ontology for Media Resources

With the advent of new ways of exploiting CH 3D models through VR and AR, we can be sure that a typical example of data enrichment would be through media files [6]. Images, videos, audios, texts are exploitable interchangeable formats that channel information also on the intangible culture linked to the assets. Media files well-known extensions can be immediately usable through common software tools like an internet browser.

Starting from this set of files, the possibility to classify the external resources have been addressed using an integration of the ontology for Media Resources 1.0, W3C standard metadata recommendation to bridge the different descriptions of media resources and provide a core set of descriptive properties also generally valuable for any file format. This recommendation defines a core set of metadata properties for media resources, along with their mappings to elements from a set of existing metadata formats [7].

The ontology is implemented in RDF/OWL, the standard language of Semantic Web also used to define the INCEPTION H-BIM ontology. The scope of Media Resource ontology is to target most of the media resources formats commonly used on the World Wide Web.

The ontology describes the most important properties and maps a set of metadata in the commonly used set of elements shared and indicated from the media files. Different software and applications could share and reuse these metadata remapping them through the Media Resources table format.

4   INCEPTION H-BIM Ontology

One of the first steps in designing the INCEPTION platform was the construction of the H-BIM ontology, an extendable open standard that uses existing ontologies as part of its architecture and tries to bridge the gap between geometry and knowledge representation.

4.1   Architectural Classification

The H-BIM ontology uses Semantic Web technologies as RDF/OWL language to export the definitions of the architectural structure of the building and use it as a knowledge layer to interlink the site elements with any structured knowledge not directly mapped on the geometry layer.

The ontology starts from a subset of the IfcOWL open standard [8] but incrementally adds the correct hierarchical specification of sections and components.

The geometrical descriptions are translated from ifcOWL/IFC towards a dedicated GEOM ontology allowing the integrated use of a library being able to convert the content into 2D/3D representations for (web-based) 3D viewers and third-party applications. This project ontology is the base to link specific properties to a specific part of the building and to retrieve all the components related to these properties. In the properties set, the reference and the link to some documents, that are resources needed to assert the origin of the modelling decisions, can be added.

To correctly map this behaviour, we need to attach a unique global identifier for every component in the model and define semantic (as opposed to geometrical), relationships between parts.

The base ontology allows linking different ontologies to enrich information related to a single component and have all the hierarchically related building parts automatically updated if the relationship is transitive.

Due to the interoperable structure, the ontology will be incremented and developed over time, but being INCEPTION focused on user needs and approaches, to assess that the "base" was correctly set, we have tested it with examples from the INCEPTION demonstration cases, analysing their context and the different users' approaches.

By collecting the results of the INCEPTION use cases, it was possible to highlight two primary requirements:

1. the need to identify a specific part of the model;
2. the need to link information or other documents to a specific component and retrieve them easily.

Those requirements define the core part of the H-BIM ontology:

A set of classifications for building components compatible with the CH needs that can be enriched with definitions, synonyms, information and documentation.

This approach clearly does not accomplish the possibilities of enrichment and all the possible relationships between different parts of the model but provides a way to interlink the information related to the building history and representation to the specific part of the model to which they are connected.

A "feature to document" granularity provides many benefits also from a usability point of view, allowing the user to explore the documentation inside the model, giving a sort of "3D structure" to a flat set of files. Other filters, like timeframe, users scope and file formats, can be applied to the document retrieval, giving a very personalised experience.

4.2   Architectural Features Names

Standard BIM does not entirely answer to the complex nomenclature of heritage buildings, so the ontology classification of the building features was integrated with a glossary to bridge the gaps in current BIM definitions and have a first hierarchical classification on Heritage buildings.

The INCEPTION approach in the selection of the nomenclature also examined classical architectural sources: a "taxonomy of monuments" according to holistic e-documentation needs is the base to understand the approach on "names". It is the starting point of a classification in Semantic Web language in order to connect terms with H-BIM ontology.

Nevertheless, the glossary was set by a top-down approach, starting from the INCEPTION demonstration cases' needs and scholars analyses and assessments, and checking the mapping with the Getty AAT [9], one of the most comprehensive sources of architectural definitions.

4.3   H-BIM Core Ontology

H-BIM ontology is the RDF/OWL ontology that inherits the main classification structure from ifcOWL and maps the components of a BIM in the Cultural Heritage field.

The starting point has been the analysis of the ifcOWL classes and properties and the collection of the semantically convenient classification useful for enrichment. Secondly, the glossary structure was hierarchically organised and integrated with the definition of the architectural component of the base H-BIM first ontology. Those are new classifications not yet included in the first ontology derived from ifcOWL: any new insertion was tested to understand if the model converted from IFC were correctly classified.

One of the results of this approach was that the complex hierarchy and the new descriptions automatically enriched models without losing the backward compatibility.

Flexibility was one of the criteria to be followed, so some of the classifications belong to different hierarchies. This process was also necessary to add a new definition or classification of one component without changing the original one but adding layers of descriptions and a deeper understanding of the building structure.

It is essential to highlight that this procedure also considers all the single project ontologies based on the H-BIM ontology schema: any of them is composed by architectural elements that are correctly defined in the H-BIM ontology framework.

Starting from the H-BIM architectural elements, an H-BIM Document ontology can be integrated to link external documentation defined in CH classifications to be used for single architectural elements enrichment.

5   File Management in the INCEPTION Platform

One of the issues to manage in linking external documents to a 3D model is that the INCEPTION platform could have several different files formats to be stored and to be related to different databases, which store the 3D geometry, the semantic definition of the architectural features or other enrichments like external websites.

Metadata and external structured descriptions of the file allow uncoupling the hardware storage organisation and the proper content of the file from the H-BIM ontology relationship.

Hence the ontology can take for granted the correspondence between the metadata description of the file and its content, giving to the INCEPTION platform the task to correctly store and visualise them. Given the standard use case of media files attached to the 3D model to enhance interactive experiences, tools to visualise and explore the content could be easily accessible by the user, enriching his experience in real time.

If the file is not available in some already managed formats, the INCEPTION platform incremental flexibility allows to develop or connect external applications to explore specific data formats, for example, files regarding structure or technical analysis like infrared outputs or measurement sets.

5.1   URL and URI

In order to be linked through a semantic relationship, files have to be provided with a URI or a URL [10].

A Uniform Resource Identifier (URI) is a compact string of characters for identifying an abstract or physical resource. The syntax of URI's goes back from Tim Berners Lee, the creator of the internet. A Uniform Resource Locator (URL) is a form of URI which expresses an address to access the referenced object using network protocols. An http URI is a URL.

When embedded within a document, a URL in its absolute form may contain a lot of information already known from the context of that document's retrieval, including the scheme, network location, and parts of the URL path. For existing Internet access protocols, it is normal to assign the encoding of the access algorithm into something concise enough to be designated as an address.

The primary component of the Resource Description Framework (RDF) data model is a semantic triple, the atomic data entity in the Semantic Web. As its name indicates, a triple is a set of three entities that codifies a statement about semantic data listing three parts: subject, predicate, object. Every part of an RDF triple is individually addressable via unique URIs, which represent entities and properties.

Once saved in the INCEPTION platform storage, the documentation file has then a URL to be reached from, that can be used to share the content through the platform, giving information about the location, the format and the scheme This URL can also be considered its unique identifier but being correlated with a web location any change of it should then be remapped to the new position.

To link external files through RDF triples, a persistent URI should be associated with the individual file, leaving the access position to different properties of the H-BIM document position.

5.2 Semantic Metadata

The link between the specific 3D model part and the file properties and location is mapped in a project-specific ontology that is not bounded to the file format or metadata. Files to map are usually already existing, and the INCEPTION approach is that the platform should not ask that formats and descriptions must be constrained to a specific standard at the upload time. Otherwise, there is the risk to avoid enriching the model due to the lack of time and resources.

Some of the metadata could be implicitly set at upload time, like the file format, size and location, other could be optionally filled getting them from the user profile, like the scope of the file or its possible utilisation.

The inclusion of ontologies like Media Resources 1.0 also gives a better mapping on possible existing metadata embedded on the file, so to save them in the semantic repository and reuse them if there is they are useful from a specific use case.

In this way, without the user having to set a massive amount of other information, at upload time we would understand much information from the context of the upload operation. The INCEPTION platform is also able to set a defined set of project specific properties, allowing to retrieve them to enhance the user experience.

6 H-BIM Document Ontology

H-BIM ontology has been extended to include an easy way to link all kind of external documentation to the whole 3D model or an interesting part of it.

The H-BIM Document ontology now includes and complete W3C Media Resource 1.0 to enrich with different digital files the single architectural element. The file can also not be of a specific media format, but the raw content is saved and retrieved from the platform, using metadata and properties to assign further external usages.

In CH field we can have a lot of peculiar formats and technical archives for data acquisition, so the W3C Media Resource was a starting point to define general properties necessary for files storage not related to common file extensions but typical from the CH domain.

To upload the file, the user also has to identify all content' scopes and significances: this allows to filter the documentation by the user needs and avoid showing a bundle of not significant information to the final user that could be different from the one who has uploaded the file.

For example, the historian can filter the uploaded documents retrieving only the ones with the scope of "CH enrichment" and significance "historical", the technician only the ones with scope "maintenance" and the administrator the one with significance "economic".

Every document can have multiple scopes and significances. Although the "scope" property seems to be the most relevant, we have chosen to add the not mandatory property "significance" for CH standards compliance (Fig. 1).

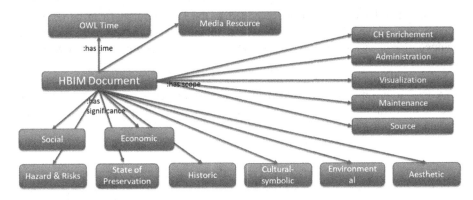

Fig. 1. Each document can set the not mandatory properties HBIM:hasScope and HBIM: hasSignificance to clarify the context of the content.

6.1 An INCEPTION Use Case: Istituto degli Innocenti

The developments achieved within INCEPTION allowed testing their validity on several real use cases, such as the documentation of the Istituto degli Innocenti in Florence.

The building was originally designed by Filippo Brunelleschi, who left after few years from the beginning of the construction. Several modifications affected the building through time, resulting in a site that partially represents the whole history of the city. The schemas below show how the workflow, made possible by the INCEP-TION developments, can enhance the knowledge of the building. From the architectural research, the reconstruction of several models in different eras was possible, also giving the chance to connect the digitised source material to each building element.

Starting from a BIM model created for historical reconstructions, it is possible to attach to each element source material as well as detailed models, a specific nomenclature (i.e., the original capitals) and a time classification (see Fig. 2).

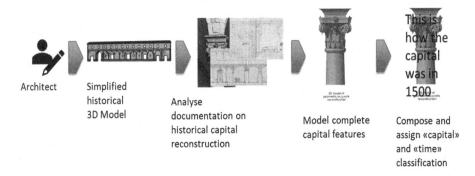

Fig. 2. How to use the H-BIM ontology to explain and manage the reconstruction of a capital at a given time of the building history.

On the other hand, the knowledge added to the H-BIM model can be accessed, explored and enjoyed by the tourist who is looking for more information on the monument that he is visiting and can be guided by exploring significant documentation attached by renowned researchers.

A user, such as a tourist or a citizen, can take advantage of the knowledge added by researchers or technical users guided by a specifically developed storytelling that, thanks to the definition of different points of interest, can highlight the documentation collected around the building (See Fig. 3).

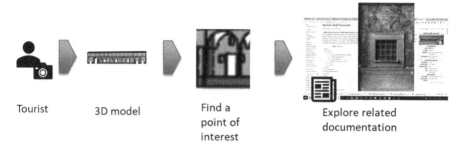

Tourist   3D model   Find a
                                    point of
                                    interest   Explore related
                                                                documentation

Fig. 3. Exploring the model with deep granularity through linked data and external documentation enrichment.

7   Conclusions

Cultural Heritage buildings' and sites' importance hugely depends on their significance on the broad spectrum of social and historical events that involved the asset through time.

IFC/ifcOWL, the open BIM standard for digital representation of physical and functional features, has to be enhanced to manage the complexity of CH assets, leading to the so-called H-BIM approach.

One of the main issues to be addressed within the H-BIM environment is the amount of external documentation related to different scopes connected to the 3D model: research, conservation, maintenance, restoration, administrative or historical issues. Although there are many examples of documentation included in software storages like databases, ontologies, we can assume that one of the main envisaged solutions is just allowing linking existing digitised files to the model.

The H-BIM ontology layer allows relating the 3D model to other knowledge sources, like external files, to a level of granularity that could be embedded in the virtual model navigation. Linking the different files to the specific architectural features allows managing required knowledge separated into different levels to be filtered depending on the user context and to be connected with each part of the building structure.

To upload and store the files, a set of semantic properties (metadata) can be extracted from the contextual information and set through the H-BIM Documentation ontology, partially derived from the W3C Media Resources. Other properties, like the scope or the significance of the information, can be added directly to the semantic repository.

Those properties are essential to retrieve the significant information accurately and to manage the file visualisation. The INCEPTION end user can thus enjoy a focused user experience, navigating the model and retrieving the information on the part of the building and the topics he is interested in. The filtering through 3D position and properties avoids overwhelming the visitor with not relevant information depending on its profile: layering the knowledge without deciding a priori which documentation can be useful but allowing the access "just in time" so to allow a genuinely "holistic" experience. Other metadata, such as size and format, are useful to embed media viewers and to give the building explorer a powerful and easy experience.

INCEPTION approach allows to incrementally enrich the model with external information but also to navigate it depending on the context, receiving the correct information at any time of the user experience.

Acknowledgements. The project is under development by a consortium of fourteen partners from ten European countries led by the Department of Architecture of the University of Ferrara. Academic partners of the Consortium, in addition to the Department of Architecture of the University of Ferrara, include the University of Ljubljana (Slovenia), the National Technical University of Athens (Greece), the Cyprus University of Technology (Cyprus), the University of Zagreb (Croatia), the research centers Consorzio Futuro in Ricerca (Italy) and Cartif (Spain). The clustering of small medium enterprises includes: DEMO Consultants BV (The Netherlands), 3L Architects (Germany), Nemoris (Italy), RDF (Bulgaria), 13BIS Consulting (France), Z + F (Germany), Vision and Business Consultants (Greece).

The INCEPTION project has been applied under the Work Programme Europe in a changing world – inclusive, innovative and reflective Societies (Call - Reflective Societies: Cultural Heritage and European Identities, Reflective-7-2014, Advanced 3D modelling for accessing and understanding European cultural assets).

This research project has received funding from the European Union's H2020 Framework Programme for research and innovation under Grant agreement no 665220.

References

1. Logothetis, S., Delinasiou, A., Stylianidis, E.: Building information modelling for cultural heritage: a review. ISPRS Ann. Photogram. Remote Sens. Spat. Inf. Sci. 2(5), 177 (2015)
2. Arayici, Y., Counsell, J., Mahdjoubi, L., Nagy, G.A., Hawas, S., Dweidar, K. (eds.): Heritage Building Information Modelling. Routledge, Abingdon (2017)
3. Pauwels, P., Bod, R., Di Mascio, D., De Meyer, R.: Integrating building information modelling and semantic web technologies for the management of built heritage information. In: Digital Heritage International Congress, vol. 1, pp. 481–488. IEEE (2013)
4. Ioannides, M., et al. (eds.): Digital Heritage. Progress in Cultural Heritage: Documentation, Preservation, and Protection. Springer, Cham (2016). https://doi.org/10.1007/978-3-319-48974-2. Proceedings of the 5th International Conference, EuroMed 2014
5. Ioannides, M., et al.: Towards monuments' holistic digital documentation: the saint neophytos enkleistriotis case study. In: Ioannides, M. (ed.) EuroMed 2016. LNCS, vol. 10058, pp. 442–473. Springer, Cham (2016). https://doi.org/10.1007/978-3-319-48496-9_36

6. Ioannides, M., Magnenat-Thalmann, N., Papagiannakis, G. (eds.): Mixed Reality and Gamification for Cultural Heritage. Springer, Cham (2017). https://doi.org/10.1007/978-3-319-49607-8
7. Brusaporci, S.: Advanced mixed heritage: a visual turn through digitality and reality of architecture. Int. J. Comput. Methods Heritage Sci. (IJCMHS) 2(1), 40–60 (2018)
8. Bonsma, P., et al.: IOP Conf. Series: Materials Science and Engineering, vol. 364 (2018)
9. López, F.J., Martin Lerones, P., Llamas, J., Gómez-García-Bermejo, J., Zalama, E.: Linking H-BIM graphical and semantic information through the Getty AAT: Practical application to the Castle of Torrelobatón. In: IOP Conference Series: Materials Science and Engineering, vol. 364 (2018)
10. Berners-Lee, T., Fielding, R., Masinter, L.: IETF (Internet Engineering Task Force). Uniform Resource Identifiers (URI): Generic Syntax (1998)

In Situ Advanced Diagnostics and Inspection by Non-destructive Techniques and UAV as Input to Numerical Model and Structural Analysis - Case Study

Vlatka Rajčić[(✉)], Mislav Stepinac, and Jure Barbalić

Faculty of Civil Engineering, University of Zagreb, Kačićeva 26, Zagreb, Croatia
vrajcic@grad.hr

Abstract. Assessment and structural health monitoring of existing timber structures has experienced huge interest in last decades. The main reasons are clear messages that sustainable development is a long term goal of the global policy which results in modifications or substitutions or extensions of existing buildings and engineering works (Kyoto protocol 1997, all further World Climate Summits) beside protection and conservation of the heritage buildings built in timber and assessment and collection of crucial data for the design and long-term behavior of new structures in timber. At the moment, European norms (Eurocodes) cover engineering principles that could be used to form the basis of assessment of structures or structural elements but basically they are concepted for the design of new structures. In this paper assessment methods for timber structures are summarized and the most common ones are briefly explained. The main focus of the paper is to present non-destructive and semi-destructive test methods for timber structures and use of unmanned vehicle for gathering the data which were inputs for numerical model and structural analysis of the structure. The whole protocol is shown on actual case study of the H2020 Project INCEPTION, Technical Museum Nikola Tesla in Zagreb, Croatia.

Keywords: Assessment · Timber · NDT · Heritage · Technical museum

1 Introduction

Structural health monitoring (SHM) is important topic for preservation of both old and new structures, thus, SHM within the context of timber structures is not adequately represented in strategic documents. Timber is often recognized as less durable material and timber structures as short-time lasting structures. While monitoring helps to continuously survey the condition of structures, non-destructive (NDT) methods aim to describe the existing condition of relevant areas of the structure [1]. There are two main areas of assessment and monitoring of timber structures: monitoring & assessment of historical timber structures and monitoring & assessment of relatively new structures erected recently as a result of significant advances and development with the field of new timber materials, timber structures and timber construction in general. The assessment of the structural health of old timber structures is different than the

© Springer Nature Switzerland AG 2019
A. Moropoulou et al. (Eds.): TMM_CH 2018, CCIS 962, pp. 359–371, 2019.
https://doi.org/10.1007/978-3-030-12960-6_24

assessment of the new timber structures, e.g. large-span structures. Therefore, advantages in technology with requirements for preservation of both historical objects and new timber structures provoked an increased interest in scientific and professional community in assessment methods for timber structures.

The need for an assessment of an existing structure can be based upon a multitude of reasons. Among the most typical are given in [2]: if errors in the planning or construction period become known, on the occasion of change of use of the building, in case of doubts about the structural safety, caused by visual damage, due to apparently inadequate serviceability an usability; because of exceptional incidents or accidental loads which might have damaged the structure; in the case of arising suspicion due to material-, construction- or system-inherent impairment of the structural safety, if a simple, initially unfounded suspicions shall be eliminated, when the remaining lifetime, determined during a previous assessment, has expired. The time and cost of structural assessment are justified by ensuring the safety, protecting of capital investments and cultural heritage.

The last decades were marked by a significant widening in the range of application of timber in structures and consequently a growing importance of the assessment of these structures. A wide variety of methods exist to assess timber structures, however, their frequency and scope, the decision making approach concerning safety and the necessary interventions are far from being agreed upon. The COST Action FP1101 which ended in year 2015 was dealing with the main problems of existing timber structures; assessment, reinforcement and monitoring of such structures. Lot of the information about the mentioned topic can be found on the website of the Action (http://www.costfp1101.eu) [3].

The main purpose of this paper is to summarize the most important assessment methods for existing timber structures. In addition, use of unmanned aerial vehicles (UAV) and photo digitization of the data which were gathered were used as an inputs for numerical model and structural analysis of the structure. The whole protocol is shown on actual case study of the H2020 Project INCEPTION, Technical Museum Nikola Tesla in Zagreb, Croatia.

2 Structures and Guidelines

Over the past years, a multitude of guidelines on how to approach the inspection and maintenance of existing timber structures have been published, however, only a few countries have published applicable code-type documents for the assessment of existing structures [2]. There is a large number of methods and guidelines for the assessment of existing timber structures but some of them are applicable only for a certain types of structures.

Historical timber structures represent an important part of the World Cultural Heritage and many of them are still in function: they must be preserved in order to guarantee their functionality and conserved for their historical value [1]. The most common method of assessment of historical timber structures is a combination of on-site inspection and non-destructive tests. Visual inspection is giving an idea about the condition of the structure in whole, identifies weak and critical zones and allows the information about the state of the structural stability and state of timber members, and respectively critical elements and joints in timber structures. Many NDT tests and models can be used in order to assess the state of conservation and the mechanical-physical properties of old timber members and joints: all the NDT tests, offer the same limits of the on-site inspection, because the data are referred to the time of realization [1]. Systematic review of criteria to be used in the assessment of load-bearing timber structures in heritage buildings is presented by document issued by CEN TC 346 Conservation of Cultural Heritage WG10 Heritage timber and Cruz et al. [4].

3   Assessment Methods

In this chapter the majority of the NDT and semi-destructive methods to assess existing timber structures are listed, and the most common ones are briefly explained. Very broad overview is given by Colla et al. [5], in the reports of the FP7 European project SMooHS (www.smoohs.eu). Dietsch and Kreuzinger [2] summarized the most common methods: visual (hands-on) inspection, tapping (sounding), mapping of cracks, measurement of environmental conditions, measurement of timber moisture content, endoscopy, penetration resistance, pull-out resistance, drill resistance, core drilling, shear tests on core samples, stress waves, X-ray, dynamic response, load tests (proof loading), strain measurement, microscopic and chemical laboratory methods, macroscopic laboratory methods—testing of specimen. Tannert et al. in [6] explained also several new techniques such as infrared thermography, glue line test, screw withdrawal, radial cores to determine compressive strength, pin pushing and surface hardness. Detailed explanations of every method can be found in [7–16]. New methods such as UAV and photo capturing of the objects were explained and presented in a way that they can be useful for structural assessment of buildings.

3.1   Visual Inspection

The simplest and most common NDT technique is visual inspection and it should be first step in assessing timber members in structure and whole structure itself. Obvious damages can be easily identified, including external damage, decay, crushed fibres, creep, or presence of severe cracks. The most common examples of damages and deterioration in structural timber elements which can be identified by visual inspection are:

(a) Poor construction details in structural timber elements
(b) Mistakes during execution of the structure
(c) Inadequate modifications to the original project
(d) Lack of maintenance and monitoring
(e) Physical-chemical-biological weathering reactions due to environmental parameters.

Both natural defects and deterioration of wood have a detrimental effect on the mechanical properties of the material. Deterioration caused by biotic attack causes a decrement of the original quality of wood, not only because of the general decrease of density but also because of the chemical alteration of the wood substance, as in the case of decay caused by rot [17]. Visual inspection has definite limitations: variability stems from differences in visual acuity and training/experience of personnel, problems with access, knowledge is limited to the exterior surface of the wood.

3.2 Ultrasonic Echo Technique

Stress wave and ultrasound methods for investigating wood are based on the propagation of compression waves through wood. The performed tests are based on the time-of-flight measurement to determine wave propagation speed. In these measurement systems, a mechanical or ultrasonic impact is used to impart a wave into a member. Piezoelectric sensors at two points on the member are used to sense passing of the wave. The time it takes for the wave to travel between sensors is measured and used to compute wave propagation speed. Longer propagation times are generally indicative of the presence of defects, deteriorated wood or wood with lower stiffness or density. Stress wave techniques are also, however, affected by other factors, including MC, wood species and growth-ring orientation [17]. The speed of propagation is directly correlated to the modulus of elasticity (MoE), but primary is correlated to the local singularities (knots, grain direction, degradation area…). When propagation velocity of the longitudinal stress wave is gained it is easy to achieve value of MOE if density of member is known. Rajčić [22, 23] proposes correlation between the ultrasound propagation velocity in a wooden element including other mechanical properties, i.e. strength of wood obtained by destructive laboratory testing. The correlation terms in [22–24] are provided for the velocity of ultrasound propagation for directions parallel and perpendicular to the grain, as derived from the "in situ" testing of very old wooden structures conducted in the scope of the FP7 project "Smart monitoring of historic structures" [22, 23]. In the context of her master's thesis, Rajčić proposes in [24] correlations obtained by neural network analysis. Data for the study were obtained by testing the network on a large number of samples investigated using both non-destructive and destructive methods.

3.3 Measurement of Timber Moisture Content

One of the most important factors affecting the performance and properties of wood is its moisture content. The amount of water present in wood can affect its weight, strength, workability, susceptibility to biological attack and dimensional stability in a particular end use. Moisture content is simply the mass of moisture present in wood divided by the mass of the wood with no moisture in it, expressed as a percentage. The dimensional changes of wood due to changes in moisture content (shrinkage, swelling) are different in the three material axes (longitudinal, tangential or radial). Shrinkage and swelling are significantly more pronounced in radial and tangential direction than in longitudinal direction. It is estimated that over 80% of the in-service problems associated with wood are in some way related to its moisture content.

Two general approaches to determine wood moisture content can be distinguished. In direct measurements, the moisture content is determined by oven-drying or water extraction, whereby both are destructive methods with respect to timber members in-situ. Indirect measurement methods use physical properties of wood which are correlated to the wood moisture content [18]. The most common moisture meters are electrical resistance meters which work on the principle that, as the moisture content of a piece timber increases, its electrical resistance decreases. Electrical resistance meters measure the conductivity between more pin electrodes that are pushed into the timber element and are calibrated to provide the user with a corresponding moisture content reading.

3.4 Drill Resistance

To detect the quality of cross-sections, decay in timber elements and to determine density of timber elements drill/penetration techniques are used. Drilling resistance is classified as quasi-non-destructive because a small diameter (1.5 mm–3 mm) hole remains in the specimen after testing. Drill resistance devices operate under the premise that resistance to penetration is correlated with material density. Drill resistance is determined by measuring the power required to cut through the material. Plotting drill resistance versus drill tip depth results in a drill-resistance profile that can be used to evaluate the internal condition of timber member and identify locations of various stages of decay.

3.5 Infrared Thermography

Infrared thermography (IRT) is a non-destructive investigation technique, which is becoming more frequently employed in civil and architectural inspections, in the diagnostic phase, in preventive maintenance or to verify the outcome of interventions. On historic structures, it allows investigating details of construction (e.g. hidden structure or masonry texture behind the plaster), damage and material decay (e.g. moisture, plaster detachment from a wall, cracks pattern evolution, temperature pattern evolution, microclimatic conditions mapping). The presence of a subsurface defect modifies the diffusion rate of thermal propagation. Infrared thermography is a contactless NDT technique able to record the distribution of surface temperatures and thus to unveil details of what is under the surface, within shallow depths, or its thermal behavior.

3.6 Digital Photography and Unmanned Aerial Vehicles (UAV)

Digital photography is the most likely candidate for alternatives to laser scanning efforts. This applies also to cameras embedded on UAV or mobile robots, commonly used for terrestrial/aerial mapping and imagery. The challenges and importance of structural damage assessment, in particular its critical role in efficient post-disaster response, have placed this discipline in the spotlight of the remote sensing community [19]. For rapid damage assessment, remote sensing has been found to be very useful, however, so far it has not reached the level of detail and accuracy of ground-based surveys. Modern science and practice is aimed at maximizing the potential of modern multi-perspective oblique imagery captured from UAVs, using both the high-resolution image data and derived 3-D point clouds, resulting in a detailed representation of all parts of a building. Severe damage could be determined directly from the 3-D point cloud data, while for the distinguishing of lower damage levels structural engineering expertise remains necessary [20]. The aim of 3-D point cloud is that experts could visually identify a number of damage features that are related to a structure.

4   Technical Museum Nikola Tesla

Technical Museum is one of the most visited Museums in Zagreb. The museum presents a powerful and distinctive scientific and educational center in the field of technical sciences. However, relatively few people know that the entire structure was designed and constructed as a timber structure and as such represents a rare existing example of European engineering concept of expo-halls timber structures with large span from the early 20 century. As such, in technical terms, it is example of European architectural heritage of building in timber. Technical Museum of Zagreb was founded on the model of existing large scientific and technical museums in the world, the common type; it is a complex museum of science and technology. The facility is under the protection of the Conservation Department in Zagreb, Protection of Cultural Heritage.

Museum was built in the 1948 as temporary structure for Zagreb International Fair. During the years this facility was used for various social and sport activities but now it is property of Technical Museum Zagreb. Designed by architect Marijan Haberle as temporary building, three building units are still in intensive use, with a very large number of annual visits. Total area of museum is 44 000 m^2 (14 000 m^2 covered space). After Zagreb Fair moved to a new location, these facilities were used for various social and sports activities and later donated to the Technical Museum.

Project of the condition assessment and reconstruction was finished in a period between 2010 and 2015 under the lead of professor Vlatka Rajčić (University of Zagreb).

Main objectives for the INCEPTION PROJECT regarding the museum were: to enhance the understanding of the history of institution, it's role in Republic of Croatia, the artistic value represented by the building and the artworks; 3d model, video presentation of the object and collection of exhibits; BIM model for condition assessment and improvement of energy efficiency in order to find solutions for thermal comfort (Fig. 1).

Layout of the main exhibition hall of the Technical Museum is designed as a circular segment of a ring (see Fig. 2).

External dimensions of the main hall are 81.27 m (east facade)/87.75 m (west facade) × 25.40 m. Total area of ground plan is about 2137 m^2. East and west facade are partly glazed and partly covered with wooden panels, while the north and south facade are entirely covered with wooden panels. The main load-bearing system is 13 truss frames with spacing of 6.8 to 7.3 m. Main timber frames are interconnected with 11 secondary trusses with spacing of 1.6 m to 3 m.

The highest elevation of the hall from the front-eastern side of the building is 19.74 m. Vertical load-bearing structure - columns are designed as four or six assembled vertical beams with "N" lattice infill. Spatial stabilization of the building was done with three horizontal transverse, four longitudinal horizontal wind

Fig. 1. Zagreb fair in 1950 (first use of the structure).

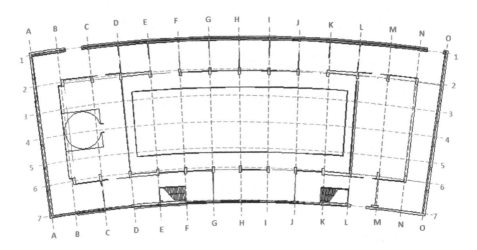

Fig. 2. Ground floor plan of the main hall.

stabilizations and two vertical wind bracings. In Fig. 3 cross section through-out the structure is given. The main frames of structure are spanning in the east – west direction.

Static analysis was conducted in order to check load bearing capacity and stiffness of the structure according to Eurocode norms. The interior space of the building consists of a ground floor area with a height from floor to ceiling of 15.00 m. Spatial stabilization of the building was done with three horizontal transverse, four longitudinal horizontal wind stabilizations and two vertical wind bracings.

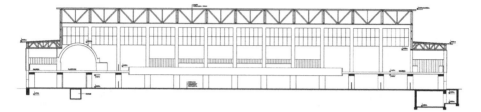

Fig. 3. Longitudinal view of the main hall.

Except the central hall, two side spaces exist. Side exhibition space is circling around the central part. Structural elements in these parts are small columns from the basement to the gallery and they support gallery floor structure which consists of the horizontal beams 36/40 cm at distance of 1.25 m. Roof structure is timber truss (Fig. 4).

Fig. 4. Secondary roof trusses (opened while reconstruction works lasted).

4.1 Visual and NDT Inspection

When the replacement of old facade (made from timber grids with glass infill) in the main exhibition hall has been scheduled, damages and excessive strains on the respective revision spots on the timber structure were found. Therefore it was decided to conduct a detailed survey of the timber structure (visual inspection, geometry recording, condition of structural elements survey and NDT determination of residual strength and stiffness). Assessment of the structure started with visual control of the structural elements. Inspection activities revealed severe damage to the large number of facade columns. The western part of the facade had direct contact of the timber columns with foundation and decay was caused by a contact with rainwater or humidity. The specimen was taken from one of the damaged columns (Fig. 5).

Fig. 5. NDT instruments (resistograph, ultrasound device and moisture-meter).

Moisture measured in the given positions varied from 6% to the maximum moisture content of 20%. Tests by acoustic ultrasound were carried out to obtain modulus of elasticity of timber structural elements perpendicular to the fibres.

In addition to these tests, several samples were taken and determined the average density of samples of 549 kg/m^3. Using the destructive testing in the laboratory with two groups of samples cut from the parts of columns which were removed and replaced because they were rotten at the foundation, it was confirmed that the average compression strength is 35 MPa. It can be concluded that original timber is soft timber of a very good quality. While assessing the structure, many defects were detected, not dangerous for the global bearing capacity, serviceability and structural stability, but this local damage can lead to local instability. Investigation work also revealed damage to the large number of columns of the west façade – the detail of the timber columns with direct contact with the ground which caused the decay of the lower part of the pillars where pillars due to damage on the system of the water drainage were constantly in contact with water. In many of the elements deep cracks were observed in longitudinal direction and in the vicinity of connectors. Examination revealed that a number of elements had visible torsion rotations of the cross section with respect to the vertical axis (Fig. 6).

Fig. 6. Columns of the main frame; decay of the columns on the western part of museum.

The last part of the assessment of the timber structure was detailed survey of timber joints and connections between elements. Lot of connections in the structure were either poorly executed or their properties decayed over time. A number of joints were executed with improper installation of fasteners so their load bearing capacity is questionable. Several joints in the tensile zones of the timber trusses have been made by just a few fasteners. Carpentry joints were generally in good condition but some were not performing in the way in which they were intended to perform. The main reasons are the changes that have occurred over time and/or initial wrong execution of timber joints.

After the complete assessment of timber structure of the Technical Museum Nikola Tesla, proposals and recommendation for reconstruction, repair and/or strengthening of the structure were given. The main reconstruction works and replacement of bad parts on main roof timber truss are briefly presented in Fig. 7.

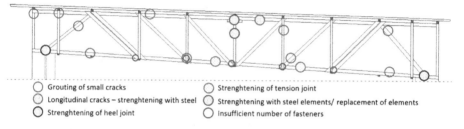

○ Grouting of small cracks
○ Longitudinal cracks – strenghtening with steel
○ Strenghtening of heel joint
○ Strenghtening of tension joint
○ Strenghtening with steel elements/ replacement of elements
○ Insufficient number of fasteners

Fig. 7. Reconstruction works obtained on the main roof truss of the Museum.

In parallel with the structural and condition assessment of the museum, a 3-D model of the museum was created using UAV and photogrammetry. Data capturing is done and post processing of the gathered data is still ongoing. High resolution photographs were taken and point cloud model was created. The taking photos procedure rule must respect that 75% of the previous photo should be on the next shot. The flight height should be maintained the same for the entire duration of the shooting. The automatic flight is generic and is good for rough volumes calculations, but the manual flight is far more productive if experienced personnel is doing the work. In the next step, photos are being processed in one of the photogrammetry programs. The photos are aligned and reference points are defined. Values of the coordinates for the corresponding points are entered and the entire planned area is georeferenced (Figs. 8, 9 and 10).

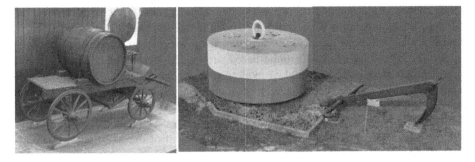

Fig. 8. 3-D digitalization of the artefacts from the Museum using 3-D point clouds and photo documentation

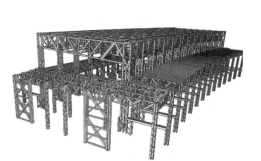

Fig. 9. Simple 3D model of the structure of the Museum.

Fig. 10. BIM model of the whole Museum with detail BIM model of complex timber structure

5   Conclusion

At the moment lot of different assessment techniques exists and are representing promising methods for a quantitative description of the current condition of timber members in timber structures. This concerns material properties like modulus of elasticity, moisture content and density as well as structural properties like dynamic characteristics, localization of inhomogeneities, cracks, and biological attack [1]. Although some methods and instruments shown to be of very high quality and are necessary for the evaluation of structures, some devices do not guarantee value for which they were designed. Nevertheless, new techniques and devices are influencing the "classical" condition assessment of structures. For rapid damage assessment, remote sensing has been found to be very useful, however, so far it has not reached the level of detail and accuracy of ground-based surveys. Modern science and practice is aimed at maximizing the potential of modern multi-perspective oblique imagery captured from UAVs, using both the high-resolution image data and derived 3-D point clouds, resulting in a detailed representation of all parts of a building.

The main focus of this paper was to present non-destructive and semi-destructive test methods which are highly in use for the assessment of old and under protection heritage objects and to show them on the special case study in Croatia, Technical Museum Nikola Tesla. More information and detailed procedure assessment can be found in [21–25].

References

1. Krause, M., et al.: Needs for further developing monitoring and NDT-methods fir timber structures. In: Proceedings of the International Conference on Structural Health Assessment of Timber Structures, pp. 89–99 (2015)
2. Dietsch, P., Kreuzinger, H.: Guideline on the assessment of timber structures: summary. Eng. Struct. **33**(11), 2983–2986 (2011)
3. D'Ayala, D., et al.: Interdisciplinary knowledge transfer and technological applications for assessment, strengthening and monitoring of timber structures. In: Proceedings of the International Conference on Structural Health Assessment of Timber Structures, pp. 49–60 (2015)
4. Cruz, H., et al.: Guidelines for on-site assessment of historic timber structures. Int. J. Archit. Herit. **9**, 277–289 (2015)
5. Colla, C., et al.: Laboratory and on-site testing activities part 3 - Historical timber elements, Report of SMooHS European FP7 project
6. Tannert, T., et al.: In situ assessment of structural timber using semi-destructive techniques. Mater. Struct. **47**(5), 767–785 (2014)
7. Kasal, B., Lear, G., Anthony, R.: Radiography. In: Kasal, B., Tannert, T. (eds.) In Situ Assessment of Structural Timber. RILEM State of the Art Reports, vol. 7, pp. 39–50. Springer, Dordrecht (2010). https://doi.org/10.1007/978-94-007-0560-9_4
8. Pošta, J., Dolejš, J., Vitek, L.: Situ non-destructive examination of timber elements. Adv. Mater. Res. **778**, 250–257 (2013)
9. Kuklik, P., Kuklikova, A.: Methods of evaluation of timber elements. Wood Res. **46**(1), 1–10 (2001)
10. Kasal, B., Anthony, R.: Advances in in situ evaluation of timber structures. Prog. Struct. Eng. Mater. **6**(2), 94–103 (2004)
11. Carpentier, O., et al.: Active and quantitative infrared thermography using frequential analysis applied to the monitoring of historic timber structures. In: Proceedings of the International Conference on Structural Health Assessment of Timber Structures, pp. 61–70 (2015)
12. Hasnikova, H., Vidensky, J., Kuklik, P.: Influence of the device structure on the outputs of semi-destructive testing of wood by Pilodyn 6J. In: Proceedings of the International Conference on Structural Health Assessment of Timber Structures, pp. 504–513 (2015)

13. Nowak, T., Hamrol-Bielecka, K., Jasienko, J.: Experimental testing of glued laminated timber members using ultrasonic and stress wave techniques. In: Proceedings of the International Conference on Structural Health Assessment of Timber Structures, pp. 523–534 (2015)

14. Sandak, J., et al.: An alternative way of determining mechanical properties of wood my measuring cutting forces. In: Proceedings of the International Conference on Structural Health Assessment of Timber Structures, pp. 543–552 (2015)

15. Yamaguchi, N., Nakao, M.: In-situ assessment method for timber based on shear strength predicted with screw withdrawals. In: Proceedings of the International Conference on Structural Health Assessment of Timber Structures, pp. 569–578 (2015)

16. Tannert, T., Kasal, B., Anthony, R.: RILEM TC 215 In-situ assessment of structural timber: report on activities and application of assessment methods. In: Proceedings of the World Conference on Timber Engineering, pp. 642–648 (2010)

17. Piazza, M., Riggio, M.: Visual strength-grading and NDT of timber in traditional structures. J. Build. Apprais. 3(4), 267–296 (2008)

18. Franke, B., et al.: Assessment and monitoring of the moisture content of timber bridges, available online

19. Rastiveis, H., Samadzadegan, F., Reinartz, P.: A fuzzy decision making system for building damage map creation using high resolution satellite imagery. Nat. Hazards Earth Syst. Sci. 13, 455–472 (2013)

20. Fernandez Galarreta, J., Kerle, N., Gerke, M.: UAV-based urban structural damage assessment using object-based image analysis and semantic reasoning. Nat. Hazards Earth Syst. Sci. 15, 1087–1101 (2015)

21. Stepinac, M., Rajčić, V., Barbalić, J.: Inspection and condition assessment of existing timber structures. Građevinar 69(9), 861–873 (2017)

22. Rajčić, V., et al.: Smart Monitoring of Historic Structures Case study 3: Palazzo Malvezzi, Report of SMooHS European FP7 project. http://www.smoohs.eu/tiki-in-dex.php?page=project%20results. Accessed 14 Feb 2012

23. Rajčić, V., Cola, C.: Correlation between destructive and four NDT techniques tests on historic timber elements. In: Proceedings of the 1st European Conference of Cultural Heritage Protection, Berlin, Germany, pp. 148–155 (2012)

24. Rajčić, V.: Neural network for wood member classification based on the results from nondestructive testings of wood samples. In: Proceedings of the 4th International Conference of Slovenian Society for Nondestructive Testing, Ljubljana, Slovenija, pp. 59–66 (1997)

25. Rajčić, V., Čizmar, D., Stepinac, M.: Reconstruction of the Technical Museum in Zagreb. Adv. Mater. Res. 778, 919–926 (2013)

Current and Potential Applications of AR/VR Technologies in Cultural Heritage. "INCEPTION Virtual Museum HAMH: A Use Case on BIM and AR/VR Modelling for the Historical Archive Museum of Hydra Greece"

Dimitrios Karadimas[1][(✉)], Leonidas Somakos[1], Dimitrios Bakalbasis[1],
Alexandros Prassas[1], Konstantina Adamopoulou[2],
and George Karadimas[1]

[1] Vision Business Consultants, 10 Antinoros Str., 11634 Athens, Greece
d.karadimas@vbc.gr
[2] Historical Archive Museum of Hydra, Hydra, Greece

Abstract. The paper is based on the activities of the EU-funded Project INCEPTION which realizes innovation in 3D modelling of Cultural Heritage (CH) through an inclusive approach for time-dynamic 3D reconstruction of artefacts, built and social environments. Specifically, the paper describes the various modern approaches of AR/VR Apps in CH and explores their potential in opening new possibilities on the way Historically or Culturally significant sites and objects can be presented using digital means. The authors argue if AR/VR technologies are today mature enough for facilitating CH adequate knowledge interpretation and interaction, while at the same time exploring the existing technologies to deploy AR/VR solutions in the field. Moreover, the paper focuses on the Building Information Modeling (BIM) applications and presents how CH information can be collected and presented in a modular way. The scope of the Historical Archive - Museum of Hydra (HAMH) demo case in Hydra, Greece, as part of project INCEPTION is presented, comprising the development of a Web/Mobile App (AR/VR), offering an innovative, interactive & enriched alternative to conventional Museum/CH Site Tour Guiding.

Keywords: Cultural · Heritage · 3D · Augmented Reality · Virtual Reality ·
Building Information Modeling

1 Introduction

Virtual and Augmented Reality (VR/AR) seem to have become nowadays the mainstream to whatever concerns Cultural Heritage (CH) preservation and promotion, as well as an increasingly important tool for the research, the communication and the popularization of CH [7].

© Springer Nature Switzerland AG 2019
A. Moropoulou et al. (Eds.): TMM_CH 2018, CCIS 962, pp. 372–381, 2019.
https://doi.org/10.1007/978-3-030-12960-6_25

It is also true that VR and AR are no longer a wishful thinking or something to work towards. They are well established and develop quite rapidly. It is also evident that the audio and visual illusion created by the currently available VR/AR hardware and software, has serious present and future potential in the realm of CH learning [6].

A significant amount of 3D reconstructions of artefacts, monuments and entire sites have already been realized, meeting the consent of both specialist and public at large [1]. Nevertheless, until recently, most of these reconstructions were essentially stationary and often missing an important factor, that of the human presence and consent. Fortunately, this has changed and today almost all relevant stakeholders and users such as historians, archivists, librarians, museum curators, are cooperating with ICT providers and VR specialists to preserve and promote their work.

2 Modern Approaches of AR/VR Apps in CH

One of the main trends in whatever concerns AR/VR in CH preservation, is the Responsive, Adaptive and Evolvable behaviours in immersive Virtual environments that capture culture and tangible/intangible heritage. This approach usually entails Multiuser Virtual Environments, high definition imaging, stereoscopic displays, sometimes interactive cinema, requiring in any instance Intelligent and High-Performance Computing.

Another trend identified more often is the application of VR and AR technologies and methodologies in Galleries, Libraries, Archives and Museums to offer Interactive Exhibits and promote the Digital Transformations of Museums with Immersive & Interactive VR and AR Apps as a narrative and/or Education activity focusing in CH [2]. The first cases of virtual museums constituted relatively simple applications running on desktop computers or were accessible through the web. Still, as technology advanced, the current trends include applications able to exploit different types of emerging technologies, Augmented Reality systems, multi-touch surfaces and haptic devices [5]. Moreover, recent advances in IT hardware, brought about reduced costs in obtaining VR gear and apparatus, that enabled and supported the general immersive visualization and interaction in virtual environments.

Still though, evaluating VR systems [1] is an exciting but rather challenging task. Relevant activities on the topic usually approach the evaluation from three specific viewpoints: a. learning effectiveness of these systems [9] b. user performance evaluation, using objective numerical capacities such as completion time, error rate and user experience and c. user enjoyment, engagement and satisfaction.

However, not taking into consideration the various difficulties of such endeavour, VR/AR applications bring significant added value in for e.g. museums as an extra and rather interesting source of information about a certain topic. Indicatively, museum visitors may be able to walk through a new environment and get a much more immersive and memorable individual experience.

As a result, the application of AR/VR technologies in museums, besides offering the option to present content and a new personal experience, can also be used to preserve exhibitions by for e.g. making a 360° video or by making a video suitable for

VR. In that way the exhibition can be preserved and most importantly be experienced many years after the actual exhibition took place.

3 VR/AR Future and Potential in CH Application

It is argued that the capacity to develop sophisticated and flexible web/mobile applications based on AR/VR is immense, considering the excess of powerful mobile computing devices for displaying such applications [9].

Current developments and increasing availability of new technologies, also offer new use-cases of VR/AR applications in a wide area of fields.

Additionally, there are various applicable techniques for the capturing and digital presentation/preservation of CH objects, such as Photograph based techniques, Spatial Models, Virtual Environment etc. However, the lack of automatic reconstruction methods followed by efficient post-processing slows down the progress in the area [12].

Today though, the world's most celebrated places have already been rebuilt in VR environments [4], while the diffusion of smartphones and tablets has already paved the way for a multitude of Mobile Augmented Reality (MAR) applications in CH, most of which are museum guides, visually augmenting physical exhibits with background or interpretive information [3].

As such, integration of research in computer vision/graphics and human-computer interface will undoubtedly overcome current limitations and restrictions as the development of ICT technologies and software is rather swift, offering the ability to bring CH on the screen of any interested web users [10].

4 Scope of the Demonstration Project for the Historical Archive-Hydra Museum (HAMH) in Greece

4.1 Project INCEPTION

The H2020 EU-funded Project INCEPTION (665220) realises innovation in 3D modelling of cultural heritage through an inclusive approach for time-dynamic 3D reconstruction of artefacts, built and social environments. It enriches the European identity through understanding of how European cultural heritage continuously evolves over long periods of time. INCEPTION's Inclusive approach comprises: time dynamics of 3D reconstruction ('forever'); addresses scientists, engineers, authorities and citizens ('for everybody'); and provides methods and tools applicable across Europe ('from everywhere').

INCEPTION solves the shortcomings of state-of-the-art 3D reconstruction by significantly enhancing the functionalities, capabilities and cost-effectiveness of instruments and deployment procedures for 3D laser survey, data acquisition and processing. It solves the accuracy and efficiency of 3D capturing by integrating

Geospatial Information, Global and Indoor Positioning Systems (GIS, GPS, IPS) both through hardware interfaces as well as software algorithms.

INCEPTION methods and tools will result in 3D models that are easily accessible for all user groups and interoperable for use by different hardware and software. It develops an open-standard Semantic Web platform for Building Information Models for Cultural Heritage (HBIM) to be implemented in user-friendly Augmented Reality (VR and AR) operable on mobile devices.

INCEPTION collaborative research and demonstration involves all disciplines (both social and technical sciences), technologies and sectors essential for creation and use of 3D models of cultural heritage. SMEs are the thrust of INCEPTION consortium that will bring the innovation into creative industries of design, manufacturing and ICT. The Consortium is fully supported by a Stakeholder Panel that represents an international organisation (UNESCO), European and national public institutions, and NGOs in all fields of cultural heritage.

4.2   Scope

As part of the EU-funded H2020 INCEPTION (665220) project, each partner has undertaken to create for its own case study, an innovative application of Enhanced Virtual/ Augmented Reality (VR/AR) for guided tours of cultural heritage sites. The Case Study[1] in Greece is the Historical Archive-Hydra Museum (HAMH). HAMH is a

Fig. 1. HAMH entrance

local branch of the State Archives and is located near the port of Hydra island. It was founded in 1918 and was housed in a building that was built at the expense of Hydra shipowner and benefactor Ghika N. Kouloura. in 1972, for construction purposes, the old building was demolished, and, in its place, the new large structure was built, which was officially inaugurated in July 1996 and has been operating since then. It is an impressive

Fig. 2. Museum exhibits

building with unique features of traditional architecture such as the stone exterior facing, which is a mainstream technique of the buildings of Hydra. Its content consists of great works of Modern Greek art of the 18th and 19th century, numerous objects - heirlooms from the war of 1821, traditional costumes, maps, navigational instruments, cannons, carved wooden parts of ships, models of shipbuilding and samples of its

---

[1] The design and implementation of the demonstration project is carried out in the framework of the European Program INCEPTION (H2020 Framework Program for Research and Innovation) with code 665220, while its implementation is undertaken by the partner VBC - Vision Business Consultants.

historical texts of the same period (over 18,000 records) as well as an impressive photographic record of the island's tradition and history (Figs. 1 and 2).

The main purpose of the Case Study is to develop an Application (AP) of Augmented (AR) and Virtual Reality (VR) Tourist Guide of the Museum to substitute the conventional tourist guide and be used as best practice in relevant Cultural Heritage Sites. The application is addressed to Scientists in History and Architecture, Cultural Heritage Researchers, Tourists/Visitors, as well as to State Services, such as the State General Archives and the Ministry of Tourism. AR/VR applications offer a unique experience by enabling the visualization of the exhibits as well as the

Fig. 3. Museum exhibits

interaction with them. The case study will focus on 3D rendering and visualization of characteristic building blocks and exhibits aiming at highlighting the cultural heritage (Fig. 3).

The categories of exhibits to be used in the projects under development implementation will be: (a) Ships/Models of Ships, (b) Weapons/Battle Objects, (c) Archives, (d) Traditional Costumes, (e) Tables, (f) Sculptures and (g) Marine Equipment.

4.3 Methodology

The Data Collection methodology follows strictly the INCEPTION (INCEPTION DAP[2]) protocol. The databases that will be created with this approach will be further elaborated using IT tools such as Autodesk Revit, Autodesk Recap, Unity 3D, etc (Fig. 4).

The following technologies will be used:

- Photogrammetry
- Laser Scanning
- Drone and DSLR HD Video and Photos
- BIM (Building Information Model).

Field work will be carried out at the Museum of Hydra with the following activities:

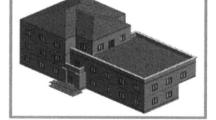

Fig. 4. 3D model of the museum

- Laser/External scanning for data creation
- Creation of data from selected exhibits
- 360° panorama HDR capture of outdoor photos and selected interior spaces of the museum
- Outdoor and high-resolution photos, as well as indoor photos of the museum, using a drone (Fig. 5).

---

[2] EU Project INCEPTION's Deliverable 2.1: Optimized data acquisition protocol.

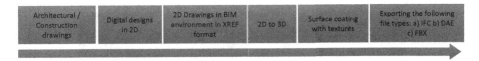

Fig. 5. Schematic representation of the work

The data collected from the above fieldwork will be used to create a 3D-BIM model of the museum's interior and exterior space through the Revit software. The model will be enriched with semantic historical and cultural significance and a museum-based Historical-BIM model will eventually be created. The model will then be exported through Revit to a suitable file format, so it is possible to use it in a Unity3D environment (usually a DAE file).

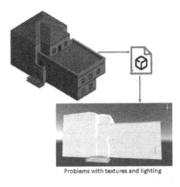

Fig. 6. Indicative export problems from Revit to Unity

When exporting the 3D model from Revit to Unity3D, various technical problems may arise, such as difficulty in capturing the texture of the surfaces of scanned objects and the way they are depicted in the two software. These problems can be resolved by additional field measurements and interventions in how data is converted from one symbol to another (Fig. 6).

Ultimately, Unity3D applications will be deployed for Android, iOS and Windows systems and devices. There is still a debate among the project researchers regarding the possibility to also store the entire Enhanced Reality application and its components so that they can be downloaded and installed in the internal storage of the device; a feature that the Partnership still assesses at this stage in the development of the project.

4.4   Expected Results

The expected results from the case study of the Hydra Museum are an Augmented (AR) and a Virtual Reality (VR) application that will be able to run on several operating systems and devices such as GearVR, Oculus Rift and eventually Microsoft's Hololens (mixed reality application) (Figs. 7 and 8).

Fig. 7. Indicative expected results from the AR App

Fig. 8. Displaying virtual text in the AR application.

The AR application will combine digital information with the user environment in real time. It will use the existing environment and overlay it with new multimedia information. When recognizing an object from an Enhanced Reality application, virtual text, images, videos, graphics, 3D models & animations can be displayed on the mobile device screen (Fig. 9).

Fig. 9. Indicative 360° photo of the museum interior space

The VR application will constitute a virtual tour of the HAMH (Virtual Museum). The guide will be able to offer users a unique experience using any computer or mobile device, as it will enable the user to navigate in a virtual environment using a VR headset. Through the application, the partners will also export High Resolution 360° photos, which can further be combined with various levels of additional content, including special audio files from narrators and historians, as well as interactive 3D models of selected objects (Figs. 10 and 11).

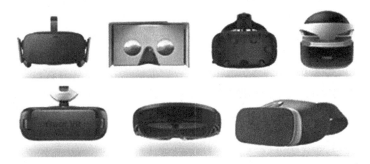

Fig. 10. Modern Virtual Reality devices (VR)

Fig. 11. Initial VR implementation stage

5 Conclusions

The paper presented in short, the potential of AR and VR Technologies applications in Cultural Heritage such as museums and archaeological sites, for information provision and enhancing the visiting experience. The authors have argued that the integration of research in computer vision/graphics and human-computer interface will undoubtedly overcome current limitations and restrictions as the development of ICT technologies and software is rather swift, offering the ability to bring CH on the screen of any interested web users. Given the current rapid ICT technologies development, it is also safe to assume that AR/VR technologies will offer significant prospects to cultural organisations for providing added value to their visitors experience by investing in developing applications for hardware owned by their visitors.

AR/VR technologies can significantly influence the customer experience, save exhibition space and contribute in a higher visitor gratification as they allow to present a picture, its creator and even the painting process itself in such ways it was unreachable before with the use of an everyday smartphone [6]. The amplified service is also a basis for product differentiation and a tool of brand building. It offers such advantages which are nowadays not commonly reachable [6].

Project Acknowledgements

- Project title: Inclusive Cultural Heritage in Europe through 3D semantic modelling
- Grant Agreement: 665220
- Duration in months: 48
- Call (part) identifier: H2020-REFLECTIVE-7-2014
- Topic: REFLECTIVE-7-2014: Advanced 3D modelling for accessing and understanding European cultural assets.

Project Partners

- UNIVERSITA DEGLI STUDI DI FERRARA - UNIFE (Italy)
- UNIVERZA V LJUBLJANI - UL (Slovenia)
- NATIONAL TECHNICAL UNIVERSITY OF ATHENS - NTUA (Greece)
- CYPRUS UNIVERSITY OF TECHNOLOGY - CUT (Cyprus)
- SVEUCILISTE U ZAGREBU GRADEVINSKI FAKULTET - UNIZAG (Croatia)
- CONSORZIO FUTURO IN RICERCA - CFR (Italy)
- FUNDACION CARTIF - CARTIF (Spain)
- DEMO CONSULTANTS BV - DMO (Netherlands)
- LENZE-LUIG 3-L-PLAN GBR - 3L (Germany)
- NEMORIS SRL - NMR (Italy)
- RDF OOD AR DI EF - RDF (Bulgaria)
- BIS CONSULTING - BIS (France)
- ZOLLER & FROHLICH GMBH - Z+F (Germany)
- DIMITRIOS KARADIMAS - VBC (Greece)

References

1. Bowman, D.A., Gabbard, J.L., Hix, D.: A survey of usability evaluation in virtual environments: classification and comparison of methods. Presence: Teleoperators Virtual Environ. **11**(4), 404–424 (2002)
2. Ch'ng, E., Cai, Y., Pan, Z., Thwaites, H.: Virtual and augmented reality in culture and heritage. Presence Special Issue on Culture and Heritage (2017)
3. Engelke, T., Keil, J., Pujol, L., Eleftheratou, S.: A digital look at physical museum exhibits: designing personalized stories with handheld augmented reality in museums. In: Conference: Digital Heritage International Congress, Marseille, vol. 2 (2013). https://doi.org/10.1109/digitalheritage.2013.6744836
4. Fakotakis, D.: Virtual reality exploration of world heritage sites: shaping the future of travel, 23 July (2018). https://www.evolving-science.com
5. Papantoniou, G., Loizides, F., Lanitis, A., Michaelides, D.: Digitization, restoration and visualization of terracotta figurines from the 'house of orpheus', nea paphos, cyprus. In: Ioannides, M., Fritsch, D., Leissner, J., Davies, R., Remondino, F., Caffo, R. (eds.) EuroMed 2012. LNCS, vol. 7616, pp. 543–550. Springer, Heidelberg (2012). https://doi.org/10.1007/978-3-642-34234-9_56
6. Attila, K., Edit, B.: Beyond reality – the possibilities of augmented reality in cultural and heritage tourism. In: 2nd International Tourism and Sport Management Conference, Debrecen, 05–06 September 2012

7. Onyesolu, M.O.: Understanding virtual reality technology: advances and applications (Chapter), InTech, March 2011 (2011). https://doi.org/10.5772/15529
8. Machidon, O., Duguleana, M., Carrozzino, M.: Virtual humans in cultural heritage ICT applications: a review. J. Cultural Herit. **33**, 249–260 (2018). https://doi.org/10.1016/j.culher.2018.01.007
9. Sylaiou, S., et al.: Evaluation of a cultural heritage augmented reality game. In: Cartographies of Mind, Soul and Knowledge (2015)
10. Wishart, J., Triggs, P.: Museum scouts: exploring how schools, museums and interactive technologies can work together to support learning. Comput. Educ. **54**(3), 669–678 (2010)
11. Wrzesien, M., Alcañiz Raya, M.: Learning in serious virtual worlds: evaluation of learning effectiveness and appeal to students in the E-junior project. Comput. Educ. **55**(1), 178–187 (2010)
12. Zara, J.: Virtual reality and cultural heritage on the web. In: Proceedings of the 7th International Conference on Computer Graphics and Artificial Intelligence (3IA 2004), Limoges, France, pp. 101–112 (2004). ISBN 2-914256-06-X

Heritage at Risk

Reconstructing the Hellenistic Heritage: Chemical Processes, Devices and Products from Illustrated Greek Manuscripts—An Interdisciplinary Approach

Dimitrios Yfantis[(✉)]

School of Chemical Engineering, National Technical University of Athens,
NTUA, Athens, Greece
dyfantis@central.ntua.gr

Abstract. After the conquest of Alexandria in Egypt by the Arabs (642 AD) the mainly empirical knowledge of chemical technology was transmitted to the West as Alchemy. Fragments of the entire knowledge reached our days in the form of later manuscripts (10th to 18th century) in several collections. Many manuscripts contain illustrations of devices for conducting chemical processes like distillation, sublimation, digestion and others. There is lack of information about the operation and the product of these apparatus. In this communication we have focused on five figures from illustrated manuscripts of St Marc Library in Venice (in book of M. Berthelot «Collection des anciens Alchimistes Grecs» 1887). By comparison of the illustrations and with aid of the Graeco-roman literature we have concluded that in a manuscript of alembic (Fig. 5) is shown probably the dry distillation of mineral sulfur for the production of pure sulfur. In mixture with others sulfur is a critical material for giving gold-like color to metallic objects.

Keywords: Hellenistic heritage · Illustrated manuscripts ·
Chemical technology · Apparatus

1 Introduction

In the capital city of Alexandria during the Hellenistic era (ca.300 BC–ca.300 AC) flourished the arts and sciences. There is no consensus among Historians for the temporal distinction or separation between the Hellenistic and Roman period. By some Historians categorically is considered as Hellenistic period from the death of Alexander the great in 243 BC to the naval battle of Actium in 54 BC or to death of Cleopatra VII last Ptolemaic Ruler of Egypt in 53 BC [1]. But speaking about "Hellenistic Heritage" is preferable to be mentioned the period ca.300 BC–ca.300 AC. The biggest achievement of the Hellenistic civilization lies in the development of sciences particularly in the 3rd century BC [2] Aristotle[1] (384–322 BC) and his peripatic School in

---

[1] Aristotle born in Stagira, Macedonia, Greek philosopher, logician and influential scientist (founder of Lyceum).

© Springer Nature Switzerland AG 2019
A. Moropoulou et al. (Eds.): TMM_CH 2018, CCIS 962, pp. 385–396, 2019.
https://doi.org/10.1007/978-3-030-12960-6_26

Athens (*Λύκειον*-Aristotle's Lyceum) has influenced strongly this development. Theophrastus[2] (ca.372–287 BC) Aristotle's successor continued his work by adding new subjects Later worthy authors such as Dioscorides[3] (ca.40–90AD) [3, 4], Vitruvius[4] (circa 100BC) [5] and Pliny[5] (100AD–180) [6] used extensively texts of Theophrastus [7, 8] King Ptolemy I following the instruction of Demetrius Phalireus[6] (350–238 BC) established Museum (Mouseion, "*Μουσείον*") a School according to the model of Aristotle. Among others an excellent library was available with thousands of papyri in which had recorded the knowledge of the Greek world. Although due to a fire which destroyed a large part of the contained works, its role in the dissemination of Greek philosophy and science was very important in late Antiquity.

2   The Sources of Chemical Technology

In particular, chemistry and chemical technology have rescued the works of a large number of scientists characterized as **"ancient alchemists"** although the word Alchemy[7] probably with partly Arab origin appears in the Latin West the 10th century BC. It is generally accepted that "two alchemists" Bolos of Mende[8] at the beginning of Hellenistic period (circa 200 BC) and Zosimos of Panopolis[9] (mentioned in Suida) about the end have contributed decisively to the development of chemical technology.

Papyri written in Greek in the 3rd–4th c.AD found in mummies at Thebes of Egypt correspond to a large extent to works of Bolos, Zosimos and Dioscorides (Leiden and Stockholm) [9].

After the conquest of Alexandria in Egypt by the Arabs (642 AD) this empirical knowledge was transmitted to the West [10] Syrian intellectuals knowing common Greek (a dialect characterized as koine) and Arabic acting as mediators have translated works into Arabic. Later many Arabic texts of Greek origin were translated into Latin.

---

[2] Theophrastus born in Eresus Lesbos influential Peripatic philosopher, pupil of Aristotle considered as the first chemist and mineralogist of the antiquity (Pliny, Vitruvius, Dioscorides and others were influenced by his writings).

[3] Dioscorides Pedanius born in Anazarbus,Cilicia, now Turkey physician and pharmacologist whose work "*De materia medica*" was the leading pharmacological text for 16 centuries in Europe.

[4] Vitruvius (Marcus Vitruvius Polio), Roman architect, engineer, and author of the famous book "*De Architectura*".

[5] Pliny the elder (Lat.*Gaius Plinius Secundus*) born in Como, Italy encyclopaedist influenced by ancient Greek authors (natural History his main work in 37 books based according Pliny on 327 Greeks and 146 Roman Authors).

[6] Demetrius Phalereus born in Phaleron near Athens, Athenian orator, statesman and philosopher. He became prominent at the court of Ptolemy I in Egypt.

[7] Instead of Alchemy the terms *chemeutike, chemeusis* correspond better to the chemical technology of the Hellenistic era.

[8] Bolos of Mende circa.200 BC Hellenized Egyptian alchemist frequently mentioned as Pseudo-Democritus influenced by Theophrastus, Author of *physica et mystica*.

[9] Zosimos of Panopolis, Egyptian alchemist of the early Hellenistic period influenced by ancient Egyptian technology.

Characteristic is the case of Graeco-Syrian Morianos (Morianes) which is mentioned as an important alchemist in the book "*Basilica Chymica*" of Oswald Croll edited 1609 as explained by Yfantis [11].

3 Development of Chemical Technology During the Hellenistic Period

The Hellenistic chemical technology has combined the knowledge of classical Greece, with the Egyptian experience of gold processing, glass manufacturing and Syrian Astronomy. Elements of Mysticism and philosophical concepts like Neoplatonism and Gnosticism are mixed with chemical technology [12].

In this way a kind of Alexandrian "Alchemy" was created. This hybrid and syncretic knowledge has led finally through many routes (Byzantium, Rome, Caliphate of Cordoba in Spain) to the later Alchemy in West [13].

Fragments of the entire knowledge reached our days in the form of later manuscripts (10th to the 18th century) in several collections. Many manuscripts contain illustrations of devices for conducting chemical processes like distillation, digestion, sublimation and others. There is lack of information about the operation and the product of these apparatus [14].

4 Illustrated Manuscripts of Chemical Devices

The practice direction of chemical knowledge is evidenced from the development of devices for conducting chemical processes.

We focused on five manuscripts of St Marc Library in Venice published in the book of Berthelot [10] «Collection des anciens Alchimistes Grecs» edited in 1887 [15].

5 Explanation of Selected Manuscripts–Comments

In Fig. 1 a distillation apparatus (alembic[11]) is shown: a material (liquid or not) is heated in a vessel ($\lambda\omega\pi\acute{\alpha}\varsigma$) by a heating source ($\kappa\alpha\acute{\upsilon}\sigma\tau\rho\alpha$) vapors are condensed ($\chi\alpha\lambda\kappa\varepsilon\acute{\iota}o\nu$) and through a pipe ($\sigma\omega\lambda\acute{\eta}\nu$) collected to a receiver. The picture demonstrates that the technique of distillation was known in Hellenistic era and refers to the concepts of Aristotle [16] Later the distillation process used for water purification was attributed to famous Arab alchemist Geber[12] [11].

---

[10] Berthelot born in Paris French physical/organic Chemist and pioneer science Historian (chemical technology, chemistry, alchemy).

[11] Alambic of partly Arabic origin (Al) and Greek ambix ($A\mu\beta\upsilon\xi$).

[12] Geber born in Iran, is the Latinized form of the name of Jabir ibn Hayyan. Geber is considered as «the father» of Arabic Alchemy with a strong influence to medieval Europe.

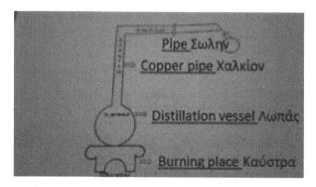

Fig. 1. Distillation device (Berthelot [15] Fig. 16 p. 250)

In Fig. 2 Two medieval distillation devices are shown from the book of Lafont [17] Medieval manuscripts have emerged from the Hellenistic period. Comparing these manuscripts with the Hellenistic ones we find that the core elements of the distillation apparatus are retained. Particularly in device of the down part we observe an evolution: a condensation pipe is placed in a tank full of water to facilitate the cooling of vapors. The reliability of medieval illustrated manuscripts is not satisfactory due to interventions in copying. Figure 2 is reported in other bibliographic source entitled manuscript of Cristobal de Paris, *Lucidarum*, 1498-9 [18].

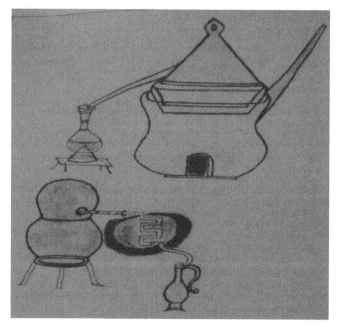

Fig. 2. Manuscript of Christophorus, *Lucidarum* 1498–1499 (Lafont [17] De l'Alchimie a Chimie p. 79)

In Fig. 3 a simple distillation device of a contemporary chemical laboratory is shown. The likeness is apparent with the Hellenistic device of Fig. 1. The modern device is originated from the German chemist Liebig (1803–1873) [19], [23] and [24].

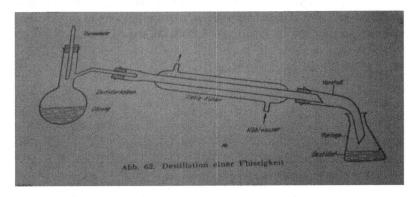

Fig. 3. Distillation device (Klement 1950 Fig. 62 p. 330)

In Fig. 4 a manuscript entitled Chrysopoiia of Cleopatra (*ΚΛΕΟΠΑΤΡΗΣ ΧΡΥΣΟΠΟΙΙΑ*) is shown containing symbols, devices and postulates. We suppose that this manuscript probably dated in 11th century belongs to Cleopatra's work referring to gold superficial processing in order to give them a hue color similar to gold. Upper left in three concentric circles are recorded secret axioms *(δόξαι)* as «εν το παν και διαυτου το παν.. **all is one and by one is the all**». In the center the symbols of mercury, silver and gold. In the bottom left is pictured the Ouroboros snake (ουροβόρος όφις) i.e. a **snake biting its own tail** with the axiom «εν το παν - all is one» attributed to the matter as entity. In these axioms is traced the influence of philosophers of classical Greece like Parmenides, Empedocles and Aristotle. At the bottom right of the figure an Alambic with two receivers (dibicos) is designed. There is no information about the product of the process. We suppose that it is something related to Gold or his imitation [20]. For political reasons practitioners the art of *chrysopoiia* (**working in gold**) in Alexandria were considered cheats (suspected fraud), were persecuted by the Roman authorities and burned their books on Diocletian era (284–30 AD). This fact is mentioned in the lemma of Suida (10th ca AC) on Chemistry *"cheimeia, cheimeusis: η του χρυσού και αργύρου κατασκευή ης τα βιβλία διερευνησάμενος ο Διοκλητιανός έκαυσε..."* Diocletian after inspection of the books for the production of silver and gold gave command to burn [21]. Therefore the *Chrysopoiia* of Cleopatra refers probably to the superficial change of the color of objects in order to give them a hue colors similar to that of precious gold and not to the chimerical attempt of **transmutation** of alchemists during the middle ages. In terms of alchemy as transmutation is meant the transformation in depth throughout the whole mass of a non- precious metal such as lead into gold. It is mentioned here that in ancient Greece the color of a metal is combined as a property of his matter, change of the color due to mixing with another metal or attack by another material could be considered as **transmutation**. The alloying of our days was confusing in ancient and medieval era. This misunderstanding is shown in the following text from of Aristotle's work *περί γενέσεως και φθοράς* "on Coming – to- be and Passing –away *De Generatione et corruptione*"..*όπερ επι τουτων συμβαινει ...ο γαρ καττίτερος ως πάθος τι ών άνευ ύλης του χαλκού σχεδόν αφανίζεται και μιχθείς άπεισι χρωματίσας μόνον* ...Tin almost disappears, as if it was property of copper without his own material, and after mixing is lost giving only the color [22].

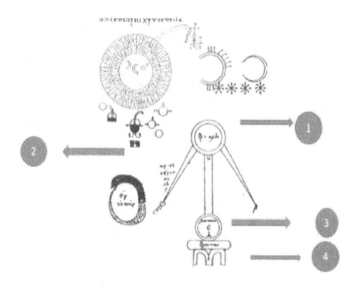

1. *φιάλη* = vessel, flask

2. *αμφίχειρος σωλήν* = thumb tube
(probably is meant a tube with reverse flow direction to the vertical tube)

3. *λωπάς* = flask, vessel

4. *φώτα* = place of flames (furnace)

Fig. 4. *Chrysopoiia* of Cleopatra (Berthelot [15] Fig. 14 p. 237)

In Fig. 5 a distillation device (Alambic) is shown. We observe a strong similarity with the alambic of Fig. 4. But two important information are given in this figure: (1) a material named apyron theion (*άπυρον θείον*) is heated (2) a material named *thion hydor*,θείον ύδωρ is collected as liquid in a vessel (*βικίον*) In Pliny's physical History these names are elucidated: ***apyron theion*** is the Greek name of sulfur mineral and means "untouched by fire" [6] ...***genera IIII: vivum quod graeci apyron vocant***...the Latin name is ***vivum sulphuris*** e.g. live sulphur and thion hydor *(θείον ύδωρ)* means liquid sulfur e.g. melt of sulfur that is collected in a small flask vessel (melting point of Sulfur 112.8°–119.2°, boiling point 444.5°). Consequently after solidification pure sulfur is produced [19, 24].

In our opinion the equipment of Fig. 5 is a device of dry distillation (alambic) for the production of sulfur that is a crucial material **for *Chrysopoiia*** or metal coloring because it reacts with many metals with mild heating. The importance of sulfur is transmitted in Geber's theory for mercury and sulfur as basic elements of transmutation and later in Latin Alchemy.

A possible application of sulfur or his compounds refers to the coloring of bronze known as artificial "black patina". This hypothesis is supported by the text of

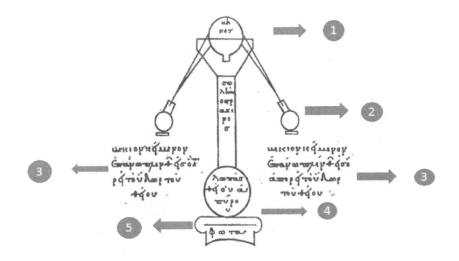

1. *βῆκος* = vessel of liquidification
2. *σωλήν οστράκινος* = clay tube

3. *βικίον κείμενον επάνω πλίνθου εἰς ο απορρεί το ύδωρ του θείου* = small flask that lies up to a brick to whom liquid Sulfur flows out

4. *Λωπάς θείου απύρου* = flask of mineral sulfur

Fig. 5. Distillation device alambic (Berthelot [15] Fig. 14 p. 247)

Plutarch[13] (45–125 AD) for coloring bronze statues at Delphi by the unknown technique of farmaxis (*φάρμαξις*) [25]. Moreover chemical analyses of bronze sculptures from a wreck in Mahdia (Tunisia) showed that during the Hellenistic period it was known the technique of creating black patinas using sulfur [26].

The dry distillation of the mineral sulfur described in Fig. 5 (dibicos alambic) has laboratory character but may be associated with two modern methods of sulfur extraction of industrial scale. The first concerns the methodology of sulfur melt production applied in Sicily. This method is not environmentally friendly due to sulfur dioxide liberation. The second known as Frasch method concerns the **in situ** sulfur melt production from ores of sulfur using high temperature steam [19, 24].

In Fig. 6 is shown a series of devices which probably correspond to the described in the lost original text. In the upper part we can read the phrase: *οι δε τύποι ούτοι* ..these types are ...In the down part is written: *ανω τα ουράνια, κάτω τα επιγηία δι αρρενος και θήλεος πληρουμενον το εργον...* Above the celestial, the earth's down; by the male and the female the work is accomplished ... Probably an allegorical axiom or a mystic formula for the work that is performed in these devices.

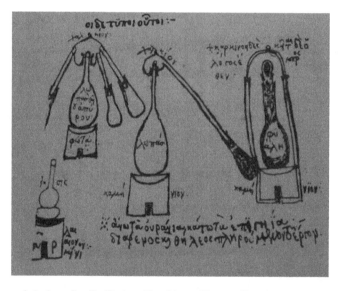

Fig. 6. Types of devices for distillation-Alambic et Vase a digestion (Berthelot [15] Fig. 37 p. 284)

Five devices are shown. On the right part a type of Alambic with three receivers called tribicos (*τρίβικος*) instead of two dibicos (*δίβικος*) as in Figs. 4 and 5. In the sphrerical vessel of distillation is written as in Fig. 4 *Λωπάς θείου απύρου* = flask or vessel of mineral sulfur.

---

[13] Plutarch born in Chaeronia, Boeotia, Greece, Biographer and Author (also was a priest in Delphi oracle).

In Fig. 7 is shown a series of devices which probably correspond to the described in the lost original text as in Fig. 6.

In the down part is shown the phrase: «**Ζωσίμου του πανοπολίτου γνησία γραφή**».

It is an original work of Zosimos of Panopolis.

Four devices are shown similar to those of Fig. 6 with some differences. On the upper left part a **tribicos** is designed as in Fig. 6 without the word *χαλκείον*. Also in the spherical vessel is written the phrase «*λωπας θείου απύρου*», as in Fig. 5. On the center a distillation with one container named *φιάλη* (vessel for collection of distillate) is designed as in Fig. 1. The condensation part (*χαλκείον*) is not developed clearly as in Fig. 6. In general the main characteristics of the distillation devices are maintained as shown in Figs. 1, 2, 3, 4, 5, 6, and 7. The terminology varies as for example in heating sources. Terms such as *φώτα* (flames), *καύστρα* (place of burning), *πυρ* (fire) and *καμήνιον* (furnace) are used according to the desired process [27].

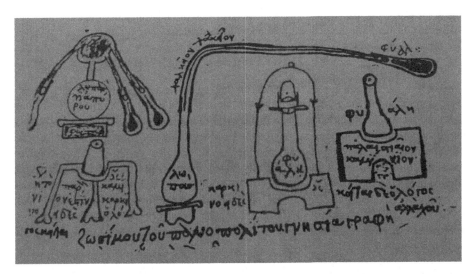

Fig. 7. Devices for distillation and digestion -Alambic et Vase a digestion (Berthelot [15] Fig. 38 p. 287).

In Fig. 8 a part of Hellenistic papyrus of the famous Leiden collection is presented with a treatise on Gold and Silver attributed to Dioscorides [12, 15]. It describes several recipes for gold and silver processing with different materials. Some terms of the manuscript are characteristic: silver gilding (*ΑΡΓΥΡΟΥ ΧΡΥCΩCΙC*), writing with gold color chrysography (*ΧΡΥΣΟΓΡΑΦΙΑ*), coloring of an object containing Gold (*ΧΡΥΣΙΟΥ ΧΡΩCΙC*), working in gold (*ΧΡΥΣΙΟΥ ΠΟΙΗΣΙΣ*) [28]. In our opinion these terms should probably be related to the *Chrysopoiia* of Cleopatra. It is noteworthy that the recipes of the Leiden Papyrus are detected as color preparation techniques for decorating surfaces of materials such as parchment, papyrus wood, marble) [29]. This tradition remains alive in the monasteries of Saint Mount (Athos) in Chalkidike (Greece) [30, 31].

Fig. 8. A Hellenistic papyrus exhibited in National Museum of Antiquities (Rijksmuseum van Oudheden), Leiden, The Netherlands (file photo of the author).

6   Conclusion

By comparison of the studied illustrations and with aid of the rich Graeco-roman literature we have concluded that in a manuscript of alambic (Fig. 5) is shown probably the dry distillation of mineral sulfur for the production of pure sulfur. In mixture with others sulfur is a critical material coloring metals e.g. for giving gold-like color to metallic objects. We believe that the illustrated manuscripts of St Marc Library in Venice belong to Cleopatra's work. The devices shown in these manuscripts are similar to medieval manuscripts. A part of papyrus of Leiden collection is related with Dioscorides and Chrysopoiia of Cleopatra Further research with an interdisciplinary approach is needed for elucidation of methods and materials developed during the Hellenistic era.

Acknowledgement. My thanks to Sychem Group of companies SA (Advanced Water Technologies) for valuable technical assistance.

References

1. Shipley, G.: The Greek World after Alexander 323–30 BC. Taylor & Francis Group/Routledge, New York/London (2005)
2. Wilcken Ulrich Griechische Geschichte im Rahmen der Altertumsgeschichte, Verlag von R. Oldenburg MUENCHEN (1962)
3. Dioscorides: *De materia medica*. Book fifth (*Διοσκουρίδου περί ύλης ιατρικής ΒΙΒΛΙΟ ΠΕΜΠΤΟ*) edition. Georgiadis, Athens (2004)
4. Dioscorides: *De materia medica*. Book fifth (*Διοσκουρίδου περί ύλης ιατρικής ΒΙΒΛΙΟ ΠΕΜΠΤΟ*) editor E. Varela edition. Zitros, Thessaloniki (2006)
5. Vitruvius: *De Architectura* books VI-X edition plethron, Athens (1998)

6. Pliny: Natural History BOOKS 33-35, translated by H. Hackham LOEB Classical library Harvard university press first published 1952 reprint 2003 (liber xxxiv, xxi100 (2003)

7. Theophrastus: On stones- on fire, περί λίθων -περί πυρός, vol. 8. Cactus edition, Athens (1998)

8. Katsaros, Th.: Chromatology of Theophrastus from Eressos: analyses, identification, and contribution to the works of cultural heritage. Ph.D., University of the Aegean, Rhodes, Greece (2009)

9. Irby-Massie, G.L., Keyser, P.T.: Greek Science of the Hellenistic Era: A Sourcebook. Psychology Press, Portland (2002)

10. De Lacy, O.L.: How Greek Science Passed to Arabs. Routledge, London (1949)

11. Yfantis, D.: L'Art psammurgique et la Chimie par M. Stephanides Mytilene 1909 a book from historical library of NTUA analysis and comments. In: Proceedings of Conference 170 Years NTUA, Athens, Greece, pp. 199–208 (2008)

12. Berthelot, M.: Les origines de l' Alchimie, Georges Steinheil, Editeur, Paris, 1885 (translated in Greek by M. Stephanides entitled " the coming –to –be of Alchemy" -Η της Αλχημείας γένεσις 1906 Athens Greece) (1885)

13. The new Encyclopaedia Britannica, Lemma Alchemy pp. 431–436, 1979 macropaedia knowledge in depth volume 1, founded 1768 15th edition

14. Canavas, C.: Distillation techniques from the classical Greek Antiquity to the late Hellenistic era. In: Proceedings of 1st International Conference Ancient Greek Technology, Thessaloniki, pp. 285–292 (1990)

15. Berthelot, M.: Collection des anciens ALCHIMISTES GRECS, Georges Steinheil, Editeur, First Edition, Paris (1887)

16. Aristotle: *Meteorologia B* - Μετεωρολογικά I, vol. 13. Cactus edition, Athens (1994). (translation in modern Greek)

17. Lafont, O.: De l'Alchimie a la Chimie Ellipses Edition Marketing (2000). (in French-Greek translation)

18. Archaeology & Arts, Alchemy in Byzantium from 1204 to 1453, pp. 23–39 fig.7 issue 108 2008

19. Klement, R.: Allgemeine und Anorganische Chemie (General and Inorganic Chemistry Wissenschaftliche Verlagsgesellschaft G.m.b.H., Stuttgart (1949). (in German)

20. Yfantis, D.: Ancient chemical technology: texts from Graeco – Roman literature an interdisciplinary approach. In: 10th International Symposium on the Conservation of Monuments in Mediterranean Basin, Athens, Greece, Chap. 57, pp. 531–537. Springer (2019)

21. Suida byzantine lexicon 10c AC, Thessaloniki edition thyrathen (2002)

22. Aristotle, on Coming – to- be and Passing – away (περί γενέσεως και φθοράς, De Generatione et corruptione). Cactus edition, Athens (1992)

23. Yfantis, D.K., Yfantis, N.: Decoration and protection of surfaces in antiquity from texts of Graeco-roman literature. In: 13th Symposium of Paints Athens, pp. 110–121 (2018). (in Greek)

24. Catakis, D.: Courses in Inorganic Chemistry edition of University of Athens, Athens (1976). (in Greek)

25. Plutarch: Delphic dialogues, vol. 2. Zitros, Thessaloniki (2002)

26. Wuensche, R.: On the coloring of the bronze head with a headband of victory in Munich. In: Multicolored gods- colors in ancient sculptures, an exhibition of the National Archaeological Museum of Athens in cooperation with Glyptothek of Munich and Goethe Institute in Athens, p. 115 (2007)

27. Varella, E.: A temperature control in ancient Greek experimental procedures. In: Proceedings of 1st International Conference Ancient Greek Technology, Thessaloniki, pp. 293–301 (1990)

28. Liddel, Scott: Greek –English Lexicon abridged edition (abridged from Liddel and Scott's Greek-English Lexicon Oxford at the Clarendon Press first edition 1891) (1891)
29. Papathanassiou, M.: The old tradition of illumination techniques. In: Proceedings of 1st International Conference Ancient Greek Technology, Thessaloniki, pp. 331–336 (1990)
30. Mastrotheodoros, D., Anagnostopoulos, D., Beltsios, K., Filippaki, E.: Glittering on the wall: gildings on Greek post – Byzantine wall paintings. Presented in 1st International Conference on Transdisciplinary Multispectral Modeling and Cooperation for the Preservation of Cultural Heritage, Athens (2018)
31. Kapetanidis, N.: Colorants and techniques used in the frescoes of Panselinos (14th century). In: 11th Symposium of Paints Athens, pp. 359–393 (2009). (in Greek)

Glittering on the Wall: Gildings on Greek Post-Byzantine Wall Paintings

Georgios P. Mastrotheodoros[1]([⊠]) [ID], Dimitrios F. Anagnostopoulos[2],
Konstantinos G. Beltsios[2], Eleni Filippaki[1], and Yannis Bassiakos[1]

[1] Institute of Nanoscience and Nanotechnology, NCSR 'Demokritos',
Aghia Paraskevi, Attiki, Greece
mastroteog@yahoo.gr,
g.mastrotheodoros@inn.demokritos.gr
[2] Materials Science and Engineering Department, University of Ioannina,
Ioannina, Greece

Abstract. Monumental painting flourished in Greek territories during the post-Byzantine period (1453 AD – early 19th century) despite the Ottoman dominance. Craftsmen working in this framework employed gilding in order to highlight specific pictorial elements (e.g. halos). Pertinent microsamples have been studied by means of several analytical techniques in the past, yet the relevant studies have mostly focused on the organic adhesives and, thus, the leaves themselves have been largely overlooked. The current study focuses on the investigation of the gold leaves, both in terms of micromorphology and composition. Samples from several post-Byzantine monuments of Epirus territory (NW Greece) were studied by employing techniques such as Optical Microscopy (OM), Scanning Electron Microscopy coupled with Energy Dispersive X-Ray Analyzer (SEM-EDX) and micro X-Ray Fluorescence (μ-XRF). A novel application of the latter technique allows for an estimation of the leaf thickness and pertinent data are compared with the results of the direct SEM thickness determination. Analytical data were evaluated in the light of a recent study that focused on the gildings of contemporary Greek portable paintings. It was thus established that post-Byzantine craftsmen were using the same gold leaves, regardless of the nature of the substrate (wall or panel). Moreover, the studied leaves turned to be of enhanced purity and reduced thickness in comparison to those found in contemporary Western European paintings.

Keywords: Gold leaf · Gilding · Mordant · OM · SEM-EDX · μ-XRF

1 Introduction

In the year 1453 AD Ottoman Turks seized Constantinople, the capital of the Byzantine Empire; this event marks the end of the long-lasting byzantine period and the beginning of the subsequent post-byzantine one (post-1453 AD – early 19th century). However, despite the Ottoman conquest most of the inhabitants of former byzantine areas retained their Christian faith, and continued practicing their ritual duties. Consequently – and in the frame of an elastic towards non-Muslim inhabitants Ottoman policy- new

© Springer Nature Switzerland AG 2019
A. Moropoulou et al. (Eds.): TMM_CH 2018, CCIS 962, pp. 397–404, 2019.
https://doi.org/10.1007/978-3-030-12960-6_27

monasteries and churches were built and embellished with wall paintings; hence the monumental ritual painting experienced a noteworthy boost [1].

Post-Byzantine painters employed very often gold leaves in their work, as in the context of Eastern Orthodox painting the shine of precious metals were thought to represent the divine glory and the heavenly space [2]. Gold is extremely ductile and it can be transformed to practically infinitely thin leaves. Indeed, in the framework of painting, gold has been almost exclusively used in the form of thin leaves. For the transformation of the solid metal (: gold) to thin leaves, a rather complex technique was employed, which was first developed possibly in Ancient Egypt during the 3rd millennium BC and remained practically the same up until recently [3]. In brief, the process may be described as follows: the metal raw material is transformed into thin sheet, cut to rectangular pieces and inserted between properly treated and usually soft, e.g. parchment or paper, sheets placed inside a pouch. Careful hammering follows and the thinning process includes various steps of new cutting and insertion between sheet-spacers [4, 5]. It has been well documented that during the Middle Ages coins very often served as the metal source for leaves manufacturing, and thanks to an information provided by Cennino Cennini it is possible to estimate that the average thickness of a high-quality medieval gold leaf was below one micron [6, 7]. Of course several alternatives to the highly-prized gold leaves were also in use, the commonest pertinent examples being silver and tin leaves, which were often covered by yellowish transparent layers in order to imitate gold [7].

Thin metal leaves were adhered to surfaces by proteinaceous or oily sizes/adhesives (alternative: "poliments"/"mordants") and two relevant terms describe the commonest application processes: "water gilding" and "oil/mordant gilding". The former term describes the process of applying metal leaves on a "protein glue plus bole (: fine iron-rich clay)" mixture, and often includes an intense final burnishing step leading to a high luster finishing [6, 8]. On the other hand, the oil gilding technique requires application of the leaves on a tacky "siccative oil plus various ingredients" adhesive, and results in a rather dull appearance as intense burnishing is not possible [7].

In the medieval and latter technical literature, recipes pertinent to the craft of gilding abound. Various manuscripts such as the early 12th century "De Diversis Artibus" [5] and the renowned "Il libro dell' Arte" by Cennino Cennini (ca. 1400) describe these processes in detail [6]. Similar directions appear also in the Greek "Hermeneia of the art of painting", a text compiled ca. 1730 by the hieromonk and painter Dionysius of Fourna [9]. The text is divided into two separate parts, a technical and an iconographical one. In the former, detailed directions pertinent to both the crafts of portable and monumental painting are included. Among them, Dionysius offers several recipes for gold leaf adhesives, and describes in detail the techniques of gilding on panel paintings (: icons), wall paintings, woodcarvings and paper, as well as methods for transforming gold leaves into powder [9].

Although several scholars have been lately engaged in the analytical study of post-byzantine paintings, the relevant works focused largely on pigment and painting technique identification [10–12]. As a matter of fact we are aware only of a single study, exclusively devoted to the identification of gildings on Greek icons [13] and few works that deal with gildings on wall paintings and focus on employed adhesives [14, 15]. Given this literature gap, a need for further investigation on post-byzantine wall paintings gildings emerges. In the framework of the current study, micro-samples from the gildings of several monuments have been subjected to analysis by means of indicated techniques. More precisely, the studied samples were collected from wall paintings that embellish the "katholika" (: main churches) of the Eleousa and Stratigopoulos/Diliou monasteries in the Ioannina lake islet, as well as the churches of St Nikolaos and Transfiguration in the Krapsi and Klimatia villages respectively. It is worth noting that all these monuments are located in the Epirus territory (NW Greece) and have been painted around mid-16th century during the early stages of the emergence of the renowned Epirus/Northwest-Greece School of iconography [16].

2   Materials and Methods

Minute samples were carefully collected (using surgical equipment) exclusively from damaged areas/losses of the wall paintings. Samples were subsequently studied under a stereoscope at magnifications 10–40x and embedded, properly oriented, into polyester resin. Upon resin curing the molded samples were subjected to grinding and polishing in order to create proper cross-sections, which were then examined and photographed below an optical microscope in the reflected visible light mode (OM, Leica, DMRXP, magnification: 20–200x). Afterwards samples were sputter-coated with conductive carbon using a carbon vaporizer (Balzers, CED 030) and thoroughly examined with a scanning electron microscope coupled with an energy dispersive X-ray analyzer (SEM-EDX, FEI, Quanta Inspect D 8334) that utilizes a built-in software for spectra acquisition and processing (EDAX, Genesis-Spectrum). Finally, the gilded surfaces of several non-embedded samples were further studied using an energy dispersive micro-XRF spectrometer (Bruker, M1-Mistral). The latter device is equipped with a high energy resolution silicon drift detector (SDD) and a micro-focus X-ray W tube (measuring conditions: 50 kV voltage and 800 μA current). The collected μ-XRF spectra were subsequently processed in order to evaluate the thickness of the gold leaves.

3   Results and Discussion

3.1   Stereoscope and OM Probing

Samples were thoroughly examined under a stereoscope, and this preliminary inspection revealed the presence of surface accumulation layers as well as extensive losses of the gold leaves (Fig. 1a). Subsequent OM probing of cross sections revealed that the samples from all four monuments exhibit the same layer sequence. In particular, a rather thick yellow preparatory layer was applied to the white plaster-substrate, a reddish layer followed and, finally, the metal leaf was applied as a top layer (Fig. 1b). Yet one shall note that despite the identical layer-sequence seen in all studied samples, there exist significant differentiations as regards the thickness of the yellow and red layers (Fig. 1b–c).

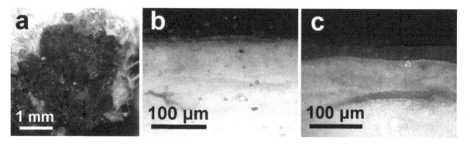

Fig. 1. (a) Sample from a gilded halo, stereoscope 16x, surface view; Eleousa monastery katholikon. (b) Cross-section of sample pictured in (a), OM, 200x. (c) Cross-section of a sample collected from a gilded halo in Stratigopoulos/Diliou monastery katholikon, OM, 200x. (Color figure online)

3.2   SEM-EDX Probing

Upon SEM-EDX examination, samples' morphological characteristics were thoroughly inspected, while the elemental compositions of the various layers, including those of the metal leaves, were estimated. The compositional features of the yellow substrate, high content of iron and an appreciable level of alumino-silicates, suggest the use of a yellow ochre-type pigment (Fig. 2a). On the other hand, the reddish layer contains- among other elements- lead at a substantial level (Fig. 2b); the presence of the latter element hints towards employment of an organic adhesive/mordant intermixed with a lead-based substance (probably a dryer). Besides, it is well known that mixtures of siccative oils with lead and other pigments were routinely used as mordants for gold leaves in the framework of post-Byzantine monumental painting [9, 14, 15]. The reddish mordant of the Eleousa sample is an exception as it contains no lead and possibly a different, non-siccative oil adhesive was employed [9, 13].

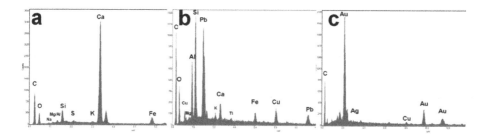

Fig. 2. Indicative EDX spectra accumulated upon analyzing the Stratigopoulos/Diliou monastery sample (Fig. 1c). (a) Spectrum from the yellow preparatory layer/substrate. (b) Spectrum from the reddish adhesive layer. (c) Metal leaf spectrum. (Color figure online)

On the other hand, several EDX analyses (conducted on folded-leaf areas) revealed that the metal leaves are composed almost exclusively of gold, as in the corresponding spectra only minor admixtures of copper and silver were occasionally detected

(Fig. 2c); indeed, on the basis of preliminary quantitative EDX results it turned out that in all instances gold content exceeds 96% (wt%). It is worth noting that very similar (in terms of composition), high quality gold leaves were extensively used in contemporary (post-byzantine) Greek icon paintings too [13]. On the contrary, recent studies have shown that in the framework of post-Renaissance western European gilding metal leaves of lower gold content were mostly used [17, 18].

Finally, upon the SEM-EDX examination of the samples' cross sections, a preliminary estimation of the gold leaves' thickness was attempted. For this purpose, the samples were inspected in higher magnifications (up to 20,000x) and the leaf thickness was directly determined by using a built-in facility of the SEM device that allows the precise measurement of dimensions (Fig. 3a–b). The pertinent results indicate that the leaves employed in all four studied wall paintings are very thin (<1 μm) (Fig. 3a–b). However it must be noted, that the SEM inspection revealed the existence of several folded areas of the gold leaves, and these must be recognized and corresponding spots should not be used for leaf thickness determination (see e.g. Fig. 3a).

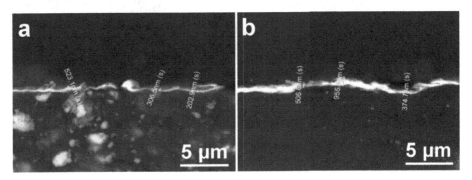

Fig. 3. Direct (SEM) determination of gold leaf thickness. (a) Sample from the Stratigopoulos/Diliou monastery katholikon (Ioannina lake islet). (b) Sample from the transfiguration of our Savior church (Klimatia village). Both photomicrographs have been taken using the BSE detector, at 12,000x magnification.

3.3   Micro-XRF Gold Leaf Thickness Determination

In order to evaluate with precision the thickness of the gold leaves, relative X-ray fluorescence intensity analysis was also performed. This type of analysis requires invariable fluorescence target area and fixed sample position, two prerequisites that are met thanks to the specifications of the employed M1-Mistral μ-XRF spectrometer. Initially a calibration curve was estimated by using a solid pure Au target as a reference, and a Monte-Carlo simulation code [19]. The validity of the calibration curve was subsequently verified through the analysis of a modern gold leaf which is of a given thickness (0.100 μm). In this case the extracted film thickness was 0.095 μm, an estimation which is in agreement within 5% of the given value.

Following verification of the accuracy of the analytical protocol, the samples in discussion were subjected to further μ-XRF analysis. The experimental results proved to be of enhanced significance as they showed clearly that the direct SEM thickness determination leads to an appreciable overestimation. Indeed, μ-XRF analysis yielded notably smaller values for the leaves thickness, as the latter were estimated to range between 0.1 and 0.3 μm. An indicative example is shown in Fig. 4, where the results from the analysis of the Eleousa sample are shown; in this very case the leaf thickness was estimated to be ∼ 0.2 μm. It is worth noting that μ-XRF results are in very good agreement with the theoretically estimated thickness of the high-quality medieval gold leaves, as this has been calculated on the basis of the relevant bibliographic references [6, 7].

The analytical results are summarized in Table 1.

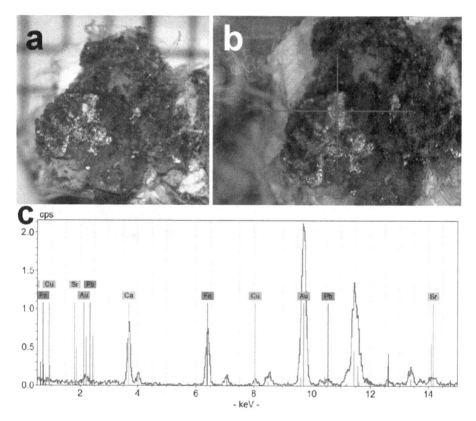

Fig. 4. μ-XRF determination of gold leaf thickness. (a) Sample from the Eleousa monastery katholikon (Ioannina lake islet), stereoscopic view (20x). (b) Same sample upon μ-XRF analysis; crosshairs mark the analyzed area. (c) The relevant XRF spectrum. Note that the μ-XRF beam spot measures 200 μm × 200 μm.

Table 1. Summary of the analytical results.

Monument	Mordant composition (major to minor, elements, EDX)	Leaf composition (elements, EDX)	Leaf thickness (μm) (SEM/μ-XRF)
Eleousa	Ca, Si, Al, Fe, Mg, Na, Cl, K, Ti	Au (minor Cu/Ag)	0.1–0.7/∼0.2
Diliou	Pb, Si, Al, Ca, Cu, Fe, Na, Mg, K	Au (minor Cu/Ag)	0.2–0.7/∼0.1
St Nikolaos	Si, Ca, Pb, Al, Fe, K, Mg	Au (minor Cu/Ag)	0.4–0.9/∼0.1
Transfiguration	Pb, Si, Ca, Al, Fe, K, Cu	Au (minor Cu/Ag)	0.3–0.8/∼0.2

4 Concluding Remarks

Samples from several post-byzantine, wall painting gilded decorations have been subjected to analytical investigation. Results have shown that in most cases the high quality gold leaves have been attached onto an organic-based, lead-containing adhesive which lies on top of a yellow ochre substrate. Differences in terms of adhesive/ preparatory layers thickness and composition were revealed, and these may imply the employment of different workshops/craftsmen in the decoration of the various monuments.

The thickness of the gold leaves was determined using both direct SEM estimation and a novel μ-XRF approach; the latter proved to be more accurate, as the SEM inspection leads to an overestimation of the leaf thickness. It is assumed that the SEM approach leads to overestimated thickness values because upon samples grinding/ polishing, minor gold leaf foldings are created, which are not easily discernible during the subsequent SEM observation (Fig. 3a–b).

Moreover, the wall painting leaves turned to be very similar in terms of composition and morphology/thickness to those used in the framework of the contemporary panel painting. This fact denotes that post-byzantine painters used the same gold leaves, regardless of the substrate they were working on (panel or wall). Finally it is also noted that the Epirus wall paintings leaves are of enhanced purity and reduced thickness in comparison to those found in contemporary Western European gildings.

Acknowledgments. This research was completed as a result of support from a Greek State Scholarships Foundation (IKY) scholarship program; the work was co-financed by the European Union (European Social Fund-ESF) and Greek national funds through the action entitled "Reinforcement of Postdoctoral Researchers", in the framework of the Operational Program "Human Resources Development Program, Education and Lifelong Learning" of the National Strategic Reference Framework (NSRF) 2014–2020 (grant number: 2016-050-0503-7689). Dr. V. Papadopoulou and Dr. K. Soueref, directors of the Arta and Ioannina Ephorates of Antiquities respectively, are sincerely thanked for their support. The personnel of the Ioannina Ephorate of Antiquities and especially the archaeologists P. Dimitrakopoulou, E. Katerini and D. Rapti, along with the conservation personnel are also thanked for their help during sampling. Special thanks are due to the General Directorate for the Restoration, Museums and Technical Works and the Directorate for Byzantine and Post-Byzantine Antiquities (both divisions of the Greek Ministry of Culture and Sports) for sampling permissions.

References

1. Chatzidakis, M.: Greek painters after the fall of constantinople (1450–1830), (in Greek), vol. 1. Institute of Historical Research/National Hellenic Research Foundation, Athens (1987)
2. Ouspensky, L.: The Theology of icon in Orthodox Church, (in Greek). Transl. S. Marinis, Armos, Athens (1998)
3. Nicholson, E.D.: The ancient craft of gold beating. Gold Bull. **12**(4), 161–166 (1979)
4. Darque-Ceretti, E., Felder, E., Aucouturier, M.: Foil and leaf gilding on cultural artifacts; forming and adhesion. Rev. Matéria **16**(1), 540–559 (2011)
5. Theophilus: On Divers Arts. Transl. J.G. Hawthorne and C.S. Smith, Dover, New York (1979)
6. Cennini, C.: The Craftsman's Handbook. Yale University Press, New Haven (1954). Transl. D.V. Thompson
7. Bomford, D., Dunkerton, J., Gordon, D., Roy, A.: Art in the Making: Italian Painting Before 1400. National Gallery, London (1990)
8. Mactaggard, P., Mactaggard, A.: Practical Gilding. Archetype, London (2005)
9. Dionysius of Fourna: The Hermeneia of the art of painting, (in Greek). Edited and published by A. Papadopoulos-Kerameus, Petersburg, 1909, reprinted by K. Spanos, Athens (1997)
10. Daniilia, S., Tsakalof, A., Bairachtari, K., Chryssoulakis, Y.: The Byzantine wall paintings from the Protaton Church on Mount Athos, Greece: tradition and science. J. Archaeol. Sci. **34**, 1971–1984 (2007)
11. Iordanidis, A., Garcia-Guinea, J., Strati, A., Gkimourtzina, A.: A comparative study of pigments from the wall paintings of two Greek Byzantine churches. Anal. Lett. **47**, 2708–2721 (2014)
12. Cheilakou, E., Troullinos, M., Koui, M.: Identification of pigments on Byzantine wall paintings from Crete (14th century AD) using non-invasive Fiber Optics Diffuse Reflectance Spectroscopy (FORS). J. Archaeol. Sci. **41**, 541–555 (2014)
13. Mastrotheodoros, G.P., Beltsios, K.G., Bassiakos, Y., Papadopoulou, V.: On the metal-leaf decorations of post-byzantine Greek icons. Archaeometry **60**(2), 269–289 (2018)
14. Katsibiri, O., Boon, J.: Investigation of the gilding techniques in two post-Byzantine wall paintings using micro-analytical techniques. Spectrochim. Acta Part B **59**, 1593–1599 (2004)
15. Katsibiri, O., Howe, R.F.: Microscopic, mass spectrometric and spectroscopic characterization of the mordants used for gilding on wall paintings from three post-Byzantine monasteries in Thessalia, Greece. Microchem. J. **94**, 83–89 (2010)
16. Acheimastou-Potamianou, M.: The Filanthropinon Monastery and the first phase of the Post-Byzantine Painting (in Greek), Second edition, Archaeologicon Deltion 31, Fund of Archaeological Proceeds, Athens (1995)
17. Sandu, I.C.A., et al.: A comparative interdisciplinary study of gilding techniques and materials in two Portuguese Baroque 'talha dourada' complexes. Estudos de conservação e restauro **4**, 47–71 (2012)
18. Bidarra, A., Coroado, J., Rocha, F.: Gold leaf analysis of three baroque altarpieces from Porto. ArcheoSciences **33**, 417–421 (2009)
19. Schoonjans, T., et al.: A general Monte Carlo simulation of energy dispersive X-ray fluorescence spectrometers—part 5: polarized radiation, stratified samples, cascade effects, M-lines. Spectrochim. Acta Part B At. Spectrosc. **70**, 10–23 (2012)

Advanced and Non-Destructive Techniques for Diagnosis, Design and Monitoring

Kinematic Analysis of Rock Instability in the Archaeological Site of Delphi Using Innovative Techniques

Kyriaki Devlioti[1], Basile Christaras[1(✉)], Vasilios Marinos[1], Konstantinos Vouvalidis[1], and Nikolaos Giannakopoulos[2]

[1] Department of Geology, Aristotle University, 54124 Thessaloniki, Greece
{kdevliot, christar, marinosv, vouval}@geo.auth.gr
[2] Department of Chemical Engineering, Aristotle University,
54124 Thessaloniki, Greece
nigianl3@hotmail.com

Abstract. The maintenance of the archaeological site of Delphi, as well as visitors and employers safety, is, directly, related to local geotechnical stability conditions. So, a first approach of understanding possible changes in slope geometry and knowledge of underlying engineering properties of the rock mass, was made, to minimize significant risks, which are associated with slope failures. This research, was conducted in the context of a PhD thesis, concerning the geotechnical risk related to the safety of archaeological sites. Laser scanning technology has been increasingly applied in geotechnical surveys, due to its high precision, high efficiency and ease of use, especially at inaccessible slopes that cannot be mapped manually. In this framework, a pilot survey was carried out at the northern rock cliff, overhanging the stadium, using LiDAR (Light detection and Ranging) technology. This research was aimed to image the rocky outcrop, produce virtual 3D computer models and collect discontinuity orientation data, in order to facilitate the geological and discontinuity mapping for its rapid evaluation of rock fall susceptibility and make a precursor mapping of the current situation in the research area, which is a typical example with high impact of rockfalls. Detailed three-dimensional models were created to distinguish the most unstable blocks, to define main rock fall source areas position, and to precisely distinguish outcropping materials and all elements at risk position. According to the results, rockfall is the most common form of landslide, as well as the most common failure mode likely to be triggered by a seismic event. As it is already known, the most important factor controlling rockfall trajectory is slope geometry. For this reason, discontinuity orientation data were collected and also their spacing and persistence on the limestone cliff were depicted. So, for investigating the existing stability conditions, kinematics of rock instability are presented.

Keywords: Delphi · LiDAR · Rockfalls · Discontinuities · Kinematic analysis

© Springer Nature Switzerland AG 2019
A. Moropoulou et al. (Eds.): TMM_CH 2018, CCIS 962, pp. 407–418, 2019.
https://doi.org/10.1007/978-3-030-12960-6_28

1   Introduction

The maintenance of the archaeological site of Delphi, as well as visitors and employers safety, is, directly, related to the local geotechnical stability conditions. So, a first approach of understanding possible changes in slope geometry and knowledge of underlying engineering properties of the rock mass was made, in order to minimize significant risks, which are associated with slope failures.

This pilot research, which was carried out at Delphi's site, was conducted in the context of a PhD thesis, concerning the geotechnical risk related to the safety of archaeological sites.

Delphi's archaeological site is of the most remarkable and imposing ones in Ancient Greece, it is designated as a world Heritage site as a site, by UNESCO and it is protected by many National and International provisions (Fig. 1) and is located, approximately, 150 km at the north-west of Athens, at the southern slopes of Parnasse Mountain, at an elevation of about 800 m, near the northern coast of the Corinthian Gulf (Fig. 2). The exposed rocks of the site belong to the Parnasse-Ghiona geotectonic zone (Jacobshagen, 1986), which builts up the mountains of Parnasse, Ghiona, Elikonas and parts of Oiti [1]. In the broader region of Delphi, and especially from the upper sections of the archaeological site, activated rockfalls are being identified from antiquity till nowadays, causing serious implications and endangering visitors and employers safety.

Fig. 1. The northern slope, above the stadium, where rockfalls occurred and a general view from Delphi's site in Google earth and geodata.

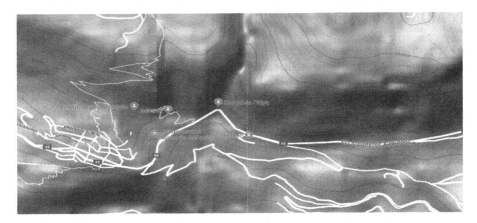

Fig. 2. Topography of the study area, from open topography data.

2   Geological and Tectonic Setting

Delphi's broader area consists mainly of the alpine formations of the Parnasse geotectonic zone, covered in places by Quaternary terrestrial loose deposits. A small area, near the coastal zone is covered by Pindos Zone Formations [4]. The geological basement is covered by neritic, medium to thick bedded limestones, dating back from Triassic to Cretaceous, followed by a sequence of flysch formations. These are alternations of brown-red siltstones, mudstones and sandstones with substantial amount of calcareous material [3]. Especially at the research area, thick-bedded limestones are being observed at the steep rock outcrop, whilst at the base of the slope, an alteration of flysch and limestone formations was depicted.

Alpine compressional tectonism caused intense deformation and fracturing of formations, resulting in folds and overthrusts formation, of upper Eocene age. This tectonism has reversed the normal sequence of the geological formations by a huge reverse fold, which resulted in places in an overthrust of the older limestones onto the younger flysch formations, as it is observed at the northern part of the archaeological site [6].

The inverted sequence of all geological formations, that are now being observed in the area, is the result of a corrugation that these formations endured, during past tectonic stages. Delphi's geological structure consists of a leaning fold, with an North-East orientation, at the inner part of which the flysch formation is depicted [5]. The latest tectonic stage in the study area is related with normal faults, with an East-West orientation. Concerning the geotechnical conditions, almost vertical discontinuities were developed which are intersected by joints, fractures and open cracks.

This pattern of fracturing results in a progressive loosening of the rock mass, which was subsequently subjected on weathering and erosion processes. This situation contributed to the development of weak zones on the rock cliff, where favorable conditions were created for limestone blocks to be detached, causing extensive rockfalls.

3 LiDAR Surveying

3D terrestrial laser scanning technology is a relatively new, but also revolutionary surveying method, due to its high accuracy, efficiency, ease of use, providing great convenience in geotechnical research.

Terrestrial scanning LiDAR (light detection and ranging) enables researchers to image the rocky outcrop, capture laser range data in a rate of thousands of individual X, Y, Z, as well as laser-intensity point clouds per second. These datasets, can be used, in conjuction with the traditionally compass measurements, in order to create 3D geological computer models and also to conduct high-precision facies characterization [2]. Particularly, all lidar instruments operate on the same premise. More specifically about its operation, a laser pulse leaves the gun, travels to a remote target, bounces off the target and finally it returns back to the detector. This two-way travel time is divided in half and multiplied by the speed of flight to calculate an accurate Z distance. Alongside, position on X and Y positions are calculated on the basis of the position of the gun, at the time when the laser pulse leaves the instrument [8]. Therefore, the survey yields a digital data set, which is a dense "point cloud", where each point is represented by a coordinate in 3D space (X, Y and Z relative to the scanner's position) and the reflected intensity of the laser beam.

For the needs of this study, an Optech Ilris 3D terrestrial (ground-based) laser scanner LiDAR was used, which was combined with field observation, in order to image the northern rock slope, at the stadium of the archaeological site of Delphi and more specifically at the northern slope of the Stadium, upwards the Sanctuary of Apollo, which is the central part of the site (Fig. 3).

Fig. 3. LiDAR equipment used for field observation, at Apollo's temple.

Point clouds that have been produced, were used to conduct the 3D computer model of the slope. In order to create the 3D representation of the outcrop strata, all high-resolution laser data were combined with the compass measurements. The field work took place in two periods, one by using the LiDAR laser and the second for field observation and traditional stratigraphic measurement. Subsequently, processing, visualization and interpretation were applied for mapping the geological surfaces in three dimensions. Specifically, all data from the terrestrial Lidar were collected and processed on a standard laptop computer with Cloud Compare v2.6.1 software (Fig. 4). Afterward, initial scans, upwards the stadium, were aligned cloud-to-cloud and merged into a single one, with a sufficient overlap between the initial scans for like-point picking, obtaining an alignment error lower than 1 (RMS < 1).

Fig. 4. At right, merge individual scans (point clouds) from the rock face of the northern slope above the stadium, after being processed, using Optech Ilris 3D laser scanner. On the left are the actual rock faces.

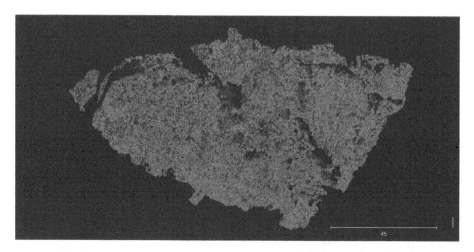

Fig. 5. 3D digital rendering of all individual scans, after alignment and merge, using Cloud Compare software.

The result of the integration of the three point clouds is illustrated in Fig. 5. As it is obvious, areas that were not imaged, are called data shadows and they are represented as black shadows in the model.

Next step of the research was the extraction of the geometrical aspects from the 3D point clouds. For this purpose, FACETS, a plugin in Cloud Compare software, was used [7], in order to extract geological planar facets from unstructured 3D point clouds (Fig. 6). More extensively, FACETS was applied into the merged 3D point cloud of the northern slope, at the archaeological site of Delphi. The outcrop faces south, with a north-south trend and it is almost 50 m high. Facet computation was implemented, a planar facet extraction was performed, their dip and dip direction was calculated and the orientations of the extracted data were reported in interactive stereogram plots.

Especially, on the one hand, using the Kd-tree algorithm, the aligned 3D cloud was subdivided into quarter cells according to the RMS value, and the result was a set of flat polygons adjusted to the original 3D point cloud. More extensively, a first clustering produced a series of planar facets, and a then, a second one, was triggered manually, where every facet was grouped into individual planes, and then all planes were all grouped, containing parallel planes with the same spatial information. In conclusion, qFACETS plugin implemented elementary planar object recognition with a minimum input value. Subsequently, these planar facets, which were recognized through the Kd-tree algorithm, grouped into planes and families.

At a second classification by their orientation, a set of 4 facets in 4 orientations was finally produced manually and well fixed with the traditional compass measurements. Finally, a graphical interface, a stereogram plot, was created into qFACETS plugin, with numerical and interactive query functionality, where the orientations were displayed, in order to explore planar objects and 3D points with normals, in 3D space (Fig. 7).

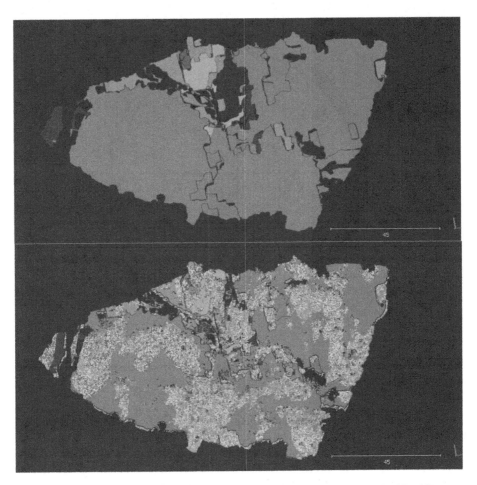

Fig. 6. Facets extracted, after classified in families, using Kd-tree algorithm in CloudCompare software. They are also depicted on the merged point cloud (at right).

4 Rockfall and Kinematic Analysis

Rockfalls are a phenomenon that represents a continuous hazard in mountainous areas and like other bedrock mass movement, can be triggered by earthquakes, or by high-intensity and long period rainfall. As it is already known, in the middle of September 2009, a prolonged rainfall activated rockfalls from the upper sections of the northern slope and fall downslope, out of the enclosure wall of the Sanctuary of Apollo north of the Portico of Attalus.

The primary reason for the failures, on the northern rocky slope, at the archaeological site of Delphi, is the disruption and the disintegration of the tectonically strained rock mass, in time, the widening and the gaps of the discontinuity sets. The factors that affect the above are the dense karstification of the limestone, which appear at the base of the slope, where the flysch formations occur, and also the water circulation at the appearance point of the two geological formations.

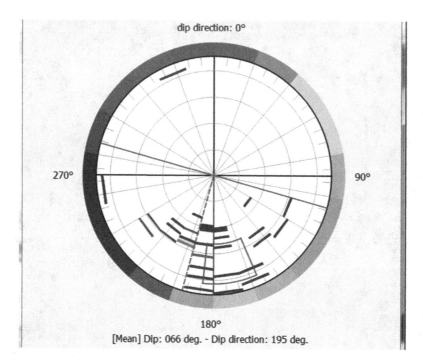

Fig. 7. Stereogram plot, where the extracted data were depicted, with the respective color ramp, in CloudCompare software.

In the research area, in order to create the morphology of the outcrop and the altitude variation of the study area, a vertical cross section from the slope till the stadium of Delphi, of a North-South orientation, was designed and then depicted on a map, in ArcGIS environment (Fig. 9). Alongside, the altitude variation was depicted on maps in Surfer software and in ArcGIS environment (Fig. 8).

After the classification and the extraction of the final planes and discontinuity sets, both by the use of LiDAR technology and compass field measurements, the dip and dip direction of each one was inserted into the Rockscience software and more specifically the Rockfall tool. According to field observations and the results from the processing of the measurement values, the predominant types if kinematic instability, in the stadium of the Delphi's archaeological site are of planar or wedge failure, and toppling of large blocks. In order to investigate the existing stability conditions and decide upon the protection measures, rockfall analysis was carried out. At first, possible future rockfall tracks were recognized, along the vertical cross section, so to calculate the related kinetic energy of the falling rocks. The falling blocks present a variety in their weight and size, therefore the indicative weight of blocks, for the simulation tests was estimated at a mean value of 20 tn and the results were depicted in diagrams (Fig. 10).

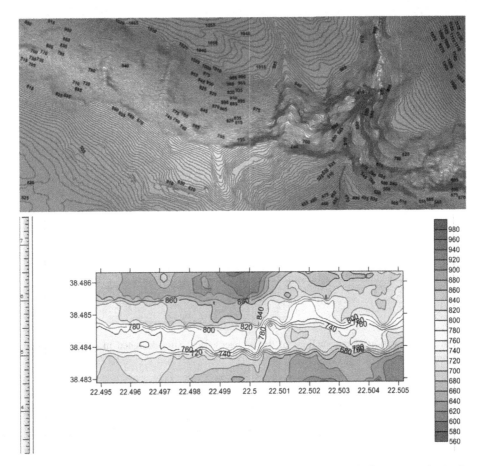

Fig. 8. Altitude variation depicted on maps, created in ArcMap (left figure) and in Surfer software (right figure).

As it obvious, rockfalls and topplings are the main failure event in the study area, while sliding over the flysch, with the aid of the water circulation, and land sliding of lateral scree were also distinguished. More extensively, according to the falling blocks height, in conjuction with the rock mass quality, two scenarios were distinguished: (a) at the upper sections of the slope, with the thick-bedded limestones of good physical and mechanical characteristics, and (b) at the lowest parts of the slope, at the limestone and flysch contact, where the rock mass appears to be more loosen and the spacing between the intersections is greater.

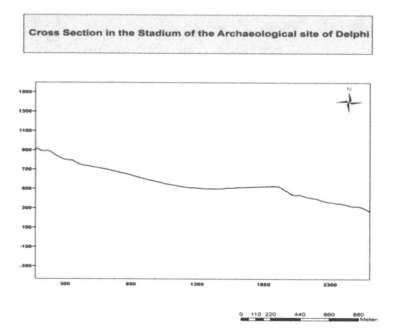

Fig. 9. Cross section in the stadium of the archaeological site, designed in ArcMap, ArcGis environmnent.

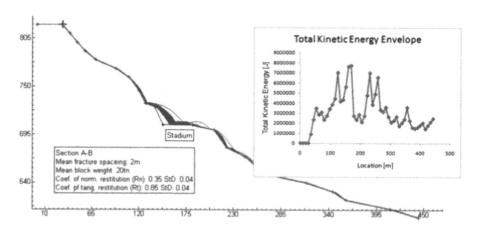

Fig. 10. Simulation of rockfalls along the vertical section, that passes through the stadium, created in rockfall software, in Rockscience. The fluctuation of the total kinetic energy along the falling block, is also depicted on the diagram.

5 Conclusions

A pilot research was carried out at the archaeological site of Delphi, in order to investigate the local geotechnical stability conditions, aiming to the maintenance of the monumental area and also visitors and employers safety. For this reason, a first approach of understanding possible changes in slope geometry and knowledge of underlying engineering properties of the rock mass, was made, in order to minimize significant risks, which are associated with slope failures.

At first, LiDAR survey carried out, in conjuction with field observation and measurements, in order to conduct high-precision facies characterization and to construct 3D geological models. These dense 3D point clouds that were acquired with the laser scanner, submitted under processing, visualization and segmentation, using the Cloud Compare software, so that they were aligned point-to-point and merged through surface reconstruction techniques and a united 3D point cloud was created (Figs. 4 and 5).

Afterwards, with the FACETS plugin, and especially the Kd-tree algorithm, the 3D cloud was subdivided into quarter cells, with a least square fitting algorithm (RMS < 1). By this procedure, planar facets were created and then clustered manually in a group containing parallel planes of all facets, that were sharing the same spatial information (Fig. 6). Finally, they were classified by their orientation and a set of 4 facets in 4 orientations was produced manually and well fixed with the traditional compass measurements.

Facet computation was implemented, a planar facet extraction was performed, their dip and dip direction was calculated and the extracted data were reported in an interactive stereogram plot, where numerical and interactive query functionality was offered, in order to see the selected objects in 3D space (Fig. 7). This interactive stereogram dialog box offers many opportunities for further computation, containing spatial characteristics about the slope and its geometrical features.

After the classification and the extraction of the final planes and discontinuity sets, both by the use of LiDAR technology and compass field measurements, a Kinematic analysis was carried out to investigate the existing stability conditions. So, the geometrical features (dip and dip direction) of each one were inserted into Rockscience software and more specifically in the Rockfall tool. It is worth noting that the falling blocks vary in size and weight and for this reason the simulation tests were performed for indicate blocks weight of 20 tn. According to field observations and the results from the processing of the measurement values, the predominant types if kinematic instability, in the stadium of the Delphi's archaeological site are of planar or wedge failure, and toppling of large blocks and the kinetic energy of 20 tn falling blocks is about 8MJ in the stadium area (Fig. 10).

According to the results of the pilot research that took place at the northern slope of the archaeological site of Delphi, it is meaningful to report the main causes of hazard and rockfalls, which are the geomorphology-outcrop of the research area, the rock mass structure and also the geometry of the discontinuity sets, the seismic effect, the compressional tectonism, the factors of weathering and erosion, as well as the easy fragmentation of the rock mass, the spacing between the discontinuities, the effect of the water persistence, humidity and frost, the water circulation in the limestone-flysch contiguity, the limestone undercut when in contact with the flysch formation.

418   K. Devlioti et al.

All the above reasons created a technicogeological environment that privilege the manifestation of instability phenomenon and rockfalls in Delphi area, factors that need to be extensively researched, with the use of innovative technological techniques.

Acknowledgements. We would like to thank the Archaeological Ephoriate of Fokida and the Archaeological site of Delphi, for their permission, facilitation and willingness in order to fulfill the field work in the stadium of the archaeological site.

References

1. Christaras, B., Vouvalidis, K.: Rockfalls occurred in the archaeological site of Delphi. In: IAEG International Congress, Auckland (2010)
2. Slob, S., Hack, R.: 3D terrestrial laser scanning as a new field measurement and monitoring technique. In: Hack, R., Azzam, R., Charlier, R. (eds.) Engineering Geology for Infrastructure Planning in Europe. LNEARTH, vol. 104, pp. 179–189. Springer, Heidelberg (2004). https:// doi.org/10.1007/978-3-540-39918-6_22
3. Celet, J.: Contribution à l'étude géologique du Parnasse-Kiona et d'une partie des régions méridionales de la Grèce continentale. Ph.D. thesis, Annales Géologiques des Pays Helléniques, pp. 1–159 (1962)
4. Piccardi, L.: Active faulting at Delphi, Greece: seismotectonic remarks and a hypothesis for the geologic environment of a myth. Geology **28**, 651–654 (2000)
5. Valkaniotis, S.: Correlation between neotectonic structures and seismicity in the broader area of Gulf (central Greece). Ph.D. thesis, Aristotle University of Thessaloniki, p. 223 (2009)
6. Marinos, P.: The archaeological site of Delphi, Greece. A site vulnerable to earthquake and landslides. In: IGCP 425 International Meeting Landslide Hazard Assessment and Mitigation for Cultural Heritage Sites and Other Locations of High Society Value, pp. 83–90. UNESCO, Paris (1999)
7. Dewez, T.J.B., Girandeau-Montaux, D., Allanic, C., Rohmer, J.: Facets: a cloudcompare plugin to extract geological planes from unstructured 3D point clouds. In: ISPRS Congress (2016)
8. Bellian, J.A., Kerans, C., Jennette, D.C.: Digital outcrop models: applications of terrestrial scanning lidar technology in stratigraphic modelling. J. Sediment. Res. **2**(75), 166–176 (2005)

Monitoring and Mapping of Deterioration Products on Cultural Heritage Monuments Using Imaging and Laser Spectroscopy

Kostas Hatzigiannakis[1]([⊠]), Kristalia Melessanaki[1],
Aggelos Philippidis[1], Olga Kokkinaki[1], Eleni Kalokairinou[1],
Panagiotis Siozos[1], Paraskevi Pouli[1], Elpida Politaki[2],
Aggeliki Psaroudaki[2], Aristides Dokoumetzidis[2],
Elissavet Katsaveli[2], Elissavet Kavoulaki[2], and Vassiliki Sithiakaki[2]

[1] Institute of Electronic Structure and Laser, Foundation for Research
and Technology-Hellas (IESL-FORTH), P.O. Box 1385, GR 71110 Heraklion,
Crete, Greece
kostas@iesl.forth.gr
[2] Ephorate of Antiquities of Heraklion, 71202 Heraklion, Greece

Abstract. Cultural Heritage (CH) outdoor monuments are susceptible to severe and extreme weather phenomena as a result of the climatic change. The prompt detection and analysis, as well as the continuous monitoring of weathering formations and deterioration products is thus crucial for their preservation and longevity.

The HERACLES project (HEritage Resilience Against CLimate Events on Site, GA 700395) aims to develop, apply and establish responsive methodologies and systems towards the mitigation of the impact caused by climate changes on the monuments and thus a multi-disciplinary research is in progress. One of the HERACLES tasks is to develop a diagnostic methodology, which will closely observe the generation and evolution of surface deterioration products on the basis of remote monitoring and mapping and in correlation with environmental data and in-situ characterization of materials.

This is implemented by means of imaging spectroscopy which, at specific wavelengths, allows the differentiation of the encrustations from stone substrates. The spectral data, recorded seasonally in-situ, on the monument, are processed and compared in order to map the spatial evolution of the pathologies. In parallel, in-situ Raman and LIBS spectroscopic measurements are performed for the chemical characterization of the studied materials, while the environmental conditions are continuously recorded. The acquired data are cross-correlated in order to elucidate the nature, understand the deterioration mechanism and determine any correlation of deterioration products with climatic changes.

The methodology is initially designed on two of the HERACLES test beds; the Palace of Knossos and the Venetian coastal fortress of Koules. Both monuments located in the city of Heraklion, Crete, Greece are representative of important historical eras, involve a number of construction materials and are subjected to various climatic conditions. This methodology, directly connected with the HERACLES ICT platform for responsive monitoring of the studied test-beds, is envisaged to be broadly implemented in the future to other CH monuments.

© Springer Nature Switzerland AG 2019
A. Moropoulou et al. (Eds.): TMM_CH 2018, CCIS 962, pp. 419–429, 2019.
https://doi.org/10.1007/978-3-030-12960-6_29

Keywords: Climate change · CH monuments ·
Monitoring of deterioration products

1   Introduction

1.1   The H2020 EU Project HERACLES

The objective of the HERACLES project, HEritage Resilience Against CLimate Events on Site, is to efficiently prevent and deal with the effects of climate change on monuments. The project approaches this aim in a holistic and multidisciplinary mode through the involvement of different expertise (end-users, industry/SMEs, scientists, engineers, conservators/restorers, social experts, as well as decision and policy makers). The core of the project is based on the development of a system exploiting an ICT platform able to collect and integrate multisource information in order to effectively provide complete and updated situational awareness and support decisions for innovative measurements improving CH resilience.

The input for the HERACLES ICT platform is collected from a wide network of global monitoring systems (i.e. satellites), with the support of local measurements from sensors installed on site at selected monuments in combination with data acquired using novel portable analytical and diagnostic instruments. The scheme also uses a number of historic data and the valuable contribution of short and long term models predicting the effects of climatic change, while it takes into account the analytical advantages offered by a highly competent network of analytical laboratories. In addition new materials for protection and cost-effective maintenance of the studied sites is also designed and investigated.

Key element in HERACLES success and performance is the design and validation of manageable methodologies and the definition of operational procedures and guidelines for risk mitigation and management. In this respect, the methodology developed for in-situ monitoring and mapping of deterioration products on CH monuments using imaging and laser spectroscopy techniques is presented as one of the tasks of the HERACLES project.

1.2   Description of the Monuments Under Study and Their Deterioration Products

The studies, described herein, refer to the investigation of a number of over-layers that can be found on the monuments' surface caused by deterioration of the original stones and/or accumulation of external material. The aim is to elucidate their nature (in correlation to the original stone substrates and their deterioration), monitor their evolution in relation to the environmental conditions and climate changes and, thus, to be able to suggest responsible preservation solutions.

A brief description of the monuments under study and their condition is presented with emphasis on their deterioration products.

The Palace of Knossos. The archaeological site of Knossos is the largest Bronze Age site on Crete and is considered Europe's oldest city. It was continuously inhabited from the Neolithic period (7000–3000 B.C.) until Roman times. The Palace of Knossos is located in the S.E. of the city of Heraklion and is the largest and the most glorious of all Minoan Palaces in Crete, covering an area of 22000 m^2. The Palace was excavated partly by Minos Kalokairinos (1878) and fully by Sir Arthur Evans (1900–1905).

The Minoan walls were constructed of rough and carved limestone and selenite (mineral gypsum) stones and the joints were filled with mortar of clay. A coating of clay and lime plaster was applied on the walls in order to create a surface for the fresco decorations. Given the sensitivity of these materials the need for preservation and restoration of the monument was obvious for Sir A. Evans in the early years of the excavation. Since 1905 most of the monument's parts were sheltered, and some restored. For the restoration, Evans reconstructed entire floors of the monument with reinforced concrete. Recently, these reconstructions are considered part of the history of the restoration of the Palace of Knossos and a monument in its own right and, thus, their conservation is considered necessary.

The Palace of Knossos is susceptible to a number of deterioration factors which, when combined with the local microclimate and the nature and structure of the walls, urge for careful planning and preservation actions. The main factors affecting the condition of the Palace of Knossos include the direct exposure to sunlight, rain, wind and atmospheric pollution, the relative humidity and temperature, which both range beyond recommended limits, and the extensive use of reinforced concrete for its restoration. In particular mineral gypsum, which is a water-soluble and extremely sensitive material, has been extensively eroded and its mechanical properties have been seriously modified.

The presence of efflorescence salts in ancient and restored masonry and roofs is one of the studied over-layers closely connected with the penetration of humidity through the jointing mortars. Furthermore, various, mainly dark-colored, crusts and layers are also investigated. Their presence is associated with exposure of the surfaces to environmental conditions, high levels of humidity and the use of cement mortar or other incompatible materials in previous restorations.

The coastal Fortress of Koules is located in the port of Heraklion facing the sea and was part of the large fortification system built by the Venetians during the 16th century in order to protect the colonial city from the Ottoman threat. There is sufficient evidence, concerning the architectural phases of the fortress, from the construction of the prime Byzantine tower, to the Venetians, the interventions of the Ottoman period as well as from the contemporary state of preservation of the monument.

It is built with large rectangular blocks of limestone and sandstone. The authentic mortar of the masonry consists of slaked lime and other aggregates (i.e. pebbles). Iron clamps also have been placed either on the internal or the external side, in order to reinforce the structure of the masonry.

The continuous and total exposure to aggressive environmental conditions such as strong north winds and sea waves has produced severe weathering on buildings' stones and mortars. Soluble salts and various types of encrustation are among the decay products, which are present on the monument's structural materials.

2   The Diagnostic Methodology for Material Characterization and Remote Monitoring and Mapping of Deterioration Products on Monuments

The diagnostic methodology developed through HERACLES for the in-situ characterization and monitoring of deterioration products on stone monuments relies on laser spectroscopy for the characterization of the deterioration products and spectral imaging for the seasonal monitoring of their evolution. The steps followed for the development of the diagnostic methodology are presented and briefly discussed below:

1. **Determination of the areas of interest.** Initially, a number of characteristic areas with intense deterioration were selected. The observed deposits were classified on the basis of their morphology and colour. Efflorescence salts as well as white and black crusts are the main objectives of this study.
2. **Material sampling for laboratory chemical analysis.** A number of samples typical of each crust type were collected by experts from different areas of the monuments under study.
3. **Development of a reference database from ex-situ analysis.** The collected samples were analysed and chemically characterized in the laboratory, via micro-Raman spectroscopy and laser-induced breakdown spectroscopy (LIBS) [1–5] in order to form a reference database for the analytical needs of the project. These two laser spectroscopic analytical techniques employ laser light to determine precisely the chemical composition of the investigated materials.

   Micro Raman spectroscopy is a broadly established method for materials analysis and also a powerful analytical tool in the context of art conservation and archaeological science [6, 7]. On the other hand, LIBS, Laser Induced Breakdown Spectroscopy, is an analytical technique that enables determination of the elemental composition of materials on the basis of the characteristic atomic emission from a micro-plasma produced by focusing a high-power laser on the (external or internal) surface of a material. LIBS technique has been used in a wide variety of analytical applications for the qualitative, semi-quantitative and quantitative analysis of cultural heritage materials [1–5]. LIBS provides very short measurement time (less than 1 s), has no particular sampling requirements and is suitable for fast screening of materials/areas.

   The two techniques are employed in combination in order to fully exploit their advantages and overcome their limitations (i.e. weak Raman signal and intense fluorescence effects, mostly caused by the presence of some organic compounds). A number of analytical measurements were performed on multiple representative spots on each sample. The experimental parameters (spot size, acquisition time, number of spectra accumulations) were carefully adjusted for the chosen spots.

   All the experimental details and measurement parameters, along with the obtained results are registered in a detailed, continuously updated, reference database.
4. **On-site chemical analysis at the areas of interest.** Following the laboratory (ex-situ) analyses and calibration of the instruments, the Raman and LIBS portable systems are moved on-site to further analyse the areas of interest.

The analytical campaigns are performed seasonally at regular intervals, and/or after major events. The in-situ analytical results are compared with the laboratory ones in order to validate the effectiveness of the instruments in providing on-site reliable measurements. The outputs of the laser-based techniques are cross correlated with imaging (multispectral) methodology data and, also, with the physical, chemical and mechanical properties of the reference materials (gypsum stone, concrete and limestone). Finally, the chemical analysis results will be matched with climatic conditions (outputs from meteorological sensors).

5. **On-site monitoring of the selected areas of interest.** In parallel, an imaging spectroscopy protocol was developed for the on-site monitoring of the evolution of the extent of deterioration over time. Imaging spectroscopy has been widely used in CH for a variety of applications including materials differentiation and stratigraphic analysis [8]. In the present study, the technique has been employed for the differentiation of the encrustations' material from the stone substrate, the mapping of the areas of interest, as well as the periodic monitoring of the evolution of these encrustations. The imaging spectroscopy protocol is implemented in two phases.

In the first phase the wavelength, which allows the optimum differentiation between the materials of the crust and the substrate, was determined. This wavelength may vary for each specific area of interest, given that the differentiation depends on the physical properties of each material. This process is briefly described herein: For each examined area spectral images are acquired in-situ. Subsequently, an ex-situ standard post-processing (image normalization and registration) process is employed [9]. A wavelength corresponding to the maximum reflectance difference between the crust and adjacent materials was determined by measuring the corresponding areas intensity values on the spectral images and by subsequent calculation of their ratios. The highest ratio value defines the wavelength where the material differentiation is optimum. For this specific wavelength a threshold is set and a 2-bit image is generated highlighting the studied crust and allowing the measurement of its extent.

The second phase involves periodic examination of each encrustation by acquiring images only at the optimum wavelength and under the same parameters. The recorded images are normalized and registered to the original image and, finally, the extent of the crust is calculated. Comparison of the periodic measurements allows monitoring of the evolution of the pathologies. The combination of the imaging data with the spectroscopic and environmental data, received from indoor and outdoor sensors, will allow the determination of the pathologies generating factors.

6. **Processing of results.** The recorded data are processed and cross-studied. During this phase the acquired data are compared with the reference database, as well as bibliographic resources, in order to identify the chemistry of the materials involved and determine their extent on the monument.

7. **Correlation of climatic changes to the monuments materials.** The identified deterioration products are correlated to the stone substrates (i.e. limestone, selenite) and the local environmental conditions in order to investigate their origin and determine the associated deterioration mechanism. In parallel, the spatial distribution of surface deterioration products and their evolution with climatic conditions over time is also estimated.

3   Experimental

3.1   Raman Spectroscopy

Laboratory Measurements (ex-situ). A mobile Raman microspectrometer (JY Horiba) was employed in the series of campaigns. The analytical description of our system is presented in a recent paper [1].

In a typical measurement, the sample or object/surface for analysis is placed under the microscope objective and the exact area to be analyzed is selected by means of a XYZ micro-positioner and the video microscope. Prior to each measurement the laser power is adjusted (with the use of neutral density filters) and spectra from standard materials (sulfur, silicon, calcite, teflon) are collected for calibration purposes. Once the area to be analyzed is selected by means of the viewing camera, the acquisition parameters (acquisition time, number of spectra accumulation) are adjusted and Raman spectra are collected.

In-situ Measurements. Raman instrument is used in situ, with its corresponding optical probe head positioned properly with respect to the area under investigation (Fig. 1a). A custom-made platform supported on a tripod enables lateral movement of the head, as well as it permits measurements from a minimum height of 60 cm up to a maximum height of 2 m from the ground.

The power delivered by the laser beam on the sample surface was set in the range of 0.5–50 mW. Typical acquisition time for each scan on the CCD was 20 s and the number of accumulation scans was approximately 5–10 consecutive scans on the same point.

3.2   Laser Induced Breakdown Spectroscopy

Laboratory Measurements (ex-situ). The portable LIBS system (LMNT II+), has been developed and constructed at FORTH-IESL [1, 2, 10] and was used in this study. A fast and reliable method has been implemented to identify main mineral(s) present on monuments material alterations. The method has been tested and exploited for rapid characterization of scale samples, such as carbonates, sulfates, oxides [11].

In-situ Measurements. The LIBS instrument is positioned close to the surface, which is analyzed, and this is achieved by mounting the LIBS probe on a monopod. This arrangement enables measurements to be performed in a height range approximately 60–200 cm from the ground (Fig. 1b).

3.3   Multispectral Imaging

All measurements were performed in-situ. For the acquisition of the spectral images the IRIS II MSI camera [9], developed at IESL-FORTH, has been employed. The camera is placed against the area of the pathology at a distance of approximately 1.2 m thus covering an area of approximately 300 × 225 mm. The area size is limited to this range because of the size of the white target (Spectralon, 300 × 300 mm) which is used as a reference reflectance material (Fig. 1c). The exact dimensions of the area as well as distances for physical reference points (coordinates) are recorded. This practice in combination with the acquisition parameters recording will allow the periodic reproduction of the measurement.

Originally, for the definition of the optimum wavelength for each pathology, a series of 28 spectral images is acquired and subsequently post-processed. Following, periodic measurements (3–6 months) are performed by capturing spectral images only at the predefined spectral wavelength and under the original positioning and acquisition parameters. The acquired images are post-processed under the original procedure. To ensure perfect matching between the area originally depicted and the area depicted at each periodic examination an extra image processing step is followed by registering the subsequent spectral images to the original one.

Based on this process, comparable periodic measurements, for each encrustation, are produced and its evolution is defined as described in Sect. 2.

Fig. 1. (a) On-site Raman measuring campaign at the Palace of Knossos, King's Megaron area, (b) On-site LIBS measuring campaign at the Palace of Knossos, King's Megaron area, (c) Imaging spectroscopy measuring at the Fortress of Koules, Room 13.

4   Discussion of Results

4.1   Efflorescence Salts

An example on the study of the seasonal evolution of efflorescence salts is presented herein. Intense presence of efflorescence salts is usually encountered on the internal surfaces of the 16th century limestone walls of the coastal Fortress of Koules. One of these walls with characteristic efflorescence salt formations (coded as HKIII-01) was selected to apply and test the methodology under development. Sampling and in-situ analysis using Raman and LIBS spectroscopy were performed in order to determine the chemical composition of the formations, while imaging spectroscopy was also used to highlight and measure their seasonal extent.

Chemical Analysis. LIBS identified the presence of strong Na atomic emission signal (Fig. 2), which is a strong indication of the presence of Na-based salts. The Raman signal, detected, was notably weak and no Raman bands were determined. It is reported in the literature that Na_2SO_4 and Na_2CO_3 give a distinct and intense Raman signal [6].

Specifically, Raman measurements on areas examined at the Palace of Knossos, throughout this project, demonstrated the presence of thenardite (Na_2SO_4). On the contrary, NaCl has no Raman signal, because it is fully polarized [12]. Therefore, it can be reliably concluded that the detected salt is NaCl.

Fig. 2. (a) The area covered with the efflorescence salts at Room 13 of Koules fortress, (b) a magnification of the area (c) LIBS emission lines from in-situ analysis of efflorescence salt, coded as HKIII-01, corresponding to Na, indicating the presence of NaCl.

Mapping/Imaging. The optimum wavelength for differentiation between this type of crust and the limestone substrate is defined at 380 nm. Four periodic measurements (summer 2017, autumn 2017, spring 2018 and summer 2018) have been performed up to now. The results are presented in Fig. 3. As can be observed, the area covered with efflorescence is larger in the summer 2017 (43.89% of the overall area) while it is significantly reduced (5.71%) in the next measurement (autumn 2017). The pathology is re-expanded, to a lower degree though, for the spring 2018 measurement (27.7%) and again reduced (16.48%) for summer 2018. The results indicate that the presence and extent of the efflorescence salts are affected by the seasonal climate changes. The specific microclimate conditions, which influence this phenomenon are currently studied on the basis of the correlation of these periodic acquisitions with the environmental monitoring data acquired from the data loggers and the meteorological stations.

4.2 Black Crust

Another typical example refers to the study of black crusts found on the surface of a 16th century limestone wall at the coastal fortress of Koules (sample area HKII01) (Fig. 4).

Chemical Analysis. Analysis of the black crust with the portable Raman instrument identified the presence of Gypsum ($CaSO_4.2H_2O$) (Fig. 5a). In Fig. 5a the recorded Raman signal presents an intense background noise, indicating the presence of highly fluorescing materials, possibly of organic/biological nature. Moreover, LIBS (Fig. 5b) identified the presence of Ca and Na. Sodium (Na) is an element that is found almost in all the investigated deposits, due to the fact that the monument is situated very close to the sea.

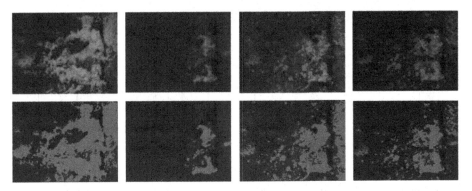

Fig. 3. The four periodic measurements for efflorescence salts. 1ˢᵗ row: Spectral images at 380 nm for (left to right) summer 2017, autumn 2017, spring 2018 and summer 2018. 2ⁿᵈ row: Highlighting of the area covered with efflorescence by applying pseudo color. (Color figure online)

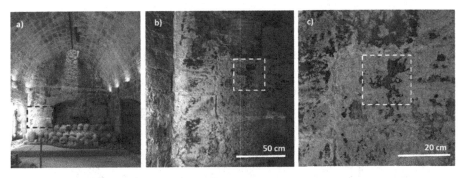

Fig. 4. (a) General view of room 18, coastal Fortress of Koules. Areas with black crust are present on the left-hand side of the rear wall (b) Closer view of the area with the black crust and (c) a magnification of the area.

Mapping/Imaging. For this type of encrustation the optimum wavelength is defined at 1000 nm. Three periodic measurements (autumn 2017, spring 2018 and summer 2018) have taken place up to now. The results are presented in Fig. 6. The area covered with black crust does not significantly vary between the periodic measurements. More specifically, the extent of the black crust was measured to cover area 31.95% of the depicted area for autumn 2017, 29.81% for spring 2018 and 32.74% for summer 2018. As expected, this crust is not significantly affected by the seasonal changes of the environmental conditions. Further periodic measurements have been scheduled while the correlation with the recorded analytical data will allow the understanding of factors, which cause the presence of these encrustations.

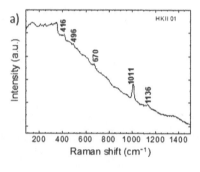
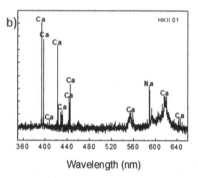

Fig. 5. (a) Raman (left) and (b) LIBS (right) spectra from black crust coded as HKII-01. Raman peaks are corresponding to Gypsum ($CaSO_4.2H_2O$) while LIBS spectrum indicated the presence of Ca and Na; Ca element corresponds to the presence of gypsum (identified by Raman spectroscopy) and Na element is due to the monument's proximity to the sea.

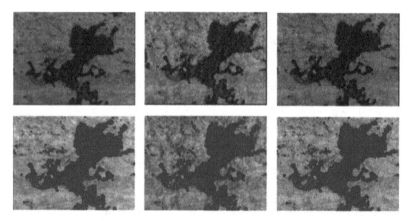

Fig. 6. The three periodic measurements for black crust. 1st row: Spectral images at 1000 nm for (left to right) autumn 2017, spring 2018 and summer 2018. 2nd row: Highlighting of the area covered with efflorescence by applying pseudo color. (Color figure online)

5   Conclusions

A diagnostic methodology to closely observe the generation and evolution of surface deterioration products on monuments is presented in this manuscript. The developed methodology, based on remote monitoring and mapping of deposits/crusts in parallel with ex-situ and in-situ characterization of materials, aims to add knowledge as regards the understanding of the deterioration mechanisms that may take place in the studied monuments and to correlate them with environmental conditions and, in long term basis, with effects associated to the climatic changes.

The individual steps of this methodology are presented and discussed with emphasis on the chemical characterization of materials on site using portable Laser Spectroscopic instruments (LIBS and Raman) developed at IESL-FORTH, as well as the actual monitoring and mapping of the spatial evolution of the pathologies. Specifically, their diagnostic potential is evaluated in comparison with a reference database created for this purpose and their monitoring effectiveness to observe changes to their extent recorded through the year is shown. These data are uploaded to the HERACLES ICT platform with the aim to enable the responsive monitoring of the monuments under study and safeguard them from effects connected with climate change.

References

1. Papliaka, Z.E., et al.: Pigments characterization in Russian icons in Greece (15th-20th CE) using portable laser spectroscopic instruments. Heritage Sci. **4**, 15 (2016)
2. Westlake, P., et al.: Studying pigments on painted plaster in Minoan, Roman and early Byzantine Crete. A multi-analytical technique approach. Anal. Bioanal. Chem. **402**, 1413–1432 (2012)
3. Giakoumaki, A., Melessanaki, K., Anglos, D.: Laser-induced breakdown spectroscopy (LIBS) in archaeological science-applications and prospects. Anal. Bioanal. Chem. **387**, 749–760 (2007)
4. Klein, S., Hildenhagen, J., Dickmann, K., Stratoudaki, T., Zafiropulos, V.: LIBS-spectroscopy for monitoring and control of the laser cleaning process of stone and medieval glass. J. Cult. Heritage **1**(Suppl. 1), S287–S292 (2000)
5. Osticioli, I., Mendes, N.F.C., Porcinai, S., Cagnini, A., Castellucci, E.: Spectroscopic analysis of works of art using a single LIBS and pulsed Raman setup. Anal. Bioanal. Chem. **394**(4), 1033–1041 (2009)
6. Madariaga, J.M., et al.: In situ analysis with portable Raman and ED-XRF spectrometers for the diagnosis of the formation of efflorescence on walls and wall paintings of the Insula IX 3 (Pompeii, Italy). J. Raman Spectrosc. **45**, 1059–1067 (2014)
7. Vandenabeele, P., Edwards, H.G.M., Moens, L.: A decade of Raman spectroscopy in art and archaeology. Chem. Rev. **107**, 675–686 (2007)
8. Liang, H.: Advances in multispectral and hyperspectral imaging for archaeology and art conservation. Appl. Phys. A **106**(2), 309–323 (2011)
9. Zacharopoulos, A., et al.: A method for the registration of spectral images of paintings and its evaluation. J. Cult. Heritage **29**, 10–18 (2018)
10. Baker, J., et al.: The height of Denier Tournois minting in Greece (1289–1313) according to new archaeometric data. Annu. Brit. Sch. Athens **112**, 267–307 (2017)
11. Siozos, P., Philippidis, A., Hadjistefanou, M., Gounarakis, C., Anglos, D.: Chemical analysis of industrial scale deposits by combined use of correlation coefficients with emission line detection of laser induced breakdown spectroscopy spectra. Spectrochim. Acta, Part B **87**, 86–91 (2013)
12. The RRUFF Project database of Raman spectra, X-ray diffraction and chemistry data for minerals. http://rruff.info. Accessed 29 Oct 2018

Radar Interferometer Application for Remote Deflection Measurements of a Slender Masonry Chimney

Georgios Livitsanos[1(✉)], Antonella Saisi[2], Dimitrios G. Aggelis[1], and Carmelo Gentile[2]

[1] Department of Mechanics of Materials and Constructions (MEMC),
Vrije Universiteit Brussel, Brussels, Belgium
{Georgios.livitsanos,daggelis}@vub.be
[2] Department of Architecture, Built Environment and Construction Engineering
(DABC), Politecnico di Milano, P.za Leonardo da Vinci, 32, 20133 Milan, Italy
{antonella.saisi,carmelo.gentile}@polimi.it

Abstract. Seismic vulnerability of many cultural heritage masonry structures has been vital the recent years especially in seismic prone countries such as Italy. Due to the frequency of the seismic motions in time and due to many limitations, concerning the available knowledge and the seismic design of these stiff masonry structures, there is the necessity of non-contact and quick ambient response recordings so as to provide a structural integrity assessment. In order to improve the knowledge about the dynamic behavior of the slender masonry structures two main different analyses are necessary. The first goal is a territorial level assessment for the determination of the seismic ground motion. The second one concerns the knowledge of the dynamic characteristic of the existing structures. Recently, radar technique has been advanced by the development of microwave interferometers which serve non-contact vibration monitoring of large structures. The main characteristic of these radar systems, is the possibility of simultaneously measuring the dynamic deflection of many points on large structures. As a result, ambient vibration measurements can provide the identification of the modal properties of a structure. This paper reports a set of on-site applications of radar interferometer technique, aiming at evaluating the capability of measuring the vibration response of a slender masonry chimney in the *Leonardo* campus of the Politecnico di Milano. The results of the investigation highlight the accuracy and the simplicity of the technique for fast dynamic response measurements as well as the capability to detect temperature effects on the fundamental frequencies on the structure.

Keywords: Ambient vibration testing · Radar interferometer ·
Modal parameters · Masonry chimney · Territorial level analysis

© Springer Nature Switzerland AG 2019
A. Moropoulou et al. (Eds.): TMM_CH 2018, CCIS 962, pp. 430–442, 2019.
https://doi.org/10.1007/978-3-030-12960-6_30

1 Introduction

As it is well known, Italy is a country with a huge number of outstanding historical monuments and Cultural Heritage buildings and a relatively seismic prone area. As a consequence, during the last decades, a great attention has been devoted to the assessment and preservation of industrial and Cultural Heritage structures. Most of these structures are nowadays included in the urban environment of the cities and surrounded by buildings and even they are considered as industrial heritage and their demolish is generally avoided, they have been subjected to poor or null maintenance operations during the years. This fact, in combination with the completely neglected seismic design to sustain gravity and wind loads, led to the necessity of an integrity evaluation. The primary goal and phase in integrity assessment of existing structures in Engineering Seismology and Geotechnical Earthquake Engineering, is to determine the seismic ground motion for the seismic design. Lately, being based on the intensity and frequency of past earthquakes, and giving specific importance on the application of special regulations of buildings in areas classified as seismic, valuable knowledge about the territorial classification is provided. This helps in the reduction of the effects of earthquakes. As a result, the seismic vulnerability of the territorial level consists a critical tool to establish priorities for future interventions. After getting this knowledge, the role of structural dynamics is the calculation of the response of the structure to a given seismic motion, in order to verify that the performance is satisfactory. However, in the case of existing structures and especially historical ones, the dynamic characteristics are difficult to be obtained due to the limited provision of the knowledge of these structures. Furthermore, there is the necessity of non-contact and non-destructive ambient response recordings so as to have an improved quick estimation about the dynamic behavior of these structures.

Dynamic testing under operational conditions or ambient vibration testing (AVT) is currently the main experimental method used for assessing the dynamic behavior of full-scale structures in service. Recently, radar technique has been advanced by the development of microwave interferometers which serve non-contact vibration monitoring of large structures. The main characteristic of these radar systems, entirely designed and developed by Italian researchers, is the possibility of simultaneously measuring the highly accuracy dynamic deflection of many points on large structures. As a result, ambient vibration measurements can provide the identification of the modal properties of a structure (natural frequencies and mode shapes). The interferometric Real-Aperture Radar (RAR) technique was introduced at the end of the 1990s as an operational tool for dynamic testing of large civil engineering structures. The combined use of radar interferometry [1] and high-resolution waveforms [2, 3] and has led to the development of the innovative radar technology which is capable of simultaneously measuring the dynamic displacement of several points on a large structure with high accuracy [4]. However, it has only been widely adopted in recent years. Many articles have focused on the use of microwave remote sensing to measure the dynamic response for a variety of structures [5, 6] such as bridges [7, 8] buildings [9], chimneys [10, 11] and ancient towers [12–16].

Masonry chimneys are structures build up many years before with only few specific information available. The first handbooks and treatises [17, 18] for the design of chimneys date back to the nineteenth century. The shape of the cross sections, the thickness of the external walls, indications for the design against the wind loads consists basic information that was provided by these sources. However, the seismic hazard design was completely neglected. On the other hand, nowadays proper design rules are imposed by specific standards. Consequently, severe collapses on masonry chimneys due to the destructive effect of the earthquakes has been frequently recorded. This paper addresses the coherent radar interferometric monitoring technique applied in a chimney located in the Leonardo campus of the Politecnico di Milano. The purpose was to evaluate the capability of radar to provide the estimation of the vibration frequency, and the amplitude of displacement for a modal analysis of the masonry chimney, solely stimulated by environmental noise such as wind. Without using artificial reflectors to increase the radar signal-to-noise ratio (SNR), results obtained using interferometric RAR are providing fast measurements of the fundamental frequency as well as an accurate estimation of the temperature effect on the structure.

2   Radar Interferometry – Structure

2.1   Working Principle

A Real Aperture Radar (RAR) is a time-of-flight ranging device, which calculates the elapsed time between the transmitted and received electromagnetic waveform of a precise frequency. An application of the RAR technique is the IBIS radar system which was designed and developed by the Italian company Ingegneria Dei Sistemi (IDS, Pisa, Italy), with various collaboration partners [4] and is commercially available in two configurations, a microwave, wide band waveform interferometer (IBIS-S) and a ground-based synthetic radar aperture (IBIS-L) system. The sensor used in this study is the IBIS-S (Image By Interferometric Survey of Structures) and is capable of high resolution dynamic displacement measurements taken simultaneously at different points of a structure. From the technical perspective, the sensor is a continuous-wave (CW) radar where the echo returning phase is used to determine the range. It is also known as stepped-frequency (SF) radar operating in the Ku frequency band [19]. This continuous stepped frequency wave serves to measure the distance from radar to the observed object. It resolves the scenario in the range direction, detecting the position in range of different targets. The magnitude of the inverse discrete Fourier transform (IDFT) of the received echoes at each time sample provides a synthetic profile which corresponds to the displacement time histories of the scattering objects in the space illuminated by the antenna beam, as a function of their relative distance from the sensor. Tall structures such as towers, are usually imposed to horizontal loads such as winds. Therefore, the prior knowledge of the motion direction is of high importance in order to have significant recorded displacement values. Due to the height of the structures, the radar positioning is being usually in the base of the structure of investigation and the calculated displacements correspond to the inclined ones along the angle of the instrument inclination with respect to the horizon. Consequently, the measurable component of the horizontal displacement d_{LOS} is given by the following equation:

$$d_{LOS} = \Delta S \frac{L}{R} = \Delta S * \cos(\theta) \tag{1}$$

where L is the distance between the radar position and the structure, R is the aslant range, and ΔS is the actual displacement by making straightforward geometric projections. From so on, the expression measured displacement, will always refer to the Line-Of-Sight component measured by the Radar, d_{LOS} of specific points with high reflectivity of the monitored structure. These points, known as targets, are strongly dependent on the geometry and the dielectric characteristic of the structure materials and usually correspond to façade structural elements intersections, resulting in "corner zones". An ideal radar range profile can be illustrated in the Fig. 1(b), where a series of targets with good electromagnetic reflectivity at different distances and different angles are being illuminated from that radar. The position of them can be identified by the range (distance from the sensor) and as a consequence the displacement time history can be detected by the transient response of these points. The remote capability of this technique up to one hundred meters to measure relative movements with submillimeter accuracy and identify the characteristics of key vibration modes without the need to enter the structure for sensors or reflectors installation is one of the main advantages. Furthermore, the easiness and the capability of measurements in a shorter time with respect to standard sensors and with negligible influence of the weather conditions makes the technique quite promising not only for validation purposes, but also for independent measurements. However, this technique causes some uncertainties and errors due to the one-dimension imaging capability, meaning that different targets are only distinguished based on their distance from the sensor. Hence, multiplicity of contributions to the same range bin, which are coming from different points in different distances from the radar without lying on the same axis, may cause measurement errors [7, 20, 21].

The apparatus is composed by a sensor module, the PC control and a power supply which consists of a 12 V battery pack capable to provide approximately 5 h autonomy.

Table 1. Main characteristics of microwave interferometer (IDS, model IBIS – D).

Parameter	Value
Maximum sampling frequency	200 Hz
Radiofrequency bandwidth	17.2 GHz
Maximum operational distance	>500 m
Displacement accuracy	<0.02 mm
Maximum range (distance) resolution	0.50 m
Max. acquisition rate	200 MHz
Weight of the entire system	12 kg
Battery autonomy	5 h

The apparatus is mounted on a tripod and it is equipped with a rotating head for adjusting the rotation measurement angle making it applicable to many different case studies. Due to the limited real-time capability of the software, a post-processing software tool named IBIs Data Viewer (IBISDV) was developed [22]. The basic function of IBISDV software is the raw data interpretation, in order to obtain the displacement history of several points in the illuminated area of the investigated structure. All the operational characteristics are summarized in Table 1. The sensor module allows detecting equivalent sub-millimeter displacements according to the standard radar-frequency letter-band nomenclature from IEEE Standard 521-1984 [19]. The elementary sampling area of the radar measurement is called radar bin corresponding to different parts of the monitored structure. The acquisition of the radar bins is conducted by the antennas, which are pyramidal horns responsible for signal transmission and receive with a half power beam-width of 0.20 rad.

2.2   Chimney Under Radar Interferometry Application

The chimney that is under investigation is a fifty-meter high masonry structure, located in Leonardo campus in the Politecnico di Milano. It is also surrounded by an 80.000-liter concrete water tank at the height of 20 m. As it is daily subjected to ambient vibration wind forces, the purpose of the study was to investigate the dynamic behavior of it. Figure 1(a) illustrates the geometric shape of the chimney. It was in use until the 90s with steam boilers for central heating throughout the university as well as for the experimental needs of the industrial Mechanics Laboratory. The main idea of the interferometer technique is to acquire the synthetic image of the scenario and the displacement time histories of the points in the scenario that are characterized by a good electromagnetic

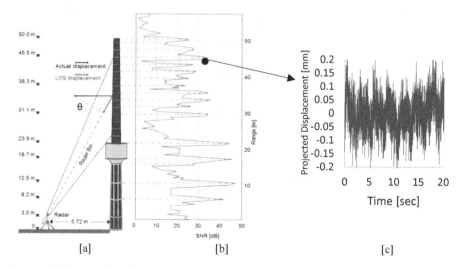

Fig. 1. (a) View of the chimney - microwave interferometer and position of the good reflecting targets detected along the height of the chimney, (b) Range profile of the scenario in the dynamic test of the chimney, (c) Displacement time histories extracted from specific reflecting targets of the radar signal profile.

reflectivity. Therefore, sufficient reflecting point are necessary. In this specific masonry structures it was observed that along the height of the chimney, steel confining rings were placed with a specific distance between them approximately about 1.8 m. These are illustrated in Fig. 1(a) between the height of 24 m and 50 m. The steel elements reflect accurately the transmitted signal and this can be validated as they coincide with the peak points that were also observed in the range profile during the acquisition (see Fig. 1b). These peaks, correspond to the deflection responses of the specific targets. The accuracy of them, is confirmed by the inspection of the displacements in time domain (Fig. 1c) as well as by the operational modal analysis (OMA).

The radar apparatus is shown in Fig. 2(a) with the provided pc for the data acquisition. An aspect of the on-site acquisition range profile is also shown in Fig. 2(b) while the confining rings which consist the main reflecting targets are depicted in Fig. 2(c). Even though the masonry chimney is symmetric in both x and y axis, measurements were conducted along three different directions in order to investigate possible identification of more than the first vibration mode (Fig. 2(d) and Table 2). The first one corresponds to the South East – North West (SE-NW) direction, the second in the South – North (S-N) and the last one in the South West – North East (SW-NE) direction. The SE-NW direction is perpendicular to SW-NE while the S-N is at 45° approximatelly. Due to the limitation of space, because of the surrounded buildings, the horizontal distance between the radar sensor and the chimney basement was varying among 5.7 to 8.6 m, for the three different measurements, resulting in rotation angles of the radar apparatus about 80° approximately (Fig. 2(d) and Table 2).

| | [a] | [b] | [c] | [d] |

Fig. 2. Deflection measurements – (a) Radar set-up, (b) Chimney illuminated area by radar, (c) Masonry chimney – steel rings-range bins (targets) in the radar illuminated area, (d) chimney top view-directions of the deflection acquisition

Table 2. Position - direction details of the radar measurements

	SE-NW	S-N	SW-NE
Radar – chimney distance [m]	8.60	5.72	8.60
Radar acquisition angle [degrees]	80.24	83.47	80.24

3 Results and Discussion

3.1 Microwave Remote Sensing – Interferometer

The modal parameters of the chimney are calculated by means of Operational Modal Analysis (OMA) technique based on "output-only" measured data under operational conditions [23]. The term "output–only", refers to the identification of the features which are only sensitive to the damage (natural frequencies, etc.) and they constitute the output of a system which is subjected to unknown inputs (ambient vibrations). The Enhanced Frequency Domain Decomposition (EFDD) technique, was used for providing the auto-spectra displacement in the frequency domain graphs and Peak Picking method for the identification of the natural frequencies and as a consequence the mode shapes. Peak Picking methodology allows the identification of the frequency using the spectral density matrices' average singular values average for all the tests carried out in the structure. The interpretation of the results was carried out by means of a commercial software, ARTeMIS Extractor. Alternatively, natural frequencies can be directly estimated by using a post-processing software tool, named IBIS Data Viewer (IBISDV) entirely developed by Italian researchers [22].

Table 3. Microwave interferometer measurements form 3 different directions according to Fig. 2(d) for fundamental frequency identification.

	SE-NW		S-N		SW-NE		
N. Frequency [Hz]	**0.74**	**0.74**	**0.76**	**0.74**	**0.76**	**0.76**	**0.74**
Date [Day]	09-05	11-05	24-05	25-05	11-05	24-05	31-05
Time [Hour]	15:45	14:45	10:45	15:30	16:15	11:50	19:00
Duration [mins]	45	45	45	45	45	45	45
Temperature °C	24.1	27.1	24.2	30.2	22.1	25.1	27
Wind Direction [°]	SE-NW 167°	NW-SE 154°	S-N 270°	SW-NE 196°	SW-NE 195°	SW-NE 205°	SW-NE 205°
Wind Speed [m/s]	4.9	5.1	3.4	5	10	4.1	5.4
Humidity [%]	50	39.2	56.5	37.3	39.5	55.5	50.8

Each measurement was arranged of 45 min duration for a more representative data acquisition. Different time periods of the day were also selected in order to investigate possible effects of the weather conditions. For each measurement, the weather data were obtained by the closest meteorological station in order to be provided with the environmental parameters such as temperature, dominant direction of the wind force, speed of the wind as well as humidity. In Table 3, different measurements from different directions are presented where finally the fundamental frequency is identified for each specific case.

Indicatively, some of the auto-spectra displacement data measured from each direction are presented in the following graphs. It is observed that in all of the cases there is consistency in the fundamental frequency identification. Approximately, all the

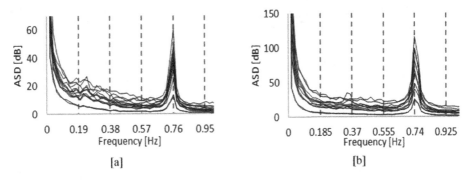

Fig. 3. Auto-spectra (ASD) of the displacement data measured by the radar interferometry - Identification of the chimney fundamental frequency from radar data (a) 24-05 (SW-NE), (b) 31-05 (SW-NE).

measurements were organized according to weather conditions. It was important the wind direction coincide with the measurements' direction in order to achieve displacement recordings higher than the accuracy (0.02 mm). Lower values are assumed as a noise. This can be also clearly illustrated in Fig. 1(c) where the measured projected displacements exceed the minimum accuracy value. Finally, the fundamental frequency was identified with an accurate consistency for every measurement. However, it was observed that the recorded values were varying between 0.74 Hz and 0.76 Hz resulting in a 2.63% difference (see Fig. 3).

This small percentage variation in the measured values cannot be attributed to the identification of another vibration mode. In literature, under experimental validation of data measured concerning the seismic period of masonry towers, Rainieri et al. proposed equations for estimating the period of the higher modes starting from the period of the corresponding fundamental mode in accordance with the formulations suggested in the Spanish seismic regulation (NSCE-02 2002). According to this, the period of the second mode is calculated as the 86% of the fundamental period resulting in a difference of 14%. This difference far exceeds the percentage of 2.63.

Furthermore, in Fig. 4, the mode shapes, calculated for the two values of the fundamental frequency, are depicted. The modal displacements are approximately in a good accordance and they clearly show that they correspond to the first vibration mode.

Consequently, the effect of the frequency variation was under the imperative need to be studied and characterized. Under this investigation, two series of one-day measurements were conducted in order to investigate this effect. Data of 45 min were recorder every hour with a time difference of 15 min from early in the morning (09:00 AM) till early in the evening (18:00 PM). In Table 4, one of the two series of measurements is presented as an indicative.

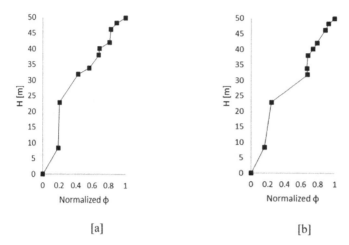

[a]   [b]

Fig. 4. Mode shapes associated to the fundament frequency (a) f = 0.74 Hz, SE-NW, (b) f = 0.76 Hz, SW-NE.

Table 4. One-day radar acquisition measurements.

SW-NE									
N. Frequency [Hz]	0.76	0.76	0.76	0.76	0.75	0.74	0.74	0.755	0.76
Time [Hour]	09:00	10:00	11:00	12:00	13:00	14:00	15:00	16:00	17:00
Duration [mins]	45	45	45	45	45	45	45	45	45
Temperature °C	25.6	27.3	28.1	29.4	30.3	30.6	29.6	30.2	26.8
Wind Direction [°]	NW-SE 158°	SW-NE 221°	S-N 267°	SW-NE 194°	SW-NE 247°	NW-SE 177°	SW-NE 236°	SW-NE 217°	SW-NE 201°
Wind Speed [m/s]	4.2	3.7	5.7	4.2	5.1	5.7	4.3	4.9	2.7
Humidity [%]	37.3	33.7	31.7	27.3	25.8	24.7	25.5	25	31

Observing the measurements during the day hours, it is obvious that there is a transition in the fundamental frequency from 0.76 Hz to 0.74 Hz. Firstly, we discard any instrumental effect because the reliability of the radar measurements was assessed with a big amount of data collected during different days, periods of each day as well as from different directions. Therefore, the repeatability and the stability of the eigenvalues enhances the assumption for reliable measurements. Furthermore, during the different measurements, there are mutations in the wind directions without the simultaneous change of the frequency, leading to the conclusion that this was not the main effect.

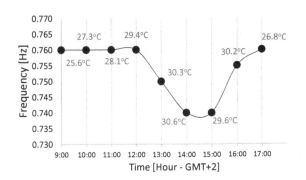

Fig. 5. Variation of identified natural frequencies due to temperature mutation during-one day measurements.

On the other hand, it is interesting to observe the progressive change of the temperature. A temperature variation between morning and evening periods of the day, can give a sensible perturbation of the target (chimney). Consequently, frequency mutations could be conceivably balanced by the temperature effect on the structure itself. The following Fig. 5 depicts the dependence of the acquired frequency on the time period of the day which is firmly connected to the temperature variations. As it is observed, the increase of the temperature downshifts the fundamental frequency value.

However, this result is contradictory according to the literature concerning masonry structures [24, 25]. Many studies have proven strong dependence of the frequency in the temperature alterations showing the simultaneous increase or decrease of them. The formula that correlates the frequency with the stiffness and the mass of the structure is the following:

$$f_n = \frac{1}{2\pi} \sqrt{\frac{k}{m}} \tag{2}$$

where *fn* represents the fundamental frequency (Hz), k the stiffness (N/m) and m the mass of the structure in (kg). It has been shown, that the increase of temperature causes closure of possible cracks due to the thermal dilatation of the materials constituting the chimney, resulting in a passive confinement for the structure while the mass remains the same. In this point, it is interesting to remind that radar measurements depend on the external surface characteristics. In this specific case, the measurements were dependent on the steel confining rings placed along the height of the structure. Consequently, increase of temperature causes thermal expansion also in the steel material. This results in relaxation and loss of tension of the rings providing also less confinement to the structure. This is an important parameter that must be also taken into account as it cannot be clarified which is the dominant effect that causes the frequency shifts. From the one hand, the lower confinement provision to the chimney from the steel rings along the height, is possible to overweight the passive confinement of the tall slender old masonry structure due to temperature. On the other hand, this effect could be dependent solely on the relaxation of the steel rings which are the targets, perturbating the reflecting echoes and masking the expected behavior. On the other hand, the small variation of the fundamental period between 1.315 and 1.35 s, does not seem to constitute a warning for the structural integrity of the structure. It finally consists a stable temperature effect of 2.6% difference among the values and not an indication of a possible degradation of strength or stiffness of the chimney. Consequently, the resistance against horizontal forces seems not to change during the two months continuous measurements.

4   Conclusions

Non-contact ambient vibration monitoring is a powerful technique for safeguarding ancient masonry structures and monuments. In these types of structures, the environmental parameters affect the dynamic characteristics in a peculiar way, which seems to be different from the cases of concrete, steel and pre-stressed concrete structures. For instance, temperature and relative humidity-driven effects on frequency changes can be observed according to previous investigations in masonry structures [26–28]. This paper describes the results of an experimental investigation which is focusing in the evaluation of the capability and the effectiveness of a coherent Real-Aperture-Radar sensor to estimate the dynamic response of a chimney, through an ambient vibration testing. The suitability of the interferometric radar is enhanced by the high level of accuracy in terms of identification of the fundamental frequency and mode shape, on the basis of a wide set of experimental data. Furthermore, environmental temperature had a strong effect in the "output-only" measured data indicating a contradictory behavior according to the literature. In masonry structures, the environmental parameters affect the dynamic characteristics in a way which seems to be different from the cases of concrete, steel and pre-stressed concrete structures. For instance, temperature and relative humidity-driven effects on frequency changes can be observed according to previous investigations in masonry structures [26–28]. This highlights the importance of the investigation as another expression of temperature effect was observed in these types of structures. Interferometric technique is always dependent on the dielectric characteristic of the target points where the radar echoes are reflected. As a result, identifying and characterizing a further parameter that is able to perturbate the results, consists also a warning that should be taken into account and renders this investigation of high importance. Finally, acquiring a quick and robust estimation of the fundamental frequency is feasible by the radar interferometer. However, validating this value by a big amount of data and continuous acquisitions provides knowledge about the structural integrity during the time. Consequently, it can be stated that a profound understanding of the chimneys' structural behavior based on this complementary approach can estimate the structural integrity and propose future strengthening interventions.

Acknowledgements. The Research Fund - Flanders (FWO) is acknowledged for funding the two months research stay in Politecnico di Milano in the Department of Architecture, Built environment and Construction engineering (DABC), (V419718 N), and the first author G.L. acknowledges FWO, for funding the four years project "AE-FracMasS: advanced Acoustic Emission analysis for Fracture mode identification in Masonry Structures" (G.0C38.15).

References

1. Henderson, F.M., Lewis, A.J.: Manual of Remote Sensing, Principles and Applications of Imaging Radar, vol. 2. Wiley, Hoboken (1998)
2. Wehner, D.R.: High Resolution Radar, 2nd edn. Artech House Inc., Norwood (1995)
3. Taylor, J.D., McEwan, T.E.: The micropower impulse radar. In: Ultra-Wideband Radar Technology, pp. 155–164 (2001)
4. Pieraccini, M., Fratini, M., Parrini, F., Macaluso, G., Atzeni, C.: High-speed CW step-frequency coherent radar for dynamic monitoring of civil engineering structures. Electron. Lett. **40**, 907–908 (2004)
5. Pieraccini, M.: Monitoring of civil infrastructures by interferometric radar: a review. Sci. World J. **2013**, 8 (2013)
6. Luzi, G., Crosetto, M., Fernández, E.: Radar interferometry for monitoring the vibration characteristics of buildings and civil structures: recent case studies in Spain. Sensors **17**, 669 (2017)
7. Gentile, C., Bernardini, G.: An interferometric radar for non-contact measurement of deflections on civil engineering structures: laboratory and full-scale tests. Struct. Infrastruct. Eng. **6**, 521–534 (2010). https://doi.org/10.1080/15732470903068557
8. Stabile, T.A., Perrone, A., Gallipoli, M.R., Ditommaso, R., Ponzo, F.C.: Dynamic survey of the Musmeci bridge by joint application of ground-based microwave radar interferometry and ambient noise standard spectral ratio techniques. IEEE Geosci. Remote Sens. Lett. **10**, 870–874 (2013)
9. Negulescu, C., et al.: Comparison of seismometer and radar measurements for the modal identification of civil engineering structures. Eng. Struct. **51**, 10–22 (2013)
10. Rödelsperger, S., Läufer, G., Gerstenecker, C., Becker, M.: Monitoring of displacements with ground-based microwave interferometry: IBIS-S and IBIS-L. J. Appl. Geodesy **4**, 41–54 (2010)
11. Gikas, V.: Ambient vibration monitoring of slender structures by microwave interferometer remote sensing. J. Appl. Geodesy **6**, 167–176 (2012)
12. Atzeni, C., Bicci, A., Dei, D., Fratini, M., Pieraccini, M.: Remote survey of the leaning tower of Pisa by interferometric sensing. IEEE Geosci. Remote Sens. Lett. **7**, 185–189 (2010)
13. Gentile, C., Saisi, A.: Dynamic testing of masonry towers using the microwave interferometry. Key Eng. Mater. **628**, 198–203 (2015)
14. Gentile, C., Saisi, A.: Radar-based vibration measurement on historic masonry towers. In: Emerging Technologies in Non-Destructive Testing V, p. 51 (2012)
15. Gentile, C., Saisi, A.: Ambient vibration testing of historic masonry towers for structural identification and damage assessment. Constr. Build. Mater. **21**, 1311–1321 (2007)
16. Luzi, G., Crosetto, M., Cuevas-González, M.: A radar-based monitoring of the Collserola tower (Barcelona). Mech. Syst. Signal Process. **49**, 234–248 (2014)
17. Breyman, G.A.: Trattato generale di costruzioni civili, voi I L. Cattaneo, L'arte Min atoria, Milano, Vallardi. 188 (1925)
18. Gouilly, A.: Théorie sur la stabilité des hautes cheminees en maçonnerie, par Al. Gouilly. J. Dejey (1876)
19. Skolnik Merrill, I.: Radar Handbook. McGraw-Hill Publishing Co. Ltd, New York (1990)
20. Gentile, C., Bernardini, G.: Radar-based measurement of deflections on bridges and large structures. Eur. J. Environ. Civ. Eng. **14**, 495–516 (2010). https://doi.org/10.1080/19648189. 2010.9693238

21. Luzi, G., Monserrat, O., Crosetto, M.: Real Aperture Radar interferometry as a tool for buildings vibration monitoring: limits and potentials from an experimental study. In: AIP Conference Proceedings, pp. 309–317 (2012)

22. Coppi, F., Gentile, C., Paolo Ricci, P.: A software tool for processing the displacement time series extracted from raw radar data. In: AIP Conference Proceedings, pp. 190–201 (2010)

23. Reynders, E.: System identification methods for (operational) modal analysis: review and comparison. Arch. Comput. Methods Eng. **19**, 51–124 (2012). https://doi.org/10.1007/s11831-012-9069-x

24. Azzara, R.M., De Roeck, G., Girardi, M., Padovani, C., Pellegrini, D., Reynders, E.: The influence of environmental parameters on the dynamic behaviour of the San Frediano bell tower in Lucca. Eng. Struct. **156**, 175–187 (2018)

25. Saisi, A., Gentile, C., Guidobaldi, M.: Post-earthquake continuous dynamic monitoring of the Gabbia Tower in Mantua, Italy. Constr. Build. Mater. **81**, 101–112 (2015)

26. Ramos, L.F., Marques, L., Lourenço, P.B., De Roeck, G., Campos-Costa, A., Roque, J.: Monitoring historical masonry structures with operational modal analysis: two case studies. Mech. Syst. Signal Process. **24**, 1291–1305 (2010). https://doi.org/10.1016/j.ymssp.2010.01.011

27. Gentile, C., Guidobaldi, M., Saisi, A.: One-year dynamic monitoring of a historic tower: damage detection under changing environment. Meccanica **51**, 2873–2889 (2016). https://doi.org/10.1007/s11012-016-0482-3

28. Cabboi, A., Gentile, C., Saisi, A.: From continuous vibration monitoring to FEM-based damage assessment: application on a stone-masonry tower. Constr. Build. Mater. **156**, 252–265 (2017). https://doi.org/10.1016/j.conbuildmat.2017.08.160

Multispectral and Hyperspectral Studies on Greek Monuments, Archaeological Objects and Paintings on Different Substrates. Achievements and Limitations

Athina Alexopoulou[1]([✉]), Agathi Anthoula Kaminari[1], and Anna Moutsatsou[2]

[1] Department of Conservation of Antiquities and Works of Art, University of West Attica, Agiou Spiridonos, 12243 Egaleo, Greece
athfrt@teiath.gr
[2] Department of Conservation, National Gallery – Alexandros Soutzos Museum, Army Park, 11525 Goudi, Greece

Abstract. The aim of the present paper is to assess the effectiveness of different protocols for hyperspectral and multispectral imaging in cultural heritage studies. In particular, spectral cubes or monochromatic images in the visible and near infrared spectrum are captured, using either reflected or transmitted light, in normal or tangent illumination conditions for micro and macro scale characteristics evaluation. Various parameters that influence the choice of the appropriate experimental setup are discussed, such as the type of the surface and substrate, the optical characteristics of the surface and the paint layers structure of the artwork. In addition, the morphology and the non-controlled environmental conditions are examined, especially when studying outdoor surfaces. Characteristic case studies are presented ranging from small archaeological objects and paintings to large indoor and outdoor wall surfaces with an emphasis on sketch and drawing detection as well as on color decoration and pigment identification. Through these case studies the authors want to show that spectral imaging techniques can offer much more possibilities than the common user can imagine provided that the application of the methods take into account the specific characteristics of the material and exploits its optical properties. Furthermore the basic criterion of the methodology presented is the optimization of primary results without post-processing of data, based on different setups and detection protocols with the constraints and specific requirements that different structures and specific conditions dictate.

Keywords: Multispectral imaging · Hyperspectral imaging · False color · Pigment identification · Stone surfaces · Marble · Hidden sketches · Modern Greek artworks · Oil paintings on paper · Sketch detection

© Springer Nature Switzerland AG 2019
A. Moropoulou et al. (Eds.): TMM_CH 2018, CCIS 962, pp. 443–461, 2019.
https://doi.org/10.1007/978-3-030-12960-6_31

1   Introduction

The authors are part of a research team active in Greece more than twenty years in the field of non-destructive examination and documentation of cultural objects and monuments of outstanding historical and artistic value. Imaging techniques have a prominent position among the applied techniques and prove to be particularly effective for a wide range of applications including diagnosis of the state of conservation, identification of materials and technical construction, indirect dating and integration into well-known construction technologies and schools [1–3]. More precisely, Visible Imaging, UV Imaging (UV Reflectance and UV Fluorescence) and IR Imaging (IR Reflectance, Transmission and False Color IR) are powerful tools towards complete documentation, study and diagnosis. The main advantages of these techniques are the non destructive character, (i.e. they do not require sampling), the quick in situ application, the entire surface inspection of the object and mapping ability, the absence of expensive consumables, as well as the ability of post-processing large volume of data. Some of the most sophisticated of these techniques, such as Infrared Reflectography have been used for decades [4] and are still been used not only for research but also as routine techniques in everyday life of museums [5]. Nevertheless, the development of advanced setups at the end of 1990 came to bring new dynamics in this knowledge field. These setups have been constructed with the reasoning of incorporating in the same equipment many different imaging structures that were applied individually in the past, increasing the number of spectral bands available and simultaneously providing spectral and spatial resolution in an extended region of visible and non visible radiation. These techniques are referred to as multispectral and hyperspectral imaging and their main feature is the full utilization of the visible range of the spectrum with the possibility of obtaining a spectral cube with a step that can reach down to 2 nm in the most modern models. However, these detectors provide the possibility of detecting in the infrared region only up to 1000 nm and are therefore differentiated from the classical infrared reflectography, which is primarily capable of taking images in the near-infrared band up to 2250 nm [6, 7]. Equivalent to infrared reflectography might be considered to be spectral imaging using modern InGaAs detectors which have a spectral sensitivity that ranges in the near infrared from 900–1700 nm [8].

The bulk of the published results regarding the multispectral and hyperspectral study of cultural heritage objects concerns portable paintings, i.e. paintings on wood and canvas [8, 9], while the available publications regarding the study paintings on paper supports, archaeological objects, but also works of particular artistic value on inorganic substrates (e.g. wall paintings, pigment decoration on stone, ceramics etc.) are clearly fewer. Furthermore, in the international literature there is mention of few examples of the application of MWIR and LWIR thermography in the Greek territory [10, 11], but not of SWIR InGaAs thermography on large scale works as the basic method of systematic and thorough diagnosis and documentation of large dimensions of internal and external surfaces and walls of historical monuments. Finally, the reported results usually refer to specific available equipment or to technical details for the improvement of equipment in a research level [12].

The originality of the present work lies in two basic and complementary axes. The first axis is the presentation of the methodology and related results from cases of application to objects and monuments for which there are no relevant references in international literature such as the systematic scanning of large stone surfaces and walls. The second axis refers to the comparative presentation and evaluation, on the end user's side, of results obtained from a series of technical and specialized infrared imaging equipment which has been applied to a range of objects and monuments, with the basic criterion of optimizing raw data, with the help of special setups and imaging protocols under different conditions (museum conditions, in situ investigations and in vitro applications) without post-processing of the data.

2 Methods and Materials

The ultimate aim of this paper is the presentation of results obtained by spectral imaging that interest mainly the end user i.e. conservator, archaeologist, conservation scientist etc. and contribute towards the condition assessment, the possible artistic value and significance of the archaeological findings and historical buildings in order to further approximate the way of expression of the artist or the constructor's technique, to enrich the reading of their stylistic features and finally to promote and incorporate them as exhibits or to further exploit and disseminate this knowledge to sensitize people in the field of cultural protection.

Part of the case studies presented were selected among a large number of recent research studies carried out in the laboratory of Physical Chemical Methods for Diagnosis and Documentation in the frame of the Non Destructive Testing activities of Advanced Research Technologies For Investigation and Conservation ARTICON Lab, at the Faculty of Applied Arts and Cultural Sciences of the University of West Attica (UWA). The other part comes from the Laboratory of Physicochemical Research, Conservation Department, National Gallery – Alexandros Soutzos Museum (EPMAS), with which ARTICON is in close collaboration. Through these case studies the authors want to show that spectral imaging techniques can offer much more possibilities than the common user can imagine provided that the application of the methods take into account the specific characteristics of the material and exploits its optical properties. Furthermore the basic criterion of the methodology presented is the optimization of primary results without post-processing of data, based on different setups and detection protocols with the constraints and specific requirements that different structures and specific conditions dictate.

Typical examples from three categories of study cases of objects, monuments and materials were selected, each requiring a different approach to both the application protocol and the type of information provided.

2.1   Description of the Instrumentation

A range of detectors sensitive to a wide spectral range including near infrared is available to the team, presenting different possibilities, advantages, and limitations. More specifically:

- Multispectral detector MuSIS 2007 was developed by IESL-FORTH, Herakleion, Greece, in late 1990s. It is equipped with a CCD detector with nominal spectral responsivity 320–1150 nm and resolution 734 × 559 pixels. The prototype imaging system belongs to the Conservation Department of the National Gallery – Alexandros Soutzos Museum, and, despite the low resolution, it has been used for IR reflectography study of more than 200 paintings on wood and canvas in spectral bands 750–950 nm and 950–1150 nm (standard illumination sources: 2 OSRAM Halogen Display/Optic Lamps 650 W with color temperature 3400 K).
- Multispectral detector MuSIS-MS (Forthphotonics Hellas S.A. now Dysis) is equipped with a 1/2″ Progressive Scan CCD (resolution 1024 × 960pixels, 8 bits) sensor, with 8 selectable spectral bands in the range of 360–1000 nm for B&W and color imaging and ability of sequential spectral image storing (spectral cube) and calculation of a full spectrum per image pixel. The imaging system belongs to the Conservation Department of the National Gallery – Alexandros Soutzos Museum (NSRF 2007–2013) and it is used for IRR and FCIR imaging study of paintings on wood, canvas and paper belonging to the museum collection (standard illumination sources: 2 OSRAM Halogen Display/Optic Lamps 650 W with colour temperature 3400 K)
- Hyperspectral detector MuSIS HS model 2009 (Forthphotonics Hellas S.A. now Dysis) is equipped with a CCD 1/200 Progressive Scan sensor (1600 × 1200 pixels, 8 bits, 15 fps) with 34 selectable spectral bands in the range of 400–1000 nm for B&W and color imaging. It has the ability of sequential spectral image storing (spectral cube) and calculation of a full spectrum per image pixel as well as False Color Infrared Imaging. The imaging system it is used for IRR, spectral cube and FCIR imaging study of icons, paintings on wood, canvas and paper, manuscripts, ceramics, stone surfaces, wall paintings and mosaics and belongs to the laboratory of Physical Chemical Methods for Diagnosis and Documentation, Department for Conservation of Antiquities and Works of Art, University of West Attica (standard illumination sources: 2 Dr. Fischer Halogen Lamps 500 W with color temperature 3200 K).
- Short wave infrared (SWIR) detector Goldeye G1 P-008 Allied vision, model 2015 is equipped with an InGaAs AVT sensor (320 × 256 pixels, 118 fps) and a Peltier cooling system. It has spectral response in the range of 900–1700 nm. It is combined with CCTV 16 mm/f1.4 and 50 mm/f1.4 KOWA lenses as well as with NIKON micro NIKKOR 60 mm f/2.8D. The imaging system it is used for SWIR thermography imaging study of icons, paintings on wood, canvas and paper, manuscripts, stone surfaces, wall paintings and belongs to laboratory of Physical Chemical Methods for Diagnosis and Documentation, Department for conservation of Antiquities and Works of Art, University of West Attica (standard illumination sources: 2 Dr. Fischer Halogen Lamps 500 W with color temperature 3200 K).

– GRUNDIG ELECTRONIC FA-76 Infrared Detector model 1986 equipped with lead sulfide-lead oxide vidicon tube, with spectral sensitivity up to 2250 nm and image resolution 625 TVlines (standard illumination sources: 2 Dr. Fischer Halogen Lamps 500 W with color temperature 3200 K). The detector is still used for in vitro special applications in the Laboratory Physical Chemical Methods for Diagnosis and Documentation, University of West Attica on a wide range of cultural objects.

2.2   Description of the Material

Application on small 3-D objects: the cases of Piraeus's lekythos, Ancient Corinth's marble pyxis and Daphne's writing tablets. The basic feature of the study of three-dimensional objects is the need to capture images in macro-photographic conditions, either due to intense curvatures or to particularly fine decorations and detailed elements such as in the cases of an Attic *Lekythos*, a writing tablet of the classical era and a Corinthian *pyxis*.

It is worth mentioning that the *lekythos* inv. 4724 and the Daphne's writing tablet inv. 7520 (see Fig. 1a, b) belongs to the important findings of the so-called Musician's Tomb of the Archaeological Museum of Piraeus [13]. More specifically, in 1981 an emergency excavation caused by road construction in Daphne, Athens, led to the uncovering of two tombs dating to 430/420BC in Odos Olgas 53. Apart from the skeletons tomb I contained four *lekythoi* (inv. 4721–4724), while in tomb II among other findings, parts of musical instruments, writing tablets and a papyrus was found. Due to the significance of the findings, a general interdisciplinary study of the objects was initiated aiming at the dating, the documentation of the current condition, the construction technique and the materials used, as well as at the acquisition of more information on decoration and the script found in the papyrus and the writing tablets. A significant part of the project was the use of non destructive testing for a thorough examination of the objects especially of the *lekythoi*, the papyrus and the writing tablets. It was the first time that hyperspectral imaging was applied to study sensitive and fragile archaeological findings of such a type [14, 15] in Greece.

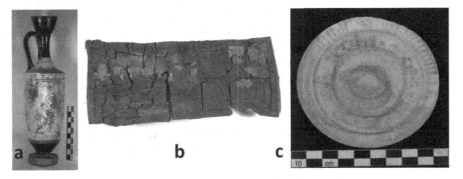

Fig. 1. (a) Piraeus's *lekythos* (inv. 4724), (b) Daphne's wooden writing tablet (inv. 7520) and (c) ancient Corinth's marble *pyxis* lid (inv. MK10905) (Color figure online)

As far as the *lekythos* inv. 4724 presented in this paper (see Fig. 1a) it is of the so-called standard shape [16] as it shows the typical design and decoration: a disc foot with a groove running around the outside near the top, a cylindrical body with a shoulder sharply set off, a long narrow neck with calyx mouth, a handle on the back between shoulder and join of neck and mouth. It presents bad condition regarding the surface, especially in the unglazed painted area. The picture field is covered by cream-white (in some places yellow-brownish discolored). In *lekythoi* inv. 4723 and 4724 traces of a figure scene and ornaments painted in matt reddish colors are hardly observed. The surface is badly damaged and the matt colors which were added after firing are not well preserved. Much of them are totally lost. After excavation, the *lekythoi* were restored, but ornament and figures were left in their fragmentary condition.

Writing wooden tablet (inv. 7452) (see Fig. 1b) has one flat side and the opposite side slightly engraved where patches of yellowish wax with traces of text can be seen. Waxed tablets were easily erasable and re-usable, and they were used for writings not intended to be kept for a long term, for example for letters and for school exercises. Writing tablets of this sort had been used for centuries in Greece and the Near East, and they continued to be used throughout antiquity.

Both lekythoi and writing tablets were studied in situ by means of spectral cube acquisition of the hyperspectral imaging system MuSIS HS in the region 420–1000 nm, with a 20 nm interval with 2 Dr. Fischer Halogen Lamps 500 W as the illumination sources in symmetrical and tangent illumination setups placed at a safe distance to avoid heating. Furthermore the object was kept in constant room temperature by air-conditioning.

Another case of spectral study of an object of archaeological interest was the study of the *pyxis* MK10905 (see Fig. 1c), dated in the last quarter of the 4th century BCE. During an excavation in Ancient Corinth an ancient cemetery was brought to light. Tomb 1 was considered to be the most important due to the vast number and high quality of the offerings it contained i.e. female figurines, silver coins, jewelry, shells and two marble *pyxides*. *Pyxis* (compass) is a small vessel with a lid usually made of marble that was meant for use by women, especially to hold their cosmetics in ancient Greece. In an early stage, conservation treatments were applied at half of the surface of the lid of *pyxis* MK10905, revealing bright colors so further study was imperative. The existence of the incrustation in areas that had not been cleaned yet made it difficult to see the decoration with the naked eye. The object was studied using the hyperspectral imaging system MuSIS HS in specific wavelengths within the region 420–1000 nm with 2 Dr. Fischer Halogen Lamps 500 W as the illumination sources in symmetrical and transmission setups. A specialized contraption was built *in situ* so as to facilitate the acquisition of transmission images in micro mode.

Application on indoor and outdoor wall surfaces - Large scale applications: the cases of wall surfaces at Horologion, Athens, and G. Halepas House – Museum, Tinos. In contrast to the case study of three-dimensional objects is the spectral imaging of large-scale surfaces such as walls or stones in monuments and historical buildings. The methodological approach to large-scale applications is determined by parameters such as the need to scan an entire surface of several square meters with images in series that satisfy the requirements of uniform conditions for capturing images, high spatial and tonal analysis and minimization of optical and geometrical deformations.

A characteristic example of a large scale spectral imaging application is the study of the stone walls of *Horologion* of Kyrristos (see Fig. 2). *Horologion* of Andronikos Kyrristos or the Tower of the Winds is located on the eastern side of the Roman Agora in Athens and is the most important building in the area. It is the oldest example of an octagonal building in the western architecture and the only one in Greece. Its total height is 13.5 m and has a diameter of about 8 m. It dates back to the 2nd half of the 1st century BC, while some scholars accept its existence already in the 2nd century BC. The monument is believed to have been raised by Andronikos Kyrristos to become the official watch on the market. Both the interior (engraved or colored sketches either in form of writing, letters and dates or in decorative elements or traces of color layers) and the exterior surfaces (state of preservation, carved patterns, texture of the stones and color decorations) of the monument were studied. The study covered a total surface of 44.1 m^2 of indoor and outdoor stone surfaces which extend in 13 m height. 4402 raw images were acquired using different set ups in visible and near infrared region (420–1700 nm), individually or in spectral cubes, in normal or raking light using both MuSIS HS system and InGaAs -Cool pix thermographic Camera and 2 Dr. Fischer Halogen Lamps 500 W as the illumination sources. The experimental result evaluation is still ongoing.

Fig. 2. The *Horologion* of Andronikos Kyrristos at the Roman Agora, Athens

It should be noted that in cases where one has to work on scaffolding, there are often limitations imposed by the narrow floor and/or the scaffold which may prevent access to certain spaces due to its vertical and horizontal bars. Restrictions also exist from the oscillation of the scaffold due to the movement of people working on it. This makes it difficult in some cases to obtain clearly focused images, while additional problems with the quality of the images taken are caused by the changing and uncontrolled intensity of the ambient radiation especially during outdoor shooting.

Another important application of spectral imaging in historical building is the case of the Halepas House-Museum in Tinos Island. Gianoulis Halepas (1851–1938) was one of the most important Greek sculptors of modern Greece. His talent and the expressiveness with which he attributed the figures throughout his lifetime are a source of inspiration for young artists. The artist's endless inspiration, the outburst of his artistic creation, forced him to express his thoughts into sketches, seeking solutions to the problems he had to deal with. Despite his unfortunate lack of the "luxury" of paper, Halepas would sketch on his father's old company records, while being at the local café or on the walls of his house, ignoring the substrate. But these sketches had the similar "bad luck" that persecuted Halepas: others were thrown away, others were burnt in the laundry fire, and others were covered with lime. However those saved are a tangible testimony of the artistic thought and an abbreviation of the creative process and inspiration of the artist.

Investigating oral testimonies and literature references according to which the artist had made drawings-compositions on the inner walls of his home, specialized scientists – researchers of the Laboratory of Advanced Research Technologies for Investigation and Conservation (ARTICON Lab), University of West Attica, at the invitation of Panormos Cultural Center "Gianoulis Halepas" of the Municipality of Tinos, studied the walls of the museum - house regarding the existence of drawings (see Fig. 3). Special equipment was transported from Athens to Tinos for the non-destructive diagnosis - documentation of the interior layers of the walls of the hall and of the two rooms of the main floor of the house, with the aim of locating drawings under the surface of lime paints, as well as for the assessment of their state of preservation. The study, approved by the Ministry of Culture, was remarkably successful, revealing sections of compositions and sketches so far unseen to visitors.

Fig. 3. Gianoulis Halepas House Museum, in Tinos Island. Part of the west inside wall surface in the main room. Visible image

The followed study methodology included the in situ application of SWIR Thermography (900–1700 nm) in order to maximize the penetrating capacity of infrared radiation to study the existence of drawings to as much depth as possible. In addition selected areas were studied by means of hyperspectral imaging system MuSIS HS of the ARTICON Lab. In all techniques, a tripod with two height extensions, with a level at the base, and two levels on its head, was used as a support base for the detectors to ensure the camera's perpendicularity with the wall area under investigation. Additionally, for a smoother image capture, a specialized camera support WalimexPro Video Rail was used which enabled the movement of the camera parallel to the surface being examined. The light sources were mounted on reflector brackets for uniform diffusion of light. Dr. Fischer 500 W and Philips 275 W halogen lamps were employed according to the prevailing environmental conditions.

The experimental conditions were determined per wall based on the specific protocols of the Non Destructive Testing activities of Advanced Research Technologies For Investigation and Conservation ARTICON Lab based on the optimum brightness and contrast ratio of the received image in relation to the type and quality of the incident lighting and the optimal magnification in relation to the analysis of the received image. The scan was carried out in a grid of a 20×20 cm window.

Application on oil paintings on canvas and paper: two case-studies from the collection of National Gallery – Alexandros Soutzos Museum, Athens. The acquisition of infrared (IR) images is especially appreciated in the case of canvas paintings because it reveals features under the pictorial layer such as underdrawings, pentimenti, etc. [17]. Multi- and Hyperspectral imaging has largely widened the application possibilities for the study of materials and painting techniques and assessment of the state of preservation of the paintings [1]. The literature concerning the study of canvas paintings refers almost exclusively to the imaging of the reflection of IR radiation in wavelengths in the near-infrared (NIR) region (760–2500 nm). Nevertheless, the optical transmittance of canvas is a feature that enables the capture even of infrared images under transmitting illumination. At the Laboratory of Physicochemical Research of the Conservation Department of the National Gallery - Alexandros Soutzos Museum, transmitted NIR imaging constitutes an inseparable part of the daily diagnostic work. One of the most complete examples of the contribution of transmitted IR imaging to the multispectral study and documentation of the canvas paintings is the examination of the painting entitled *Return to the Village* (1952) by the Greek painter Theofrastos Triantafyllides (see Fig. 4) [18]. For the capture of IR images, a MuSIS^{TM} 2007 multispectral imager was used. For the IR reflectograms, two OSRAM Halogen Display/Optic Lamps (color temperature of 3400 K) are symmetrically placed in front of the painting. In order to record the transmitted IR radiation, one light source is placed in the back side of the painting at a safe distance to avoid heating and in such a position where the lighting would be constricted into the bounds of the canvas substrate.

Fig. 4. Visible reflectance image before restoration of *Return to the Village* by Theofrastos Triantafyllides, oil on canvas, 106 × 156 cm (P. 2640)

Finally, during the research project entitled "Oil paintings on paper support: documentation of the preservation state using multispectral imaging and chemical analysis. Determination of evaluation criteria, conservation treatment proposals" realized in the framework of Archimedes III Research Funding Program, the Nikolaos Gyzis sketch *Sewing workshop* (see Fig. 5) was studied using a non destructive methodology [19]. The methodology comprised classical ultraviolet fluorescence photography, hyperspectral imaging in ultraviolet, visible and near infrared and false color recording. The hyperspectral camera MuSIS HS with a 20 nm step interval and tungsten light sources (2X500 W) for illuminating the objects, was used both for hyperspectral imaging and false colour infrared recording.

Fig. 5. N. Gyzis, *Sewing workshop*, oil and ink on paper (P. 3434)

3   Comparative Presentation of the Results

3.1   Application on Small 3-D Objects: Ceramics, Stone and Wax Substrates

As mentioned above, the basic feature of three-dimensional study objects such as ceramic and stone objects, as well as three-dimensional objects of organic materials, is the need to capture images in macro-photographic conditions, either because of intense curvatures or particularly sophisticated decorations and details.

Fig. 6.   Detail of lekythos inv. 4723 at 600 nm

Thus in the case of *lekythoi*, image acquisitions in 440 nm, 600 nm (see Fig. 6) and 1000 nm (see Fig. 7) and in false color mode helped in distinguishing the scenes on the surface from scratches and damages, and highlighting the initial drawing. This was possible because of the high contrast exhibited by the grey-black pigment on the white surface and the discrimination of red pigments from reddish toned patches on the surface coming from the burial surroundings. The matt colors almost disappeared in the monochromatic infrared radiation of 1000 nm due to their red tones and the small thickness. As it can be observed on *lekythos* inv. 4723 (see Fig. 6), a woman in three quarter view is standing on the right, her head in profile with her hair tied up. As it can be assumed according to the yellow tones that the traces of the pigment present in false color infrared recording, she wears a *peplos* with red folds. Opposite, on the left side, a young man is standing in profile. He leans on a staff, lightly bending forward, and stretches his right hand. Nothing remained of his *himation*. In *lekythos* inv. 4724 (see Fig. 7) on the right side of the acanthus monument a woman with short hair kneels in profile. She raises her right hand in a gesture of mourning; her left arm is stretched forward. Her garment, with red folds, is similar to that on inv. 4723. A red *tainia* hangs above her. The fragmentary red garment on the left, appearing with yellow tones in the FCIR belongs, according to infrared photos to a female figure. In conclusion the

Fig. 7. Detail of lekythos 4724 at 1000 nm

lekythoi show in two-figure-scenes the so-called 'visit to the tomb'. That is: a grave *stele* is framed by two human figures. Considering all this information Simon and Wehgartner [16] assert their chronology should be between 430 and 425 B.C.

In the case of Daphne's writing tablets, the recording of the reflectance of the surface in deep blue (420 nm) as well as in near infrared (1000 nm) indicates the use of a very good quality wood, as it is very smooth, homogenous and without prominent grain. However, letters can be fully highlighted using raking near infrared light in macro mode (see Fig. 8). As they are engraved their optical behavior differentiates them from the surrounding wax and their morphology appears clearly. A neat well-formed writing is recorded microscopically small as letter size is approximately 1.5 mm of height. Apart from scattered letters, lines of text are also observed. The amount of the text written must have been considerable as the tablet appears to have had about eighteen lines of writing fitted into its vertical space of 5.3 cm, with perhaps 90 characters to the line. According to West [20] the script is clearly the work of a practiced hand and the Ionic alphabet already used in public and epigraphic texts in Attica in the second half of the fifth century is recognizable. Concerning the infrared picture of the Fig. 7 West also quotes: "In the second line it is tempting to recognize an

Fig. 8. Writing tablet inv. 7420, detail. Infrared Image at 1000 nm acquired in macro mode using raking light

allusion to the magnificent verse spoken by Herakles in the Hesiodic Wedding of Keyx when he arrived uninvited at the wedding feast "αὐτόματοι δ' ἀγαθοὶ ἀγαθῶν ἐπὶ δαῖτας ἴενται", "good men come to good men's banquets of their own accord". It was a famous line, alluded to by several classical writers". Writing tablet contains texts with indications of poetic diction that contribute to the hypothesis that the deceased was a musician and probably a poet.

Correspondingly, *pyxis* MK10905 was studied *in situ* with reflected illumination with the hyperspectral imaging detector MuSIS HS at wavelengths of the visible and near infrared spectra. The images clearly highlight the portion of the color decoration that has been revealed but this setup can not provide more information about what is happening under the layer of deposits. However, the method which proved to be particularly effective in this case because of the crystalline structure and the relative transparency of the material is the recording of transmitted illumination images. In the transmitting illumination and especially in visible monochromatic radiation, the layer of salts, also translucent in the transmitting light, creates the necessary contrast so that decorative elements under the coats are revealed more intensely than the one in the cleaned area (see Fig. 9a–c). The use of different wavelengths for the examination of the colored decorations provided a better understanding of the variety of decoration themes employed for the *pyxis*.

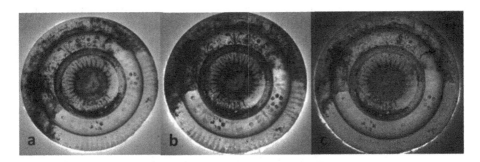

Fig. 9. *Pyxis* MK10905, transmitted light (a) VIS, (b) 560 nm and (c) 1000 nm

In the case of the *pyxis*, infrared does not particularly help to reveal the decoration, on the one hand because of the nature of the pigments, namely the red tint they present and on the other, the high transparency at 1000 nm the colors have due to their low pigment density and their small thickness rendering them almost invisible to the infrared. In contrast, infrared reflection reveals the homogeneity and purity of the marble, while the transmitting light in macro-photographic conditions emphasizes its crystalline nature.

3.2   Application on Indoor and Outdoor Wall Surfaces - Large Scale Applications: Detection of Colored Decoration, Hidden Sketches and Surface Morphology

As discussed above, the methodological approach to large-scale applications is determined by parameters such as the need to scan an entire surface of several square meters with grid images to meet the requirements of uniform imaging conditions, high spatial and tonal resolution and minimization of optical and geometric distortions.

Regarding the study of the wall surfaces of the *Horologion* it is worth mentioning that the indoor stone surfaces present an intense roughness, which differs from stone to stone according to the type and degree of the surface treatment (representative carving lines or coating preparation). The roughness is clearly highlighted using infrared raking light (see Figs. 10 and 11) at 1000 nm. Because of this morphology and the lack of transparency of the stone the acquired images gave poor results in the infrared spectra both in monochromatic acquisition at 800 and 1000 nm, as well as in the SWIR sensibility region 900–1700 nm of the InGaAs detector, comparing to that observed by naked eye. Thus a comparative evaluation is necessary of both monochromatic and SWIR acquisitions in order to extract the maximum of information. However, in the outdoor surfaces, infrared due to the smooth texture of them can reveal the high quality of white marble stone used (see Fig. 12) and at the same time contribute to record the surface with high clearness despite the presence of the yellowish *patina*. On the other hand FCIR proved to be for once more a valuable tool for the pigment identification, as it happens in the case of detection of traces of Egyptian blue according to the violet false color they present in the infrared (see Fig. 13a–c).

Fig. 10. Stone 326, raking light, 1000 nm

In contrast to *Horologion,* in the case of Halepas House – Museum the InGaAs detector proved to be the most powerful tool, even at low resolution, for revealing the hidden sketches. Even though a large number of successive lime layers was observed (which in some cases rose up to thirteen) the sketches' detection became possible because of the high penetrating ability of the SWIR radiation that the InGaAs camera can detect (up to 1700 nm). This is critical for cases like the one presented here (see Fig. 14) where the total thickness of the lime layers is about than 1 mm or greater.

Fig. 11. Stone 345, raking light, 1000 nm

Fig. 12. Stone 489, 1000 nm

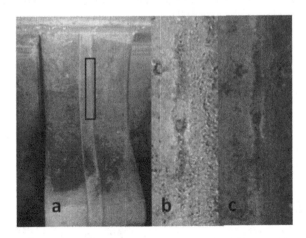

Fig. 13. Stone 187. (a) Visible image. traces of color are observed at the surface (marked area), (b) traces of blue color detected on the marked area. Visible macrophotography, (c) blue color identified as egyptian blue according to FCIR image (Color figure online)

Fig. 14. Gianoulis Halepas House Museum, in Tinos Island. Part of the west inside wall surface in the main room. SWIR image (900–1700 nm)

3.3   Application on Paintings on Canvas and Paper Support: Drawing and Pentimenti Detection, Material Characterization

As mentioned above, the optical transmittance of canvas is a feature that enables the capture even of infrared images under transmitting illumination. The cases studied suggest that transmitted NIR imaging even with the use of an imaging device with limited spectral sensitivity and low resolution may provide significant information regarding the underdrawing and underpainting in canvas paintings, in cases where reflected IR images of the same spectral band captured by the same imaging device present constraints [18]. In the case of Theofrastos Triantafyllides' painting reflected IR image in the 950–1150 nm region (see Fig. 15a) provides information only related to an underlying image playing a guitar on the right, while the image of the transmitted IR radiation of the same spectral region depicts more underlying forms such as two more human figures, architectural elements, and a glass or jug, as well as extensive under-drawing executed by both dry and wet media (see Fig. 15b).

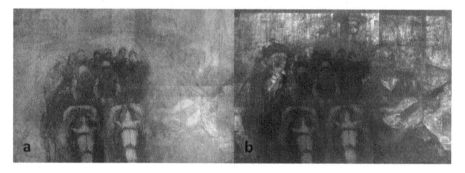

Fig. 15. Theofrastos Triantafyllides, *Return to the Village* (a) mosaic of IR reflection images and (b) mosaic of IR transmission images

In the case of Nikolaos Gyzis' oil sketch, the results of hyperspectral imaging confirmed the importance and usefulness of the non-destructive methods for sketches on paper. These methods proved to be very effective in tracking and distinguishing data produced from different depths (see Fig. 16a–d) with greater precision and resolution compared to classic reflectography. Several images yielded more information, such as those taken at 600 nm, a yellow radiation within the visible, as it provides the best possible resolution and contrast concerning papers yellowed because of ageing, or other similar surfaces. Their study in near infrared (1000 nm) gives information on the state of preservation within the material, the way the work of art was executed and, depending on the degree of reflectance of the materials used, indications about their chemical composition. Furthermore, if hyperspectral imaging is combined with other techniques such as raking light, the final image is enriched by information not only coming from the depth but also due to the texture of the surface.

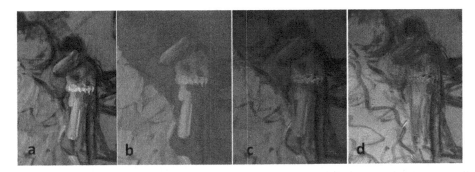

Fig. 16. Nikolaos Gyzis, *Sewing workshop*, detail, (a) VIS reflectance imaging and (b) hyperspectral imaging at 370 nm (UV), (c) 600 nm (VIS) and (d) 1000 nm (NIR) (Color figure online)

4   Conclusions - Achievements and Limitations

In the case of three-dimensional small objects of archaeological interest, the morphology and the very small thickness of the color decoration impose spectral acquisitions in macro mode with or without the combination of raking lighting, where the predominant source of information is not so much the behavior of the material in the infrared, but mainly its response to specific wavelengths and especially the visible spectrum. The macro technique proves to be very effective in cases where the object's micromorphology is characteristic for the material identification, or where unveiling of particularly detailed features and features related to the construction technique and the interpretation of the object is required.

The results clearly demonstrate that in the case of masonry decorations, the need to scan large surfaces of several square meters such as room walls or stone interior or exterior surfaces of buildings with a satisfactory resolution in a short amount of time, requires systematic capture of images in sequence with such an acquisition condition set up that will lead to images of the same tone range and of the same quality. In this case it is possible that the software-assisted capture conditions, in addition to the original calibration of the device that remains constant for all the shots, are modified at a microscale level so as to accommodate the possibly different visual behavior of the surface depending on the morphology and local optical properties. The information is related both to the behavior of the surface in the visible and to the penetrating ability of infrared radiation in different layers, if any. In addition, there is the requirement to limit geometric deformations in order to allow the mosaicing of images. Special equipment - detector movement guides and rails, etc. are also necessary. A common problem is scaffolds, limited space and vibrations, as well as the lighting conditions that may prevail in the building's interior or the effect of ambient light on external shots. A typical feature of these cases is the very large acquisition volume.

Characteristic of portable paintings on organic substrates, such as those that constitute the collection of the National Gallery - Alexandros Soutzos Museum, is the existence of colored stratification with layers of several tens or hundreds of microns (μm), which are of the type of emulsion i.e. dispersion of particles in a medium. In this case, conditions for enhancing penetrating irradiation inside the layers are required, as well as recording and highlighting the elements coming from the depth of the stratigraphy. High resolution analysis and image capture with increased contrast are desirable. Advantage is the ability to shoot in a studio or to design a space suitable for the application of *in situ* methods. It is also an advantage the ability to photograph in 2D dimensions.

Finally, it is worth mentioning that the function of spectral cube acquisition offered by certain type detectors is a very important feature because it is the most suitable tool to determine the appropriate wave length to record elements of interest with the best optical quality regarding the reflectance properties, contrast and grey scale tones. This wavelength varies in each case.

Acknowledgments. Warm thanks are due to conservator Agni-Vasileia Terlixi, to conservation scientist Dr. Eleni Kouloumpi, to curator Zina Kaloudi and conservator Christina Karadima for their cooperation on the study of *Return to the village* (EPMAS, No. P2640), as well as to Lecturer Dr. Alexios-Nikolaos Stefanis and Maria Chatzidaki for their contribution to Gianoulis Halepas House - Museum wall sketches' study (ARTICON Lab).

References

1. Fisher, C., Kakoulli, I.: Multispectral and hyperspectral imaging technologies in conservation: current research and potential applications. Rev. Conserv. **7**, 3–16 (2006)
2. Liang, H.: Advances in multispectral and hyperspectral imaging for archaeology and art conservation. Appl. Phys. A **106**, 309–323 (2012)
3. Grinzato, E.: IR thermography applied to the cultural heritage conservation. In: Proceedings of 18th World Conference on Nondestructive Testing, South African Institute for Nondestructive Testing (SAINT), Lynnwood Ridge (2012)
4. Van Asperen de Boer, J.R.J.: Reflectography of paintings using an infra-red vidicon television system. Stud. Conserv. **14**, 96–118 (1969)
5. Bonford, D. (ed.): Art in the making, underdrawings in renaissance paintings. National Gallery Company, London (2002)

6. Walmsey, E., Fletcher, C., Delaney, J.: Improved visualization of underdrawings with solid-state detectors operating in the infrared. Stud. Cons. **39**(4), 217–231 (1994)

7. Alexopoulou, A., Kaminari, A.: Study and documentation of an icon of "Saint George" by Angelos using infrared reflectography" Icons by the hand of Angelos. In: Milanou, K., Vourvopoulou, C., Vranopoulou, L., Kalliga, A.E. (eds.) The Painting Method of a Fifteenth-Century Cretan Painter, pp. 151–161. Benaki Museum, Athens (2010)

8. Ambrosini, D., et al.: Integrated reflectography and thermography for wooden paintings diagnostics. J. Cult. Heritage **11**, 196–204 (2010)

9. Pelagotti, A., Del Mastio, A., De Rosa, A., Piva, A.: Multispectral imaging of paintings, a way to material identification. IEEE Sig. Process. Mag. 27–36 (2008)

10. Moropoulou, A., Avdelidis, N.P.: Emissivity measurements on historic building materials using dual-wavelength infrared thermography. In: Proceedings of SPIE, pp. 224–228 (2001)

11. Kordatos, E.Z., Exarchos, D.A., Stavrakos, C., Moropoulou, A., Matikas, T.E.: Infrared thermographic inspection of murals and characterization of degradation in historic monuments. Constr. Build. Mat. **48**, 1261–1265 (2012)

12. Ribés, A., Schmitt, F., Pillay, R., Lahanier, C.: Calibration and spectral reconstruction for CRISATEL: an art painting multispectral acquisition system. J. Imaging Sci. Technol. **49**(6), 563–573 (2005)

13. Pöhlmann, E.: Excavation, dating and content of two tombs in Daphne, Odos Olgas 53, Athens. Greek Roman Musical Stud. **1**, 7–24 (2013)

14. Alexopoulou, A., Kaminari, A.: Multispectral imaging documentation of the findings of Tomb I and II at Daphne. Greek Roman Musical Stud. **1**, 25–60 (2013)

15. Alexopoulou, A., Kaminari, A., Panagopoulos, A.: Multispectral imaging assisted by image processing: a useful tool for the study of ancient writing and sketches on different substrates. In: 5th International Conference on NDT of HSNT-IC MINDT. CD-ROM, Athens (2013)

16. Simon, E., Wehgartner, I.: The white *lekythoi* and the dating of Tomb I. Greek Roman Musical Stud. **1**, 61–71 (2013)

17. Daffara, C., Fontana, R., Pezzati, L.: Infrared reflectography. In: Pinna, D., Galeotti, M., Mazzeo, R. (eds.) Scientific Examination for the Investigation of Paintings. A Handbook for Conservator-Restorers, p. 172. Centro Di della Edifimi srl, Florence (2009)

18. Moutsatsou, A.P., Skapoula, D., Doulgeridis, M.: The contribution of transmitted infrared imaging to non-invasive study of canvas paintings at the National Gallery – Alexandros Soutzos Museum, Greece. e-Conservation 22 (2011)

19. Alexopoulou, A., et al.: Oil paintings on paper supports: investigating the state of preservation of the paper support using NDT and microanalysis technique. In: AIPnD Art 2014-11th International Conference on Non-Destructive Investigations and Microanalysis for the Diagnostics and Conservation of Cultural and Environmental Heritage. CD-ROM, Madrid (2014)

20. West, M.: The writing tablets and papyrus from Tomb II in Daphni. Greek Roman Musical Stud. **1**, 73–92 (2013)

Infrared Hyperspectral Spectroscopic Mapping Imaging from 800 to 5000 nm. A Step Forward in the Field of Infrared "Imaging"

Stamatios Amanatiadis[✉], Georgios Apostolidis, and Georgios Karagiannis

Ormylia Foundation, Ormylia, Chalkidiki, Greece
{amanatiadis,g.apostolidis,g.karagiannis}@artdiagnosis.gr,
http://www.artdiagnosis.gr

Abstract. The purpose of this work is the development of a method for the acquisition of multispectral images at the infrared region on cultural heritage artworks. The infrared light is able to penetrate into deeper, to the surface, layers, especially at the mid and far infrared spectrum. To this end, Fourier-transform Infrared spectrophotometer, is utilized for the acquisition of multispectral data via a diffuse reflectance integration sphere to improve the quality of the detected signal. The integration sphere is mounted on a mechanical system to achieve a precise mapping of a region of interest. Then, The acquired data are combined to form the requested multispectral mapping imaging of the artwork. Advanced signal processing techniques are utilized on the spatial and spectral measurements to de-noise and enhance the imaging. Finally, the multispectral mapping reveals the sub-surface details of different inner layers.

Keywords: FTIR · Hyperspectral · Mapping imaging ·
Integration sphere

1 Introduction

The non-invasive study of cultural heritage artworks is of significant importance due to the feature extraction that can reveal, especially through hyperspectral imaging [1,2]. Several non-destructive techniques have been developed, such as acoustic microscopy [3,4], infrared [5–7], terahertz [8,9] and X-ray [10] imaging that are able to penetrate into deeper, to the surface, layers. This property provides the extraction of the sub-surface details of an artwork, revealing possible deterioration of the inner structure, thus assisting the conservation, or even hidden templates and covered artworks. Concerning the infrared spectrum, some additional features are derived, specifically chemical substance of the utilized pigments via their infrared "fingerprint" [11,12]. To this way, it is possible to

© Springer Nature Switzerland AG 2019
A. Moropoulou et al. (Eds.): TMM_CH 2018, CCIS 962, pp. 462–471, 2019.
https://doi.org/10.1007/978-3-030-12960-6_32

identify the unique characteristics of an individual artist [13]. These very important facts are enabling the development of devices that acquire multispectral data and the modeling of algorithms for their optimal processing.

To this end, the third generations of infrared spectrophotometers, namely the Fourier-Transform Infrared (FTIR) spectrophotometer, is utilized for the acquisition of multispectral data. This device covers the majority of the infrared frequencies and its basic principle depend on the Michelson interferometer that produces a modulated wide-band signal. The latter interacts to the sample that is under investigation reaching to a single photo-detector for Fourier-transform processing that leads to the infrared spectrum. Moreover, diffuse reflectance integration spheres can be utilized to improve the quality of the detected signal. However, a single measurement of an artwork is not able to reveal spatial sub-surface details. Consequently, a mechanical system is designed to achieve a precise mapping on a region of interest.

The data, acquired through the method that is described above, are combined to form the requested multispectral mapping imaging of the artwork. Furthermore, advanced signal processing techniques are utilized on the spatial and spectral measurements to de-noise and enhance the imaging. Finally, the spectrum is separated into multiple bands that correspond to different inner layers since the penetration depth of the electromagnetic radiation depends on the frequency. Therefore, a three-dimensional structure of the sub-surface details of the considered artwork is reconstructed via the developed multispectral mapping imaging system.

2   Acquisition of Infrared Reflectance Spectra

Fourier-transform infrared (FTIR) spectroscopy is a useful tool for identifying a variety of inorganic and organic compounds, based on their selective absorption of radiation in the mid-infrared region of electromagnetic spectrum. In FTIR spectroscopy, an emitted wideband infrared (IR) electromagnetic wave is absorbed by, transmitted through and reflected on the structure under investigation [14]. Thus, FTIR is one of the fastest operating measurement methods, especially when compared to other monochromatic IR techniques.

The heart of an FTIR spectrometer is the interferometer and one of the main parameters utilized to assess its performance and retrieve the wideband IR electromagnetic wave. Generally, an interferometer is an optical structure that serves to separate light bundles (waves), to move them spatially against each other and to superimpose them again. The simplest form of an interferometer is the Michelson type [15], with flat mirrors, as shown in Fig. 1. In order to experience low absorption at mid-IR wavelengths, KBr mirrors are usually utilized [16]. The light from the infrared source travels to a beam splitter that is ideally 50% reflective. Therefore 50% of the light is directed to a stationary mirror and 50% travels to the moving mirror. The light returning from both mirrors is recombined at the beam splitter. As the moving mirror travels back and forth various wavelengths of light go in and out of phase. By recording the signal observed by the detector at regular, precise intervals the raw data for the interferogram is generated. This is then Fourier transformed into the desired spectrum.

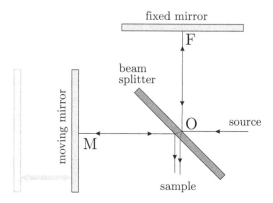

Fig. 1. A simple illustration of the Michelson interferometer and the path lengths.

As the wavelengths of the IR radiation are comparable to molecular bonds, IR waves interact with them (absorption, reflection), so the resulted spectra are indicative to molecular structures, i.e. an FTIR spectrum indicates chemical properties. An important region which is rich in information is the so-called IR fingerprint (400–$1500\,\mathrm{cm}^{-1}$). In this work FTIR spectroscopy in reflectance mode is used, i.e. the reflected IR wave is acquired and the acquisition time of a single point measurement is approximately $1\,\mathrm{s}$.

Moreover, the probe for in situ measurement of the diffuse reflectance radiation from the object is assembled using an integration sphere (also known as Ulbricht sphere) of gold surface [17]. The integration sphere is the best approach as far as wavelength range and signal/noise ratio are concerned in order to provide the possibility for the user to perform in situ measurements. The spot size of the probe is a circle with a diameter of approximately $2\,\mathrm{mm}$. Important factors for designing the integration sphere are the throughput, the sensitivity and the sphere size.

3   De-Noising Single Reflectance Scans

The acquired signals in real-life measurements are degraded due to various noising factors and especially thermal noise that is generated by the thermal agitation of the charge carriers inside an electrical conductor regardless the voltage. The more sensitive spectral regions are near the limits of the studied regime where the illuminating light power is significantly decreased; consequently the signal-to-noise ratio is severely reduced. A straightforward approach to eliminate the noise is the acquisition of various ideal measurements and apply averaging to the final signal. To this end, the basic property of white noise is exploited, namely its zero mean value, and the reflectance spectrum is "clearer". However, the acquisition time is increased significantly and in the case of a long series of measurements, and the averaging ones have to be maintained at acceptable levels (usually 8–32).

For this reason, smoothing techniques are introduced in order to enhance the acquired signal and specifically the moving average method. Considering a reflectance spectrum r, the smoothed r_s at any point n is derived

$$r_s(n) = \frac{f(0)r(n) + f(1)[r(n-1) + r(n+1)] + f(2)[r(n-2) + r(n+2)]}{f(0) + 2\sum_{m=1}^{\infty} f(m)} +$$

$$\cdots + \frac{f(m)[r(n-m) + r(n+m)]}{f(0) + 2\sum_{m=1}^{\infty} f(m)}, \quad (1)$$

where $f(m)$ is a weighting function. In the simplest case, this function is 1 for a specific number of points and 0 for the rest

$$r_s(n) = \frac{r(n) + r(n+1) + r(n-1) + \cdots + r(n-m) + r(n+m)}{1 + 2m}. \quad (2)$$

Nevertheless, this simple approach is degrading regions where abrupt variations are appearing at the spectrum, such the absorption at the IR "fingerprint". To this way, another weighted function is proposed, that is able to smooth the spectrum retaining its features, though. The proposed function is based on the normal distribution and it is derived

$$f(m) = e^{-\frac{m^2}{2\sigma^2}}, \quad (3)$$

and each point is mainly influenced by its neighbouring ones by the factor σ^2. Note, that the function values, less than 0.1%, are neglected to reduce computational cost. For values of factor σ^2 that are near zero, the smoothed spectrum is approximating the original one, while increasing it the spectrum is smoother. Typical values for σ^2 are between 1–20 and in Fig. 2 are depicted the smoothed reflectance spectra for different σ^2 values.

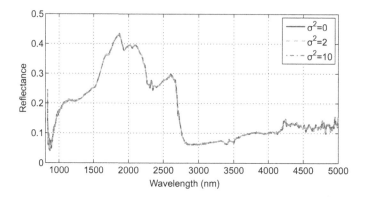

Fig. 2. Various smoothed reflectance spectra through moving average technique using the normal distribution weighting function.

4   Mapping Imaging and Visualisation Improvement

The acquisition of reflectance spectra is providing the identification of materials, such as pigments, and in the case of artworks it is interesting to extract hyperspectral images in a region of interest (ROI). For this reason a precise motorized mechanical positioning system is necessary to facilitate the mapping of the selected ROI and automate the procedure, as illustrated in Fig. 3a. In our work, the hyperspectral mapping is realized on the experimental icon "Descent from the cross", that is created at the hagiography lab of the Sacred Convent of the Annunciation in Ormylia, Halkidiki, Greece, Fig. 3b. This icon is painted over the image of "Saint James" and a blended illustration is depicted in Fig. 3c.

Fig. 3. (a) The mechanical device that enables the precise mapping of a ROI. (b) The experimental icon "Descent from the cross". A ROI is highlighted by the yellow rectangle. (c) The image of "Saint James", blended over the "Descent from the Cross". (Color figure online)

Our point measurements are illuminated via the IR-Cube provided by Bruker Optics, that emits the radiation from 900 nm–5000 nm and the diffuse reflected radiation is received using a Mercury-Cadmium-Telluride (MCT) detector (800 nm–4500 nm) that is installed on the integration sphere. Although, there is a variety of sensors that can be used [18], an MCT detector is chosen, because its sensitivity is optimal in the specific spectral area. Note that the studied spectral region is not including the infrared "fingerprint" one due to the limitations of utilized instruments. Nevertheless, the proposed methods for the extraction of hyperspectral mapping images can be applied straightforwardly for any other more wideband FTIR instrument. Additionally, the hyperspectral mapping imaging using the IR-Cube is, also, providing an initial view of the sub-surface details of an artwork.

The schematic process for the acquisition of hyperspectral mapping imaging is depicted in Fig. 4. The integration sphere is placed at the desired points through the mechanical system and the application of a focused IR beam allows the effective absorption mapping of a large area through raster scanning. The main role of FTIR mapping is the investigation of material distribution in a region of interest (ROI). The spatial resolution of the mapping depends on the one hand on the moving resolution of the fine positioning system and on the other hand on the size of the illuminating spot that is selected 2 mm, as aforementioned. However, as the resolution in a selected ROI is increased, so does the acquisition time; thus the total number of points are limited. For example, in our study the ROI is divided in 42×42 points that result in a total number of 1764 measurements. Then, the resolution is increased using advanced interpolation methods.

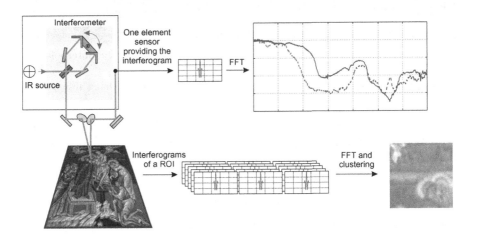

Fig. 4. A schematic depiction of the FTIR mapping procedure.

The most common method to visualize an image is the nearest neighbour one, where the values of the intermediate points are identical to the nearest one. Despite its simplicity, the image output is very poor, as depicted in Fig. 5a. For this reason, the bilinear interpolation is used, which is approximating any intermediate point by applying linear interpolation in two dimensions. Specifically, considering known values at four points $p_1(x_1, y_1), p_2(x_1, y_2), p_3(x_2, y_1)$ and $p_4(x_2, y_2)$, the unknown one $p(x, y)$ is

$$r(x, y) \approx \frac{y_2 - y}{y_2 - y_1} r(x, y_1) + \frac{y - y_1}{y_2 - y_1} r(x, y_2), \tag{4}$$

where

$$r(x, y_1) \approx \frac{x_2 - x}{x_2 - x_1} r(p_1) + \frac{x - x_1}{x_2 - x_1} r(p_3), \tag{5a}$$

$$r(x, y_2) \approx \frac{x_2 - x}{x_2 - x_1} r(p_2) + \frac{x - x_1}{x_2 - x_1} r(p_4). \tag{5b}$$

The final bilinearly interpolated image is significantly improved as shown in Fig. 5b.

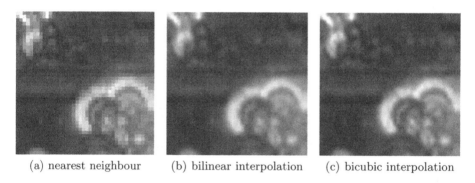

(a) nearest neighbour   (b) bilinear interpolation   (c) bicubic interpolation

Fig. 5. Spectral mapping images at the region 800 nm–1500 nm using interpolation algorithms.

An even better enhancement is achieved through the bicubic interpolation method that eliminates many interpolation artifacts. In this approach, the unknown value is approximated through

$$r(x, y) = \sum_{i=0}^{3} \sum_{j=0}^{3} \alpha_{ij} x^i y^j. \tag{6}$$

where the α_{ij} coefficients are calculated via the known points $r(p_1), r(p_2), r(p_3)$ and $r(p_4)$ and their derivatives r_x, r_y and r_{xy}. Obviously, the derivatives are present in a mapping imaging measurement; consequently a finite difference

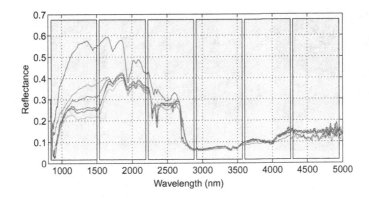

Fig. 6. Various indicative reflectance spectra of the ROI in Fig. 3b and separation into spectral regions.

scheme can be utilized. The output of the spectral image using bicubic interpolation is illustrated in Fig. 5c, where the improvements are mainly at the regions of abrupt changes, such as near the "halo".

5   Extraction of Multispectral Images

The proposed system, composed of the precise motorized positioning system, the FTIR spectrophotometer and the integration sphere as the diffuse reflectance detector, the acquisition of reflectance spectra of an artwork's ROI is feasible. Various indicative of the latter, for our studied experimental icon in Fig. 3b, are demonstrated in Fig. 6 and the combination of them is leading in a hyperspectral mapping cube, since the spectrum is continuous.

Furthermore, multispectral images are acquired through the integration of the hyperspectral mapping data at specific wavelength regions, such as those in Fig. 6. These images are providing information for the inner surface structure of artwork due to the ability of the electromagnetic energy to penetrate in deeper layers as the wavelength increases.

Multispectral images are depicted in Fig. 7 and at the near infrared region (Fig. 7a, b) the features of the "Descent from the cross" artwork are very "clear". As the wavelength increases, the image of "Saint James" is revealed, basically the halo and at 4300 nm–5000 nm regime hair are distinguished. It is worth to mention the red pigment of the angel at the top-left corner is intensively reflective, especially at the mid infrared frequencies.

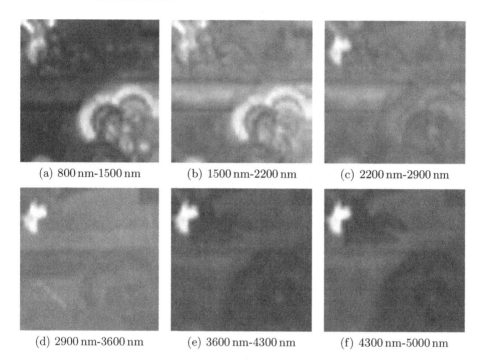

(a) 800 nm-1500 nm   (b) 1500 nm-2200 nm   (c) 2200 nm-2900 nm

(d) 2900 nm-3600 nm   (e) 3600 nm-4300 nm   (f) 4300 nm-5000 nm

Fig. 7. Multispectral images extracted via the hyperspectral mapping imaging method.

6   Conclusions

In this work a methodology to acquire hyperspectral spectroscopic mapping infrared images on cultural heritage artworks have been presented. The system consists of a precise mechanical motorized arm where the infrared sensor and illumination source are mounted. Several point measurements are acquired from a specified region of interest and they are combined to form an hyperspectral cube. Advanced signal processing algorithms are applied concerning the spatial and spectral data in order to enhance the final image quality. Finally, multispectral images can be obtained straightforwrdly via the integration of the desired spectral regions of the hyperspectral cube, revealing the sub-surface details of an artwork.

Acknowledgement. This work is part of Scan4Reco project that has received funding from the European Union Horizon 2020 Framework Programme for Research and Innovation under grant agreement no 665091.

References

1. Kubik, M.: Hyperspectral imaging: a new technique for the non-invasive study of artworks. In: Physical Techniques in the Study of Art, Archaeology and Cultural Heritage, vol. 2, pp. 199–259. Elsevier (2007)
2. Legrand, S., et al.: Examination of historical paintings by state-of-the-art hyperspectral imaging methods: from scanning infra-red spectroscopy to computed X-ray laminography. Springer Heritage Sci. **2**(1), 13 (2014)
3. Maev, R.G., Green, R., Siddiolo, A.: Review of advanced acoustical imaging techniques for nondestructive evaluation of art objects. Res. Nondestr. Eval. **17**(4), 191–204 (2006)
4. Karagiannis, G., et al.: Three-dimensional nondestructive "sampling" of art objects using acoustic microscopy and time-frequency analysis. IEEE Trans. Instrum. Meas. **60**(9), 3082–3109 (2011)
5. Karagiannis, G., et al.: Processing of UV/VIS/nIR/mIR diffuse reflectance spectra and acoustic microscopy echo graphs for stratigraphy determination, using neural networks and wavelet transform. IEEE ICTTA, pp. 1–7 (2008)
6. Sarmiento, A., et al.: Classification and identification of organic binding media in artworks by means of Fourier transform infrared spectroscopy and principal component analysis. Springer Anal. Bioanal. Chem. **399**(10), 3601–3611 (2011)
7. Attas, M., et al.: Near-infrared spectroscopic imaging in art conservation: investigation of drawing constituents. Elsevier J. Cult. Heritage **4**(2), 127–136 (2003)
8. Fukunaga, K., Hosako, I.: Innovative non-invasive analysis techniques for cultural heritage using terahertz technology. C. R. Phys. **11**(7–8), 519–526 (2010)
9. Filippidis, G., et al.: Nonlinear imaging and THz diagnostic tools in the service of cultural heritage. Springer Appl. Phys. A **106**(2), 257–263 (2012)
10. Zielińska, A., et al.: X-ray fluorescence imaging system for fast mapping of pigment distributions in cultural heritage paintings. IOP J. Instrum. **8**(10), P10011 (2013)
11. Vahur, S., Teearu, A., Leito, I.: ATR-FT-IR spectroscopy in the region of 550–230 cm- 1 for identification of inorganic pigments. Spectrochim. Acta Part A Mol. Biomol. Spectrosc. **75**(3), 1061–1072 (2010)
12. Cosentino, A.: Identification of pigments by multispectral imaging; a flowchart method. Springer Heritage Sci. **2**(1), 8 (2014)
13. Polak, A., et al.: Hyperspectral imaging combined with data classification techniques as an aid for artwork authentication. J. Cultural Heritage **26**, 1–11 (2017)
14. Griffiths, P.R., De Haseth, J.A.: Fourier Transform Infrared Spectrometry, vol. 171. Wiley, Hoboken (2007)
15. Hariharan, P.: Basics of Interferometry. Academic Press, San Diego (2010)
16. Battle, G.C., Connolly, T., Keesee, A.M.: Laser Window and Mirror Materials. Springer, Boston (2012)
17. Goebel, D.G.: Generalized integrating-sphere theory. Opt. Soc. Am. Appl. Opt. **6**(1), 125–128 (1967)
18. Burns, D.A., Ciurczak, E.W.: Handbook of Near-Infrared Analysis. CRC Press, Boca Raton (2007)

Fusion of the Infrared Imaging and the Ultrasound Techniques to Enhance the Sub-surface Characterization

Stamatios Amanatiadis[✉], Georgios Apostolidis, and Georgios Karagiannis

Ormylia Foundation, Ormylia, Chalkidiki, Greece
{amanatiadis,g.apostolidis,g.karagiannis}@artdiagnosis.gr
http://www.artdiagnosis.gr

Abstract. In this paper, the techniques of high-frequency ultrasound and infrared imaging are combined to enhance the sub-surface characterization of cultural heritage artworks. Initially, these two different modalities are studied independently focusing on the extraction of an art object's stratigraphy through acoustic microscopy and the distinction of materials, such as pigments, via their infrared fingerprint. Moreover, post-processing procedures are utilized separately for each technique to maximize the information of the acquired data. Then, robust registration methods are presented and applied on the images in order to align them spatially facilitating their fusion. Finally, the entire process is summarized in a block diagram and the fused images are presented, revealing the enhanced perspective of the artwork's sub-surface details.

Keywords: Fusion · Ultrasound · IR imaging ·
Non-destructive evaluation

1 Introduction

The details, hidden under the surface of an artwork, are of great importance since they can reveal possible deterioration, thus assisting the conservation techniques, or even unveil hidden features. Several modalities have been designed for this purpose, such as the acoustic microscopy [1], optical coherence tomography (OCT) [2] and X-ray tomography [3], the infrared [4,5] and terahertz [6,7] imaging and others that their primer objective is the non-destructive evaluation of the sub-surface details. Specifically, the acoustic microscopy is based on the propagation of the acoustic waves, differentiating it from other modalities, into deeper than the surface layers, while infrared imaging is able to detect accurately different materials, such as the artwork's pigments. Thus, the purpose of this paper is to study the combination of these two methods via the fusion of the acquired data that can enhance further the quality of the final image.

© Springer Nature Switzerland AG 2019
A. Moropoulou et al. (Eds.): TMM_CH 2018, CCIS 962, pp. 472–481, 2019.
https://doi.org/10.1007/978-3-030-12960-6_33

Initially, the acoustic microscopy, especially high-frequency ultrasounds that are of interest in this work, are produced via piezoelectric transducers and the echoes that result from the reflections on the different layers of the artwork are collected through the same device. Then, advanced signal processing techniques, such as the Hilbert transformation, are utilized to extract the inner structure of the artwork through the reflection echoes. Note, that the ultrasonic waves require special matching materials, such as ultrasound gels, that must be selected appropriately to avoid the degradation of the artwork.

Moreover, the infrared (IR) electromagnetic energy has the ability to penetrate into deeper, to the surface, layers, especially at the mid and far IR regime. To this end, the artwork can be illuminated through IR light and images at several bands, that correspond to different layers, can be acquired through the combination of a focal plane array camera and the appropriate filters. However, multispectral images are also extracted via the spectrophotometer mapping of a region of interest. Although, the image resolution is degraded using this method, the separation into different sub-surface layers is significantly better; consequently it is preferred.

Summarizing, the described methods are representing the three-dimensional inner structure of the artwork. The combination of these imaging techniques can enhance further the quality of the acquired images, though. To this end, advanced registration techniques are used to match into a specific region of interest. Moreover, the 3D sub-surface models of the two methods are fused to improve the sub-surface details of the artwork that is under consideration.

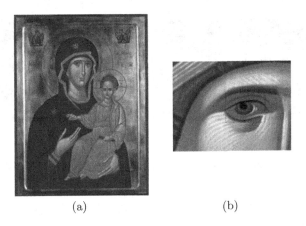

(a)   (b)

Fig. 1. (a) Visual image of a clone of the Afytou Mother of God icon and (b) the selected ROI for the application the ultrasound/FTIR stratigraphic registration technique.

2   Acquisition of Ultrasound and Infrared Images

Initially, the acoustic microscopy and the infrared spectroscopy modalities are studied independently. The data acquisition, exploiting the features of these methods, is performed on a clone of the Afytou Mother of God icon, illustrated in Fig. 1a. Specifically, the region of interest (ROI) for the scannings is located on the "eye" of the icon, Fig. 1b.

2.1   Acoustic Microscopy

Acoustic microscopy is based on emitting and receiving high frequency (>10 MHz) ultrasonic waves, i.e. short-time pulses [8–11]. The received wave consists of the backscattered reflections, or echoes, of the emitted wave which are resulted by the micro structures on and under the surface of an object. The resulted signal consists of a few time-delayed pulses whose delay, or also called the time-of flight, is associated with the distance the pulse travelled after it reflected from a micro-structure. This signal is commonly called either as the echograph or the amplitude scan (A-scan).

When an acoustic microscope acquires a series of A-scans from several spots in a region of interest (ROI), then the internal structure of the ROI can be reconstructed, hence revealing the stratigraphy of the object [1,12]. A raster scanning in a ROI result a set of equally spaced A-scans which are merged and processed to produce either cross-sectional tomographic images, B-scans (Brightness scans) or planar ones, C-scans. Also, from the A-scans of a raster scanning 3D images can be produced. In general, these images can be used in measuring the roughness of the material, acquiring the stratigraphic information, identifying of any structural defects and disclosing enclosed findings for the object under study.

(a)   (b)   (c)

Fig. 2. Ultrasound scanning of ROI or the "eye" area of the test icon.

For the purposes of this paper, the investigated stratigraphies, i.e. multi-layered painting structures, consist of layers with thicknesses of a few tens of micrometres (>10 μm). To satisfy such requirement, the developed ultrasonic system emits ultra-short wavelength pulses (>100 MHz) as the shorter the wavelength the better the better the spatial resolution is. It is worthwhile to mention that such technology is on the limits of the current state-of-the-art.

An adequate measurement consists of a scanning of a ROI with dimensions of 5 mm × 5 mm and with moving step sizes of 20 μm × 20 μm. The acquisition of ROI in Fig. 1b is depicted in Fig. 2. One extremely important concern about the ultrasonic propagation is the coupling of the transducer to the object under investigation. This problem arises from the fact that the acoustic waves cannot propagate adequately throughout gases, namely the air, so coupling materials need to be applied onto the artwork. The selection of these materials is crucial since they can possibly harm the surface and advanced approaches are used for their composition. In our problem, where a protective varnish layer is absent, the coupling gel is an organic one based on alcohol and the effect on the artwork is negligible. However, in the majority of cases, where a protective varnish layer is placed, simple hydrogel can be applied and the non-destructive nature of the method is maintained. The acquisition time at either case must not exceed the total of an hour.

2.2   Multispectral Infrared Mapping Imaging

Fourier-transform infrared (FTIR) spectroscopy is a useful tool for identifying a variety of inorganic and organic compounds, based on their selective absorption of radiation in the mid-infrared region of electromagnetic spectrum. In FTIR spectroscopy, an emitted wideband infrared (IR) electromagnetic wave is absorbed by, transmitted through and reflected on the structure under investigation [13]. Thus, FTIR is one of the fastest operating measurement methods, especially when compared to other monochromatic IR techniques. In this paper FTIR spectroscopy in reflectance mode is used, i.e. the reflected IR wave will be acquired. The wideband IR wave is realized using an interferometer (usually the Michelson one [14]), so the acquired IR wave is an interferogram whose Fourier Transform results the final spectrum. As the wavelengths of the IR radiation are comparable to molecular bonds, IR waves interact with them (absorption, reflection), so the resulted spectra are indicative to molecular structures, i.e. an FTIR spectrum indicates chemical properties. FTIR spectroscopy is a rapid and sensitive method with instrumentation that allows for numerous sampling techniques. A variety of digital signal processing techniques may be applied for the evaluation and quantification of spectral features. FTIR can be used in all applications where a dispersive spectrometer was used in the past. In addition, the improved sensitivity and speed have opened up new areas of application.

The application of a focused IR beam allows the effective absorption mapping of a large area through raster scanning. The main role of FTIR mapping is the investigation of material (chemical properties) distribution in a ROI. In IR spectroscopy the variety of molecular bonds that can be investigated depend on the spectral region (the bandwidth) of the device. The spectral region is mainly related to the interferometer's characteristics. An important region which is rich in information is the so-called IR fingerprint (400–$1500\,\mathrm{cm}^{-1}$). The spatial resolution of the mapping depends on the one hand on the moving resolution of the fine positioning system and on the other hand on the size of the illuminating spot. Several indicative reflection spectra of the test icon ROI (Fig. 1b) are demonstrated in Fig. 3. The inlet image is the final IR mapping at a specific spectral region ($2.87\,\mu\mathrm{m}$–$5\,\mu\mathrm{m}$).

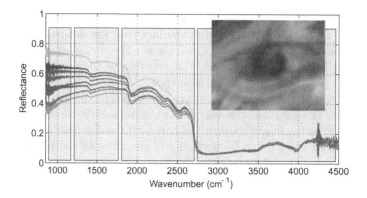

Fig. 3. Reflection spectra of the ROI or the "eye" area of the test icon and the acquired image at $2.87\,\mu m$–$5\,\mu m$ (inlet).

3 Registration

After a certain ROI is selected the scanned data can be further processed to enhance the image quality. Specifically, the ultrasound results can undergo a Hilbert transformation processing, in case of producing a "clearer" image for registration, since the "true" signal is extracted. Correspondingly, the stratigraphic images are extracted for different wavelengths and combined qualitatively to produce a multimodal description of the inner structure. This procedure is described in Fig. 4 and it is indicated that the process that follows is registration.

Image registration is the procedure of transforming different sets of data into one coordinate system. Data are usually two or multiple images, taken from different modalities/sensors, time steps, or viewpoints [15, 16]. Registration is necessary for comparing images acquired from these different measurement types, and is a pre-processing step for further image analysis, like fusion and multichannel restoration. In this paper, we focus on multimodal analysis or, specifically, on registration of images from different modalities/sensors that is of great importance for artworks, like paintings or icons, because it indicates multiple features on a location of the sample under study, like overpainting or element/molecule detection. A valuable step to facilitate the registration process is the scanning of the ROI a calibrated mechanism as depicted in Fig. 5. However, fine image alignment is, also, required.

For the case of two-image alignment, one is set as stationary setting the global coordinate system, commonly named fixed, and the other one to be placed upon the first one is called moving. At the heart of image alignment techniques is a simple 3×3 matrix called homography H. This matrix has to map the points of one image (x_1, y_1) to the corresponding of the second (x_2, y_2) as

$$\begin{bmatrix} x_1 \\ y_1 \\ 1 \end{bmatrix} = H \begin{bmatrix} x_2 \\ y_2 \\ 1 \end{bmatrix} = \begin{bmatrix} h_{11} & h_{12} & h_{13} \\ h_{21} & h_{22} & h_{23} \\ h_{31} & h_{32} & h_{33} \end{bmatrix} \begin{bmatrix} x_2 \\ y_2 \\ 1 \end{bmatrix}. \tag{1}$$

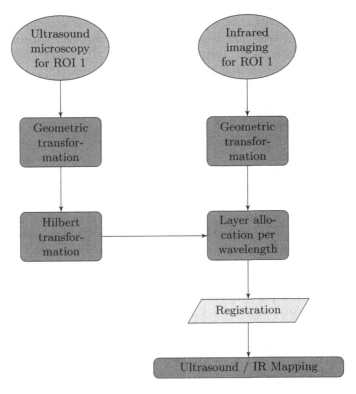

Fig. 4. Flow chart describing the stratigraphic registration process via ultrasound and IR mapping of a ROI.

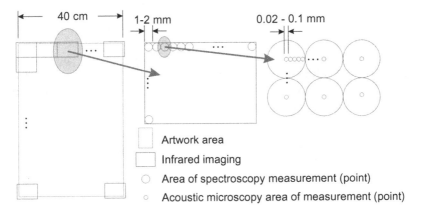

Fig. 5. Scanning areas and fidelity for different onboard modalities via an accurate mechanical system.

The homography matrix is generated by registration techniques that are categorized as area-based and feature-based. The former, sometimes called correlation-based methods or template matching [17], exploit for matching directly image intensities, without any structural analysis [18]. Orthogonal or circular ROIs are initially set and a 2D cross-correlation procedure is performed

$$C(k,l) = \sum_{m=0}^{M-1} \sum_{n=0}^{N-1} X(m,n)\bar{H}(m-k,n-l), \tag{2}$$

where X is a $M \times N$ matrix, that corresponds to one image, and $\bar{H}$ is the $P \times Q$ counterpart of the second one, as complex conjugate. The final correlation matrix C is of dimensions $(M + P - 1) \times (N + Q - 1)$ and the window pairs for which the maximum is achieved are set as the corresponding ones. Then, according to the cross-correlation matrix values the homography matrix is calculated and the registration process is completed. Although the correlation-based registration can exactly align mutually translated images only, it can also be successfully applied when slight rotation and scaling are present.

On the other hand, feature-based techniques are following a two step procedure, the feature detection and the feature matching. The first step, namely the detection, is performed via the popular algorithm SIFT (Scale-Invariant Feature Transform) [19], which is known for its robustness. Moreover, some of the feature matching algorithms are outgrowths of traditional techniques for performing manual image registration, in which an operator chooses corresponding control points in images. When the number of control points exceeds the minimum required to define the appropriate transformation model, iterative algorithms like RANSAC (RANdom SAmple Consensus) [20] can be used to robustly estimate the parameters of a particular transformation type (e.g. affine) for registration of the images.

For the case of 3D transformation of the images, as in the paper, the procedure is similar instead of the part of the homography matrix that is, now, 4×4. This matrix performs the rotation given by $R(\alpha, \beta, \gamma)$ for the corresponding yaw, pitch and roll angles, followed by a translation given by x_t, y_t, z_t, thus it has six degrees of freedom. The corresponding form of the 3D homogeneous transformation matrix is given by

$$H = \begin{bmatrix} \cos\alpha\cos\beta & \cos\alpha\sin\beta\sin\gamma - \sin\alpha\cos\gamma & \cos\alpha\sin\beta\cos\gamma + \sin\alpha\sin\gamma & x_t \\ \sin\alpha\cos\beta & \sin\alpha\sin\beta\sin\gamma - \cos\alpha\cos\gamma & \sin\alpha\sin\beta\cos\gamma - \cos\alpha\sin\gamma & y_t \\ -\sin\beta & \cos\beta\sin\gamma & \cos\beta\cos\gamma & z_t \\ 0 & 0 & 0 & 1 \end{bmatrix}. \tag{3}$$

Concerning our study of icon in Fig. 1, the registration technique output is presented in Fig. 6a for the acoustic microscopy image. The transformation matrix for this sample is derived as

$$H = \begin{bmatrix} 0.90 & 0.24 & -0.36 & 321.66 \\ 0.37 & -0.01 & 0.93 & -65 \\ 0.22 & -0.97 & -0.10 & 850 \\ 0 & 0 & 0 & 1 \end{bmatrix}. \tag{4}$$

Additionally, the registration technique is, also, applied for the IR mapping imaging and the output is demonstrated in Fig. 6b. The transformation matrix for IR mapping imaging is derived

$$H = \begin{bmatrix} 1 & 0 & 0 & -1.29 \\ 0 & 0 & -1 & 207.94 \\ 0 & 1 & 0 & 1071.24 \\ 0 & 0 & 0 & 1 \end{bmatrix}. \tag{5}$$

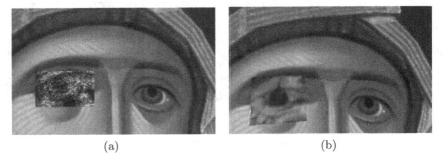

(a)   (b)

Fig. 6. Image registration for distinctive detail ROI (a) visual and ultrasound images, (b) visual and image via FTIR mapping.

4   Fusion

The process of fusion between two registered images or volumes acquired via different modalities is not a trivial one. In many cases, the details of each image are not adequate and their fusion is necessary to create a new on of enhanced details. At these cases, the images are processed in order to indicate these details, located mainly at higher frequencies. To this aim, advanced transformations, such as the wavelet and contourlet ones, are applied and the fused image is reconstructed through the maximum point value of each image [21–24]. On other cases, where the major concern is to enhance the whole image instead of the details, the process is realized at the original images. The fused image can either select the maximum or the mean value of the prototypes [25,26].

In our problem the enhancement is provided through the blending of the images obtained from the different modalities. Specifically, the ultrasonic modality offers a very fine model of the artwork and its sub-surface characteristics but there is not any clue about the utilized materials, Fig. 6a. That is the main reason that grayscale level is used for visualization. On the other hand, the FTIR mapping imaging is the ideal modality to characterize the materials, but the quality is somewhat coarse and the sub-surface details are limited. A pseudocolor mapping is used for it in order to indicate the difference between the materials, Fig. 6b. Consequently, once the control points are known after the mapping process, data from different modalities are registered for different depths. Moreover, the volumes are blended, namely the half percent of image intensity is used and the fused image is a sum of the different modalities. Note that the utilization of different colormaps is crucial for the correct and enhanced visualization. In Fig. 7 the registration of two stratigraphic images from IR and ultrasound mapping is presented, transformed using the previous matrices. It is observed that the "iris" detail is matched in the pictures and following that we can attempt to deduce substance under-surface characteristics via the qualitatively fused image.

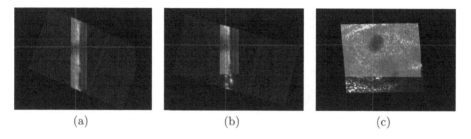

(a)   (b)   (c)

Fig. 7. Fusion of the FTIR mapping image of ROI with the respective ultrasound one.

Acknowledgement. This work is part of Scan4Reco project that has received funding from the European Union Horizon 2020 Framework Programme for Research and Innovation under grant agreement no 665091.

References

1. Karagiannis, G., et al.: Three-dimensional nondestructive "sampling" of art objects using acoustic microscopy and time-frequency analysis. IEEE Trans. Instrum. Meas. **60**(9), 3082–3109 (2011)
2. Targowski, P., Iwanicka, M.: Optical coherence tomography: its role in the non-invasive structural examination and conservation of cultural heritage objects—a review. Appl. Phys. A **106**(2), 265–277 (2012)
3. Zielińska, A., et al.: X-ray fluorescence imaging system for fast mapping of pigment distributions in cultural heritage paintings. J. Instrum. **8**(10), P10011 (2013)
4. Sarmiento, A., et al.: Classification and identification of organic binding media in artworks by means of Fourier transform infrared spectroscopy and principal component analysis. Anal. Bioanal. Chem. **399**(10), 3601–3611 (2011)

5. Attas, M., et al.: Near-infrared spectroscopic imaging in art conservation: investigation of drawing constituents. J. Cult. Heritage **4**(2), 127–136 (2003)
6. Fukunaga, K., Hosako, I.: Innovative non-invasive analysis techniques for cultural heritage using terahertz technology. C.R. Phys. **11**(7–8), 519–526 (2010)
7. Filippidis, G., et al.: Nonlinear imaging and THz diagnostic tools in the service of Cultural Heritage. Appl. Phys. A **106**(2), 257–263 (2012)
8. Briggs, A.: Advances in Acoustic Microscopy, vol. 1. Springer, New York (2013)
9. Rose, J.L.: Ultrasonic Guided Waves in Solid Media. Cambridge University Press, Cambridge (2014)
10. Cheeke, J., David, N.: Fundamentals and Applications of Ultrasonic Waves. CRC Press, Boca Raton (2012)
11. Yu, Z., Boseck, S.: Scanning acoustic microscopy and its applications to material characterization. Rev. Mod. Phys. **67**(4), 863 (1995)
12. Karagiannis, G., et al.: Processing of UV/VIS/nIR/mIR diffuse reflectance spectra and acoustic microscopy echo graphs for stratigraphy determination, using neural networks and wavelet transform. In: IEEE ICTTA, pp. 1–7 (2008)
13. Griffiths, P.R., De Haseth, J.A.: Fourier Transform Infrared Spectrometry, vol. 171. Wiley, Hoboken (2007)
14. Hariharan, P.: Basics of Interferometry. Academic Press, San Diego (2010)
15. Brown, L.G.: A survey of image registration techniques. ACM Comput. Surv. (CSUR) **24**(4), 325–376 (1992)
16. Zitova, B., Flusser, J.: Image registration methods: a survey. Image Vis. Comput. **21**(11), 977–1000 (2003)
17. Barnea, D.I., Silverman, H.F.: A class of algorithms for fast digital image registration. IEEE Trans. Comput. **21**, 179–186 (1972)
18. Althof, R.J., Wind, M.G.J., Dobbins, J.T.: A rapid and automatic image registration algorithm with subpixel accuracy. IEEE Trans. Med. Imaging **16**, 308–316 (1997)
19. Lowe, D.G.: Object recognition from local scale-invariant features. In: The Proceedings of the Seventh IEEE International Conference on Computer Vision, vol. 2, pp. 1150–1157 (1999)
20. Fischler, M.A., Bolles, R.C.: Random sample consensus: a paradigm for model fitting with applications to image analysis and automated cartography. Commun. ACM **24**(6), 381–395 (1981)
21. Rockinger, O.: Image sequence fusion using a shift-invariant wavelet transform. In: IEEE International Conference on Image Processing 1997, vol. 3, pp. 288–291 (1997)
22. Nikolov, S., Hill, P., Bull, D., Canagarajah, N.: Wavelets for image fusion. In: Petrosian, A.A., Meyer, F.G. (eds.) Wavelets in Signal and Image Analysis. Computational Imaging and Vision, vol. 19, pp. 213–241. Springer, Dordrecht (2001). https://doi.org/10.1007/978-94-015-9715-9_8
23. Liu, K., Guo, L., Chen, J.: Contourlet transform for image fusion using cycle spinning. BIAI J. Syst. Eng. Electron. **22**(2), 353–357 (2011)
24. Xiao-Bo, Q., Jing-Wen, Y., Hong-Zhi, X., Zi-Qian, Z.: Image fusion algorithm based on spatial frequency-motivated pulse coupled neural networks in nonsubsampled contourlet transform domain. Acta Automatica Sinica **34**(12), 1508–1514 (2008)
25. Pohl, C., Van Genderen, J.L.: Review article multisensor image fusion in remote sensing: concepts, methods and applications. Int. J. Remote Sens. **19**(15), 823–854 (1998)
26. Xydeas, C.S., Petrovic, V.: Objective image fusion performance measure. IEEE Electron. Lett. **36**(4), 308–309 (2000)

Moisture Climate Monitoring in Confined Spaces Using Percolation Sensors

Helge Pfeiffer[1](✉) ⓘ, Charlotte Van Steen[2], Els Verstrynge[2], and Martine Wevers[1]

[1] Department of Materials Engineering, KU Leuven, Leuven, Belgium
{helge.pfeiffer,martine.wevers}@kuleuven.be
[2] Department of Civil Engineering, KU Leuven, Leuven, Belgium
{charlotte.vansteen,els.verstrynge}@kuleuven.be

Abstract. It is a well-known fact that inappropriate levels of moisture can harm different kinds of structures, especially also in cultural heritage. This does not only regard hygroscopic materials but all assemblies where small quantities of water can change the physical properties of materials and structures. Damage modes range from crack formation due to swelling pressure, migration of harmful agents such as salts, up to corrosion and microbial damage. A specific concern is not only a certain moisture level as such, but also the periodic moisture variations as they can even be related to mechanical fatigue processes when alternating hydration pressures apply. Therefore, moisture monitoring is in many cases unavoidable and a multitude of commercial and experimental set-ups exist ranging from simple point measurements using hygrometers up to sophisticated time-domain reflectometry that provides humidity profiles at long ranges.

Inspired by an analogous problem in aircraft structures suffering e.g. from moisture ingress in floor structures, a sensing material to detect the moisture climate was developed that is based on the percolation effect. Here, an electrical conductive material is changing into an isolator when a certain humidity level is reached. The material presented is based on hygroscopic poly-vinyl alcohol and electrically conductive titanium carbonitrides. The resulting material shows a characteristic change of the resistance over many orders of magnitude when a humidity level of 80% is reached which can be fine-tuned depending on the actual application. The sensing material is then applied on a cable with a small diameter and in this way, a versatile sensor that can be used especially in confined spaces is obtained. Data transmission can start with simple read-out by a multimeter, but also a network of wireless nodes were tested as well as RFID technology.

The developed sensor PercoSens® is for the moment being applied in chemical installations and on an experimental level also in concrete structures enabling extended sensing ranges and a high independency on interfering with environmental parameters. Very recent tests showed the potential of this sensor to indicate water ingress in damaged building structures. For the latter, a sensor was embedded in concrete and after imitated crack formation, the subsequent water ingress was determined quickly.

Keywords: Moisture · Monitoring · Water damage · Percolation sensors

© Springer Nature Switzerland AG 2019
A. Moropoulou et al. (Eds.): TMM_CH 2018, CCIS 962, pp. 482–493, 2019.
https://doi.org/10.1007/978-3-030-12960-6_34

1 Introduction

1.1 General Introduction

It is a well-known fact that inappropriate levels of moisture can harm different kinds of structures, especially also in cultural heritage. Therefore, a lot of research activities are dedicated to this important subject [1]. Avoidance of harmful moisture does not only regard hygroscopic materials but all assemblies where small quantities of water can change the physical properties of materials and structures.

The damage modes range from crack formation due to swelling pressure, freezing damage, migration of harmful agents such as salts, up to corrosion and microbial damage. A specific concern is not only a certain moisture level as such, but also periodic moisture variations as they can even be related to mechanical fatigue processes when alternating hydration pressures apply. Therefore, moisture monitoring is in many cases unavoidable and a multitude of commercial and experimental set-ups exist ranging from simple point measurements using hygrometers up to sophisticated time-domain reflectometry that provides humidity profiles at long ranges.

Inspired by analogous problems in aircraft structures suffering e.g. from moisture ingress in floor structures, a sensing material was developed that is based on the percolation effect. A peculiarity is that it is essentially only operating at one certain humidity level, but the sensor response proceeds over a very big range. Here, an electrical conductive material is changing into an isolator when a certain humidity level is reached. The developed sensor PercoSens® is now also applied in chemical installations and on an experimental level at concrete structures enabling extended sensing ranges and a high independency on interfering environmental parameters. Very recent tests showed the potential of this sensor to indicate water ingress in building structures. For that purpose, a sensor was embedded into concrete and after imitated crack formation, the subsequent water ingress was determined quickly.

1.2 Percolation and Hydration Pressure

Percolation Conductivity. The functionality of the presented sensor material is based on the collapse of a percolation network [2] in electrically conducting composites after contact with water. Those composites usually consist of conductive particles that are embedded in an insulating, hygroscopic matrix. After the ingress of liquids, the swelling of the matrix will cause a characteristic sharp increase of the electrical resistance that arises from the interruption of the conductive pathways. The increase of the electrical resistance usually proceeds on an exponential scale, and it can easily be monitored by standard ohmmeters.

This is shown in Fig. 1 where the interruption of percolation conductivity is schematically presented. Such as further explained below as a matrix material the hygroscopic PVA polymer was selected, and it was mixed with electrically conductive TiCN powder (a ceramic compound, see also Fig. 2).

Hydration and Hydration Pressure. One of the important interactions of water in a confined hygroscopic system is the swelling pressure when a reservoir of pure water is entering spaces where the water activity is lowered. Inner pressures are created that can

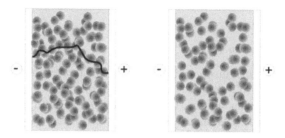

Fig. 1. Interruption of percolation conductivity within a swelling matrix (left), i.e. conductive pathways are disturbed after the ingress of liquids [3].

be the cause of osmotic-kind of hydrostatic pressure contributing to the destruction of materials. Here, the sensing material is, as explained in detail below, designed in an analogous way and the "swelling" pressure created after water ingress is used to interrupt the electrical percolation networks. Mathematically, for mechanically removing the water again against reference water, a hydration pressure P_h would be required that can be expressed by the following Eq. [4], with R, T, V_w as the molar gas constant, the absolute temperature and the molar volume of water, respectively.

$$P_h = -\left(\frac{RT}{V_W}\right) \ln\left(\frac{RH}{100}\right) \tag{1}$$

One can calculate that e.g. a hydration pressure of $P_h = 300$ MPa (3 kbar) is required to remove water hydrated at a humidity of RH = 11% from such a material. This hydration pressure is a possibility to illustrate the inner pressure acting inside the matrices. One can derive at which relative humidity, RH, the interruption of the percolation network occurs. It is in the range of RH = 75% and according to (1), this corresponds to a hydration pressure of $P_h = 40$ MPa.

The pressure can finally be understood as the mechanical pressure required bringing back the percolation material into the conductive state. And from the sorption isotherm [5], the corresponding water content can be derived which is given by a weight concentration of water of $c_{wH2O} = 13\%$.

1.3   Selected Methods for Moisture Monitoring

In essence, every phenomenon that is sensitive to water interactions could in principle be a target application for measuring moisture. Practical constraints and cost implication finally decide on the method to be chosen. A very specific aspect is the standardization of moisture measurements to enable comparability and proper moisture handling [6].

The most simple devices are hygrometers, such as the traditional hair tension hygrometers that are since decades deposited in museums and other buildings where moisture content needs to be monitored. Modern hygrometers use a variety of physical principles whereby the ability for digital data transfer is more and more important. Capacitive sensors are robust and measure the influence of moisture in the dielectric properties of materials. If the resistivity is concerned, we talk about resistive sensors. Furthermore, also thermal and gravimetric methods are used. Humidity measuring is in general more complex than temperature sensing as most of the humidity gauges are also sensitive to the temperature and require calibration. Other methods look e.g. even at the temperature distribution at historical buildings to discover the condensation risks. Furthermore, thermography is also able to detect wetness under isolations during temperature change in the evening or the morning due to its high heat capacity due to the presence of water. Also spectroscopic methods are applied, such as nuclear magnetic resonance that is able to detect relaxation times of water in a non-destructive manner for analyzing local moisture content in paint of historical paintings [7].

In the context of large scale monitoring of cultural heritage, moisture and wetness monitoring is more and more important [8]. An important aspect is monitoring the efficiency of dehumification campaigns, such as in the St. Mattey Abbey in Genoa [9]. Another example regards monitoring of the Odda's Chapel in Deerhurst, Gloucestershire where high humidity levels indicated the need for better protection of historical stone masonry [10]. Therefore, this kind of testing is also part of more fundamental research under artificial weathering conditions, such as investigated within the European project EFFESUS [11]. Another example regards the monitoring of moisture in wood which is extremely important possible by e.g. measuring the electrical conductivity of that material, whereby data transmission is supported by wireless sensor networks [12].

The advantage of the current proposed sensor in the context of large scale monitoring is the simple threshold-based working principle, the possibility for sensor integration in various materials and structures, and the very low production cost, even for large gauge lengths.

2 Methods

2.1 Composites with Percolation Behaviour

To adapt the idea of hydration respectively swelling pressure for sensing purposes, one can make use of hygroscopic synthetic polymers. In the current paper, the authors chose polyvinyl alcohol (PVA). It contains long hydrocarbon chains, and the required hygroscopic functionality arises from the –OH group. It has a typical sorption isotherm, but at lower humidities, there is almost no water uptake as the water increase is triggered by the so-called glass transition occurring at higher humidities [13].

Fig. 2. Electromicroscopy picture (SEM, scale 10 and 5 µm) of electrically conducting TiCN particles in a hygroscopic PVA matrix visible by the black background [13].

As a conductive material, a blend of titanium carbonitrides were used and mixed in appropriate ratios with the PVA. The resulting material shows a characteristic change of the electrical resistance over many orders of magnitude when a humidity level of 80% is reached (Fig. 3), which could be fine-tuned depending on the actual application.

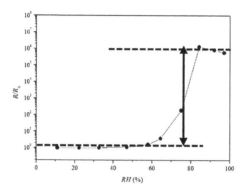

Fig. 3. Electrical resistance of TiCN/PVA mixtures as a function of the humidity, RH (bottom). The humidity and water content at which the percolation threshold, used for liquid sensing, proceeds can be determined. The remaining electrical conductivity arises from the water network still present in the composites.

This shows that the sensor is in fact a threshold sensor giving a clear response when in a certain area, humidity levels are beyond RH = 80%. The response after deposition of aqueous liquids is shown in Fig. 4.

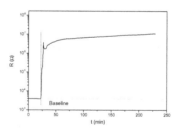

Fig. 4. Time-dependent behaviour when the sensor is in direct contact with aqueous liquids. Also note here the logarithmic scale. This method represents a kind of "threshold-sensing".

The sensing material is then applied on a cable with a small diameter and in this way, a versatile sensor to be embedded at confined spaces is obtained (Fig. 5).

A specific, in our case useful property is that only at this point, at RH = 80%, a reliable detection of liquid is possible, and that signal is many orders of magnitude above baseline variations making this measuring principle very useful for practical applications if measuring the whole range of humidity is not necessary.

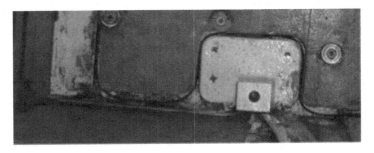

Fig. 5. Sensing cables offer interesting options for implementing in tiny, elongated spaces, here as an example shown for the floor structures of an aircraft Boeing 737 [14].

2.2 Data Reading Out and Data Transmission

Multimeter. The most simple option for testing purposes is reading out by a standard ohmmeter (multimeter) and to manually transfer the data into spreadsheets. The nature of the measurements does not require high accuracy as relevant changes usually proceed on a logarithmic scale (Fig. 4).

Microlink® System. Another system tested was a wireless sensor network (WSN) provided by Microstrain® [15]. It enabled the reading out and graphical presentation of three sensor nodes. The distance for transmission in free air is up to 1 km depending on potential obstacles. Here, a wireless protocol operating under a frequency of 2.4 GHz was used.

RFID-Tags. Radio-frequency identification technology, originally designed for counting items, is also suitable to transfer digitized data from sensors. Their potential frequency range is very broad and covers 125 kHz up to 5.8 GHz. They are e.g. applied for temperature monitoring of heat-sensitive goods. Also monitoring of humidity and other parameter is possible and commercial solutions are available. We used RFID systems from FARSENS®.

A limiting condition is always the relatively small distance between sender and receiver. It is approximately 1 m for the devices tested, but their range can be tailored and it is to expect that in future, more options will be available. A big advantage is the independency of power sources as real passive tags obtain their operational energy by the sender that transmits the respective power to the sensing instrument. Also solutions with battery-driven RFID devices and data storage are available. Finally, it is already established to implement RFID systems in smartphones, so it is thinkable to read-out sensor states of hidden sensor just by walking around in different zones under investigation and to check the data via an app or to transfer it to a desktop application.

3 Selected Results

3.1 Monitoring of Ingressing Water in Concrete Structures

Water detecting sensors based on the percolation type have been embedded in several concrete blocks during concrete casting. To verify if these sensors can uphold their functionality inside a concrete block after hardening and drying of the concrete, tests were performed. It should be noted that the blocks were designed to be rather low quality, porous concrete to ease testing the sensor functionality.

Undamaged Porous Concrete. Figure 7 shows the results of the effect of water absorption on the resistance of the percolation sensor and on the mass of the corresponding concrete block. The concrete block is first placed in a plastic box with water (Fig. 6), followed by a drying step in the oven. One can see a clear increase of mass during water absorption and after a certain moment, a huge change of resistance of the percolation sensor.

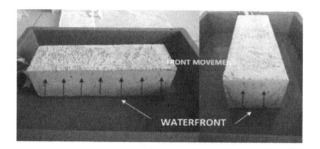

Fig. 6. Waterfront movement during the tests. Water finds its way upwards through the pores of the concrete block eventually reaching the sensors which are located in mid-center. The sensor connections are situated at the left side.

Water ingress in porous building materials can thus easily be monitored. After drying in the oven, mass of water was reduced but the sensor stayed at a relatively high level for a while – it is to expect that water captured in capillaries prevents fast drying of the sensor material.

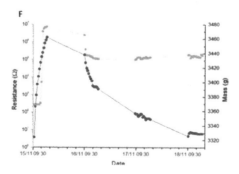

Fig. 7. Effect of water absorption and subsequent evaporation in time on resistance for a selected sensor. Absorption of water leads to the increase of mass and after a certain time also to the increase in resistance. Accelerated evaporation by heat leads to a decrease in mass and resistance as water is being removed from the system.

Induced Crack Formation. The influence of an induced crack (see concrete block in Fig. 8, Shimadzu, capacity 100 kN, not shown) on sensor resistance and mass increase is illustrated in Fig. 9. The moment of crack formation due to the applied force in the three-point bending test is clearly visible.

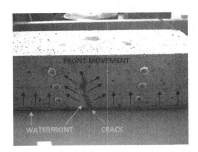

Fig. 8. Illustration of the waterfront movement during the test on the cracked concrete block. Water easily finds its way upwards through the crack of the concrete block, rapidly reaching the sensors which are located in mid-center.

After inducing the crack by a dedicated load facility (Shimadzu, capacity 100 kN), the concrete block is weighed and then placed in a plastic box containing water. Simultaneously, the resistance of the sensor is continuously monitored. Each half hour, the uptake of water is determined by weighing the concrete block and also, the waterfront level is marked.

In Fig. 9, the effect of water absorption on resistance and mass is given. The induced crack accelerated the time to uptake the water by the concrete block. Besides the uptake of water by the pores of the block, the water also entered the block through the crack, resulting in a fast propagation of water towards the center of the concrete

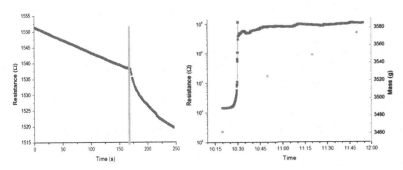

Fig. 9. Effect of crack formation and water absorption in time on resistance for a percolation sensor. Load and crack formation first changes the slope almost linearly (left), The blue line indicates the moment when the crack reached the sensor; subsequent submerging in water leads to the increase in mass and after a limited amount of time with respect to the other experiments also to a huge increase in resistance (right). (Color figure online)

block. This leads to a faster contact between the water and the sensor and thus to a rapid increase in resistance compared to previous tests.

The presence of a crack can thus be detected by these sensors. Logically, the extra water absorption possibility also results in a higher water absorption rate, indicated by the higher mass increase with time.

3.2 Water Accumulation in Pipe Isolations in the Chemical Industry

More elongated, extended sensing cables (25 m) were installed for detecting enhanced moisture levels in rockwool pipeline insulations within a chemical plant to check upscaling that concept to very big structures. The electrical resistance data was measured and transmitted by wireless sensor networks transmitting the signal (ENV-Link-mini-LXRS, LORD MicroStrain) to a distant receiver (WSDA-Base-101-LXRS) and coupled into an internet connection (Fig. 10). Currently, the sampling interval is set to 10 min and data are in principal analysed in a control unit 80 km far away from the pipeline. The system is also tested and partially running with a wireless mesh and RFID facilities to enable customisation according to maintenance needs.

Fig. 10. Installation of percolation-based moisture gauges in pipeline isolations at BASF Antwerp and scheme of the data flow from the sensors to the remote host at KU Leuven [15].

Over several months, the sensor performance was tested and analysed versus local weather and climate information such as the amount of rain fall and humidity. In this way, faulty insulation were clearly identified as checked while re-opening the respective sections later on. A typical example for these wetness events is shown in Fig. 11 presenting a sensor response as a function of time, also note the logarithmic scale that supresses baseline variations. Finally, after a couple of heavy rain events, the pipeline system became wet. After short periods of drying, subsequent heavy weather events made that the sensors finally became "saturated" as the entire isolation was wetted on larger scale. Essentially, the switch between two values clearly shows the concept of "threshold sensing".

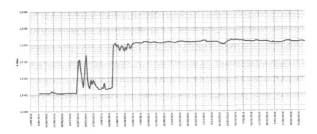

Fig. 11. Typical data (logarithmic presentation of electrical resistance versus time) for wetted pipeline isolations in a chemical plant.

3.3   Moisture in Floor Structures and Insulation Blanket of Aircraft

Another working example was the monitoring of corrosive liquids present in the floor structures of operational aircraft (Boeing 737-500, Boeing 747-400). The detection of water provided interesting options for preventing corrosion already at very early stages [14]. In that specific example, elongated wire-sensors were embedded into diverse floor structures under galleys, lavatories and entrance areas, and the system is in service for more than 7 years. The read out of sensor data is performed after approximately 100 flight hours by simple ohmmeters and it has been proven that this sampling interval is highly sufficient.

A characteristic curve is shown in Fig. 12 with clear indications on harmful "wetting events" in the floor under the galley area, most probably due to the faulty sealings. Note, wetness appears firstly temporally as "peaks", affected areas dry afterwards, and the sensor is hygroscopic acting as water buffer.

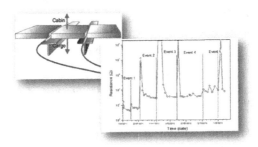

Fig. 12. Wire sensor for detecting "wetness events" in floor structures of a Boeing 737-530.

4   Summary and Motivation for Cultural Heritage

(1) The response time of the developed sensors is relatively long, but this is acceptable in the case of relatively long inspection intervals.

(2) Signals arising from leakage usually range orders of magnitude above baseline variations due to the percolation effect.

(3) Due to the absorption via the vapour phase, also liquids not in contact, but just close to the sensor will be detected, given that the vapour pressure is high enough. The sensor combines functionality of a humidity and a leakage sensor.

(4) The sensors behave partially as a buffer, i.e. they absorb the first leaking amounts of harmful water.

(5) Using threshold sensing, quantifying of liquids as well the of the accurate humidity is not possible, but this is in routine operations not a major drawback.

(6) No in-situ electronics is required because harmful liquids are always absorbed, and remain trapped for a sufficient time.

(7) The temperature dependence of the response time as well as the humidity threshold is governed by the glass transition of PVA [13].

(8) Standard ohmmeters are sufficient for reading-out of data. Wireless sensor networks (WSN) however provide more comfort.

(9) The technology presented is able to monitor moisture in circumstances that are similar to environmental conditions suited for human beings. This makes them also appropriate for the use in cultural heritage. Especially the option to cover larger areas for a very affordable price is interesting.

Acknowledgements. Part of the research leading to these results has received funding from the European Community's Seventh Framework Programme [FP7/2007–2013] under grant agreement n°212912 "AISHA II". We thank Jurgen Perremans for performing the tests and Johan Vanhulst for the technical assistance and design of the measuring set-up.

References

1. Rosina, E., Sansonetti, A., Ludwig, N.: Moisture: the problem that any conservator faced in his professional life. J. Cult. Heritage **31**, S1–S2 (2018)

2. Essam, J.W.: Percolation theory. Rep. Prog. Phys. **43**(7), 833–912 (1980)

3. Pfeiffer, H., et al.: Structural health monitoring using percolation sensors – new user cases from operational airliners and chemical plants. In: International Workshop on Structural Health Monitoring. DesTech, Stanford (2013)

4. Leneveu, D.M., Rand, R.P., Parsegian, V.A.: Measurement of forces between lecithin bilayers. Nature **259**(5544), 601–603 (1976)

5. Pfeiffer, H., et al.: Liquid detection in confined aircraft structures based on lyotropic percolation thresholds. Sens. Actuators B-Chem. **161**(1), 791–798 (2012)

6. Camuffo, D.: Standardization activity in the evaluation of moisture content. J. Cult. Heritage **31**, S10–S14 (2018)

7. Senni, L., et al.: A portable NMR sensor for moisture monitoring of wooden works of art, particularly of paintings on wood. Wood Sci. Technol. **43**(1), 167–180 (2009)

8. D'Ayala, D., Aktas, Y.D.: Moisture dynamics in the masonry fabric of historic buildings subjected to wind-driven rain and flooding. Build. Environ. **104**, 208–220 (2016)

9. Vecchiattini, R.: Moisture monitoring experience in the old town of Genoa (Italy). J. Cult. Heritage **31**, S71–S81 (2018)

10. Erkal, A., D'Ayala, D., Stephenson, V.: Evaluation of environmental impact on historical stone masonry through on-site monitoring appraisal. Q. J. Eng. Geol. Hydrogeol. **46**, 2012-060 (2013)

11. Frick, J., et al.: Moisture monitoring during an artificial weathering test of a cultural heritage compatible insulation plaster. In: 2016 19th World Conference on Non-destructive Testing (2016). Munich: ndt.net
12. Arakistain, I., Miguel Abascal, J., Munne, O.: Wireless sensor network technology for moisture monitoring of wood (2013)
13. Pfeiffer, H., et al.: Liquid detection in confined aircraft structures based on lyotropic percolation thresholds. Sens. Actuators, B: Chem. Sens. Mater. **161**, 791–798 (2012)
14. Pfeiffer, H.: Structural health monitoring makes sense. LHT Connection - The Lufthansa Technik Group Magazine (2012)
15. Pfeiffer, H., et al.: Leakage monitoring using percolation sensors for revealing structural damage in engineering structures. Struct. Control Health Monitor. **21**(6), 1030–1042 (2014)

Author Index

Adamopoulos, George I-141
Adamopoulou, Konstantina II-372
Aggelis, Dimitrios G. II-430
Alexakis, Emmanouil I-69
Alexandrakis, George I-385
Alexopoulou, Athina II-443
Amanatiadis, Stamatios II-462, II-472
Anagnostopoulos, Christos-Nikolaos I-250
Anagnostopoulos, Dimitrios F. II-397
Andrikou, Dimitra II-22
Andriotakis, George II-299
Androulaki, Theano I-128
Angelaki, Georgia I-184
Antonopoulos, Antonios I-119
Apostolidis, Georgios II-462, II-472
Apostolopoulou, Maria I-3, I-513
Arabatzis, Ioannis II-104
Argyropoulos, Ioannis I-513, II-200
Astaras, Konstantinos II-3
Asteris, Panagiotis G. I-513, II-200
Athanasiou, Stefanos I-29
Athina-Georgia, Alexopoulou I-500

Bakalbasis, Dimitrios II-372
Balodimou, Maria I-273
Barbalić, Jure II-359
Bartoli, Gianni I-487
Bassiakos, Yannis II-397
Batis, George I-58
Beltsios, Konstantinos G. II-397
Bertolin, Chiara I-402
Betti, Michele I-487
Boniotti, Cristina II-289
Bonsma, Iveta II-336, II-347
Bonsma, Peter II-336, II-347
Burnham, Bonnie II-275

Cantini, Lorenzo I-319
Cavaleri, Liborio II-200
Christaras, Basile II-407
Chrysaeidis, Leonidas II-3

Chrysochou, Nasso II-88
Couvelas, Agnes II-117

Daflou, Eleni I-58
de Vries, Pieter I-263
Delinikolas, Nikolaos II-59
Della Torre, Stefano I-319
Devlioti, Kyriaki II-407
Di Giulio, Roberto II-325
Dokoumetzidis, Aristides II-419
Doulamis, Anastasios I-376
Douvika, Maria G. I-513
Drakaki, Maria A. I-329
Drosopoulos, Georgios A. II-143
Drosou, Anastasios I-141

Efesiou, Irene I-273
Elmaloglou, Julia I-184

Fellas, Argyris II-177
Ferrari, Federico II-336, II-347
Filippaki, Eleni II-397
Floros, Christos I-427
Fudos, Ioannis I-141

G. Anagnostakis, Aristidis I-541
Gagnon, Alexandre S. I-402
Gavela, Stamatia I-309
Gentile, Carmelo II-430
Georgopoulos, Andreas I-3, I-69
Gheraldi, Francesca II-104
Giannakopoulos, Dimitrios II-299
Giannakopoulos, Nikolaos II-407
Giarleli, Mariliza II-131

Hatzigiannakis, Kostas II-419
Hellmuth, René I-232
Hughes, John I-402

Iadanza, Ernesto II-336, II-347
Ioannidis, Charalabos I-337

Ioannidis, Panagiotis   I-171
Ioannou, Byron   I-285

Jacob, Daniela   I-353
Julia, Vobiri   I-449

Kalentzi, Katerina   I-119
Kalokairinou, Eleni   II-419
Kaminari, Agathi Anthoula   II-443
Kampanis, Nikolaos   I-385
Kanellopoulou, Dimitra G.   I-456
Kapassa, Evgenia   I-69
Kapridaki, Chrysi   II-104
Karadimas, Dimitrios   II-372
Karadimas, George   II-372
Karagiannis, Georgios   II-462, II-472
Karampinis, Leonidas   I-150
Karapidis, Alexander   I-232
Katsaveli, Elissavet   II-419
Kavoulaki, Elissavet   II-419
Kokkinaki, Olga   II-419
Kontogiannis, Evangelos D.   II-36
Korres, Manolis   I-3
Kotova, Lola   I-353
Koutsoukos, Petros G.   I-456
Kozyrakis, Georgios V.   I-385
Kyriazis, Dimosthenis   I-69, I-376

Lambropoulos, Kyriakos   I-376
Lampropoulos, Kyriakos C.   I-3
Lampropoulou, Antonia   I-273
Leissner, Johanna   I-353, I-402
Liolios, Angelos A.   II-188
Livitsanos, Georgios   II-430
Loli, Arian   I-402
Lourenço, Paulo B.   II-200

Maietti, Federica   II-325, II-336, II-347
Maistrou, Eleni   I-273
Maistrou, Helen   I-552
Makris, Dimitrios   I-150
Makropoulos, Constantin   I-232
Malakasioti, Angeliki   I-171
Malaperdas, George   I-222, I-532
Maniatakis, Charilaos A.   I-44, II-225
Maravelaki, Pagona-Noni   II-104
Mariettaki, Athena-Panagiota   I-119
Marinos, Vasilios   II-407
Martín Lerones, Pedro   II-336, II-347

Mastrotheodoros, Georgios P.   II-397
Mathis, Moritz   I-353
Mavrokostidou, Maria   I-171
Mavromati, Eleftheria   II-3
Medici, Marco   II-336, II-347
Melessanaki, Kristalia   II-419
Miaoulis, Georgios   I-337
Mikolajewicz, Uwe   I-353
Miltiadou-Fezans, Androniki   II-59
Moioli, Rossella   I-319
Moraitis, Konstantinos   I-105
Moropoulos, Nikolaos   I-78, II-299
Moropoulou, Antonia   I-3, I-44, I-58, I-69,
    I-78, I-263, I-273, I-376, I-513, II-299
Moutafidou, Anastasia   I-141
Moutsatsou, Anna   II-443
Mouzakis, Charalambos   I-3

Oikonomopoulou, Apostolia   II-131

Padeletti, Giuseppina   I-360
Palamara, Eleni   I-222, I-532
Panagiotidis, Vayia V.   I-222
Panagiotis, Ilias   I-500
Papadimitriou, Alcestis   I-337
Pappa, Dimitra   I-232
Parthenios, Panagiotis   I-128
Pelekanos, Marios   I-285, II-157
Pfeiffer, Helge   II-482
Phakwago, Jan   II-143
Philippidis, Aggelos   II-419
Philokyprou, Maria   I-473
Piaia, Emanuele   II-325
Pitsilis, Vassilis   I-232
Politaki, Elpida   II-419
Pouli, Paraskevi   II-419
Poulios, Ioannis   II-262
Prassas, Alexandros   II-372
Prepis, Alkiviadis   II-239
Psaltakis, Dimitrios-Ioannis   I-119
Psaroudaki, Aggeliki   II-419
Psychogyios, Dimitris   I-552

Rajčić, Vlatka   II-359
Rakanta, Eleni   I-58
Rodrigues, Hugo   II-200

Saisi, Antonella   II-430
Salemi, Niki   II-131

Sesana, Elena   I-402
Siountri, Konstantina   I-250, II-78
Siozos, Panagiotis   II-419
Sithiakaki, Vassiliki   II-419
Skentou, Athanasia   I-513
Skourtis, Michail   I-337
Soile, Sofia   I-337
Somakos, Leonidas   II-372
Sotiropoulou, Anastasia   I-309
Spyrakos, Constantine C.   I-44, II-225
Spyrakos, Constantine   I-3
Stamatiou, Eleni   II-3
Stavroulaki, Maria E.   II-143
Stavroulakis, Georgios E.   II-143
Stefanou, Joseph   I-250
Stepinac, Mislav   II-359

Thomas, Job   II-200
Toniolo, Lucia   II-104
Touloupou, Marios   I-69
Triantafyllopoulou, Konstantia   I-203
Tucci, Grazia   I-487
Tziranis, Ioannis   I-337
Tzovaras, Dimitrios   I-141

Vagena, Evangelia   I-250
Valantou, Vasiliki   I-222

Van Steen, Charlotte   II-482
Varum, Humberto   II-200
Vavouraki, Aikaterini I.   I-456
Vergados, Dimitrios D.   I-250
Verstrynge, Els   II-482
Vlachou, Melina Aikaterini   I-150
Vlahoulis, Themistoklis   II-131
Voulodimos, Athanasios   I-337
Vouvalidis, Konstantinos   II-407
Vryonis, Panagiotis   I-532

Wehner, Florian   I-232
Wevers, Martine   II-482
Wilhelm, Stephan   I-232

Xipnitou, Maria   I-337
Xydia, Stephania   I-184

Yfantis, Dimitrios   II-385

Zacharias, Nikolaos   I-222
Zacharias, Nikos   I-532
Zacharopoulou, Angeliki   I-58
Ziri, Anna Elisabetta   II-336, II-347
Zoe, Georgiadou   I-500

Printed in the United States
By Bookmasters